Growth and Change in Post-socialist Cities of Central Europe

This book presents multidimensional socio-economic transformations taking place in the post-socialist cities located in selected countries of the Central European region.

The analysis includes case studies from the Eastern part of Germany (Chemnitz, Leipzig), Poland (Łódź, Kielce, Katowice conurbation, and peripheral urban centres from Eastern Poland), Slovakia (Bratislava, Nitra), the Czech Republic (Olomouc, Brno), and from Hungary (Pécs). The analysed urban areas have undergone far-reaching political and socio-economic changes in the last 30 years. These changes began with the collapse of communism and the centrally planned economy system in the region of Central Europe. The beginning of this period, often referred to as post-socialist transformation, dates back to 1989. The consequence of the aforementioned political processes was the multifaceted socio-economic and demographic changes that significantly affected urban areas in Central Europe. This book presents an attempt to summarize the main long-term processes of changes taking place in these urban areas and to identify contemporary and future trends in their socio-economic development.

The book will be valuable to undergraduate and postgraduate students in human geography, urban studies, economy, and city marketing, especially those with an interest in Central Europe.

Waldemar Cudny is an Associate Professor at The University of Lodz, Faculty of Geographical Sciences (Poland). He specialises in urban, tourism geography, and event studies. His publications include such books as *City Branding and Promotion: The Strategic Approach* (Routledge, 2019) and *Urban Events, Place branding and Promotion. Place Event Marketing* (Routledge, 2020).

Josef Kunc Ph.D., is an Associate Professor at Masaryk University, Faculty of Economics and Administration, Brno, Czech Republic. He specialises in the transformation of the current socio-economic and spatial environment of cities and regions, consumer preferences, shopping behaviour, and retail management as well as special interest tourism.

Routledge Contemporary Perspectives on Urban Growth, Innovation and Change

Series edited by Sharmistha Bagchi-Sen, Professor, Department of Geography and Department of Global Gender and Sexuality Studies, State University of New York-Buffalo, Buffalo, NY, USA and **Waldemar Cudny**, Associate Professor, Working at the University of Łódź, Poland.

Urban transformation affects various aspects of the physical, social, and economic spaces. This series contains monographs and edited collections that provide theoretically informed and interdisciplinary insights on the factors, patterns, processes and outcomes that facilitate or hinder urban development and transformation. Books within the series offer international and comparative perspectives from cities around the world, exploring how 'new life' may be brought to cities, and what the cities of future may look like.

Topics within the series may include: urban immigration and management, gender, sustainability and eco-cities, smart cities, technological developments and the impact on industry and on urban societies, cultural production and consumption in cities (including tourism, events and festivals), the marketing and branding of cities, and the role of various actors and policy makers in the planning and management of changing urban spaces.

If you are interested in submitting a proposal to the series please contact Faye Leerink, Commissioning Editor, faye.leerink@tandf.co.uk.

Urban Events, Place Branding and Promotion
Place Event Marketing
Edited by *Waldemar Cudny*

Place Event Marketing in the Asia Pacific Region
Branding and Promotion in Cities
Edited by *Waldemar Cudny*

Post-socialist Shrinking Cities
Edited by *Chung-Tong Wu, Maria Gunko, Tadeusz Stryjakiewicz and Kai Zhou*

Growth and Change in Post-socialist Cities of Central Europe
Edited by *Waldemar Cudny and Josef Kunc*

Interstitial Spaces of Urban Sprawl
Geographies of the Post-Suburban City in Chile
By *Cristian Silva*

Growth and Change in Post-socialist Cities of Central Europe

Edited by Waldemar Cudny and Josef Kunc

Routledge
Taylor & Francis Group

LONDON AND NEW YORK

First published 2022
by Routledge
2 Park Square, Milton Park, Abingdon, Oxon OX14 4RN

and by Routledge
605 Third Avenue, New York, NY 10158

Routledge is an imprint of the Taylor & Francis Group, an informa business

British Library Cataloguing-in-Publication Data
A catalogue record for this book is available from the British Library

Library of Congress Cataloging-in-Publication Data
Names: Cudny, Waldemar, editor. | Kunc, Josef, editor. Title: Growth and change in post-socialist cities of Central Europe / edited by Waldemar Cudny and Josef Kunc. Description: Milton Park, Abingdon, Oxon ; New York, NY : Routledge, 2022. | Includes bibliographical references and index. Identifiers: LCCN 2021033119 (print) | LCCN 2021033120 (ebook) | ISBN 9780367484477 (hardback) | ISBN 9781032132532 (paperback) | ISBN 9781003039792 (ebook)
Subjects: LCSH: Cities and towns–Europe, Central. | Post-communism–Europe, Central. | Europe, Central–Economic conditions. Classification: LCC HT145. C36 G76 2022 (print) | LCC HT145.C36 (ebook) | DDC 307.760943–dc23
LC record available at https://lccn.loc.gov/2021033119LC ebook record available at https://lccn.loc.gov/2021033120

ISBN: 978-0-367-48447-7 (hbk)
ISBN: 978-1-032-13253-2 (pbk)
ISBN: 978-1-003-03979-2 (ebk)

DOI: 10.4324/9781003039792

Typeset in Bembo
by SPi Technologies India Pvt Ltd (Straive)

Contents

List of Contributors   vii

1   Conceptualising urban transition in Central Europe   1
    WALDEMAR CUDNY, JOSEF KUNC AND IRENA DYBSKA-JAKÓBKIEWICZ

2   Post-socialist urban change and its spatial patterns: the
    case of Nitra   15
    VLADIMÍR IRA AND MARTIN BOLTIŽIAR

3   The university as the creative hub: the case of the city
    of Olomouc after 1989   30
    ZDENĚK SZCZYRBA, IRENA SMOLOVÁ, MARTIN JUREK AND DAVID FIEDOR

4   In the shadow of Karl Marx: the case of Chemnitz and its
    multiple transitions   48
    BIRGIT GLORIUS

5   Young people's life plans and their impact on the demographic
    future of a shrinking city: Kielce case study   67
    MIROSŁAW MULARCZYK AND WIOLETTA KAMIŃSKA

6   Manufacturing in the post-industrial city: the role of a
    "Hidden Sector" in the development of Pécs, Hungary   94
    GÁBOR LUX

7   Socio-economic development in Bratislava during
    post-socialism   113
    PAVOL KOREC AND SLAVOMÍR ONDOŠ

8   Brno in transition: from industrial legacy towards modern
    urban environment   136
    JOSEF KUNC AND PETR TONEV

9 Challenges and problems of re-growth: the case of Leipzig
   (Eastern Germany)   158
   DIETER RINK, MARCO BONTJE, ANNEGRET HAASE, SIGRUN KABISCH AND
   MANUEL WOLFF

10 Łódź: a multidimensional transition from an industrial
    center to a post-socialist city   178
    JOLANTA JAKÓBCZYK-GRYSZKIEWICZ

11 The socio-economic transformation of the Katowice
    conurbation in Poland   195
    ROBERT KRZYSZTOFIK

12 Population ageing processes in towns and cities situated in
    peripheral areas: an example of urban centres in
    Eastern Poland   217
    WIOLETTA KAMIŃSKA AND MIROSŁAW MULARCZYK

13 Hallmark features of post-socialist urban development in
    Central Europe   242
    JOSEF KUNC AND WALDEMAR CUDNY

    Index   254

Contributors

Martin Boltižiar, Ph.D., is Slovak geographer and landscape ecologist. He is vice-president of the Slovak Geographical Society and member of the Slovak National Geographical Committee. He is research worker at the Institute of Landscape Ecology, Slovak Academy of Sciences and Professor of Geography at the Constantine the Philosopher University in Nitra. He is specializing in physical geography, landscape ecology, and geoinformatics. His main research interest is land-use/land cover changes. He has participated in over 60 scientific projects and published more than 300 scientific papers and 17 monographs. He is a member of editorial boards of eight scientific journals.

Marco Bontje is senior lecturer in urban geography at the Department of Geography, Planning and International Development Studies of the University of Amsterdam. Within the research institute AISSR, he is member of the research group Urban Geographies. Specialising mostly in urban geography, his research has crossed disciplinary boundaries towards urban and regional planning, and population, economic and environmental geography. Research topics so far include the effectiveness of Dutch national urbanisation policy (PhD thesis); sustainable development of city-edge and (post-)suburban business locations; spatial, social, and economic conditions for competitive creative knowledge cities; and urban and regional shrinkage and regrowth.

Waldemar Cudny, Ph.D., associate professor in the University of Łódź, Faculty of Geographical Sciences, Poland. He is a human geographer specializing in urban development, festival tourism, and special interest tourisms i.e. car tourism. Waldemar Cudny is the author of over 80 research publications including books, edited volumes, and articles. His latest research includes analysis of the role of festivals in the development of urban spaces, car tourism. He is also the author of a complex monograph entitled *City Branding and Promotion. The Strategic Approach* and an edited volume *Urban Events, Place Branding and Promotion. Place Event Marketing* (published by Routledge).

Irena Dybska-Jakóbkiewicz, Ph.D., in Earth Sciences (geography), is an assistant professor in the Institute of Geography and Environmental Sciences of the Jan Kochanowski University in Kielce. Her academic interests focus on the problems of teaching geography, regional education, and the geography of perception. Lately she has done research into the perception of urban space.

David Fiedor is junior lecturer at the Department of Geography of the Faculty of Science of the Palacký University in Olomouc. He is concerned with issues of human geography, partly GIS, and statistical methods. His publication portfolio consists of several original articles also registered in the Web of Science or Scopus.

Birgit Glorius is a professor of human geography with a focus on European migration research at Chemnitz University of Technology. Her research interests and the majority of publications are in the fields of international migration an integration, with recent projects on the reception and integration of asylum seekers and refugees in Europe, notably on questions of local governance and local reception cultures. Further research areas are demographic change, geographies of education, and (im)mobility and regional development; most of her research is carried out in Eastern Germany, Poland, Bulgaria, and the Western Balkans.

Annegret Haase is an urban sociologist and works as a senior scientist at the Helmholtz Centre for Environmental Research – UFZ in Leipzig, Germany, at the Department of Urban and Environmental Sociology. Her research foci are sustainable urban development, urban transformations, social-spatial and socio-environmental processes and goal conflicts in cities, land use, green spaces, governance, inequalities, urban diversity and migration, urban shrinkage and regrowth, neighbourhood development and participation.

Vladimír Ira, Ph.D., Czecho-Slovak geographer, is currently senior research worker at the Institute of Geography, SAS in Bratislava and Professor of Geography at Universities in Bratislava and České Budějovice. He graduated from the Comenius University in Bratislava. He formerly held the position of Director of the Institute of Geography, SAS (2006–2016). He serves as chairperson of Slovak National Committee of the IGU. He is editor-in-chief of Geografický časopis. His earlier works have concentrated on urban and behavioural geography. He has also contributed to theoretical writing on sustainability and quality of life, time-space behaviour, and geographical thought.

Jolanta Jakóbczyk-Gryszkiewicz is a professor at the Faculty of Geographical Sciences at the University of Łódź, Poland. Her research is connected with settlement geography. Her doctorate presented links between the cities of Łódź urban region. Her habilitation thesis addressed issues of suburbanization in Poland's three largest cities. She received the title of full professor in 2011 publishing a study on the diversity of land prices in Łódź and its suburban zone. Professor Gryszkiewicz has authored numerous scientific publications. Her recent works are devoted to urbanization in suburban zones, gentrification in large Polish cities, and migrations to Poland from outside the EU.

Martin Jurek is an assistant at the Department of Geography in the Faculty of Science of the Palacký University in Olomouc. His research interest encompasses issues of geography of education and to some extent also physical geography. His publication portfolio consists of several original articles published in international journals registered in the Web of Science and Scopus databases.

Sigrun Kabisch is head of the Department of Urban and Environmental Sociology at the Helmholtz Centre for Environmental Research – UFZ in Leipzig, Germany. Additionally, she is a professor of Urban Geography at the University of Leipzig. She acts as chair of the Scientific Advisory Board of the Joint Programming Initiative (JPI) Urban Europe. Her main research fields are urban transformations, urban demographic change as well as multiple urban risks. In 2018, she co-edited the volume "Urban transformations: sustainable development through resource efficiency, quality of life and resilience" (Springer) which represents her interdisciplinary approaches towards sustainable urban development.

Wioletta Kamińska is a professor at the Institute of Geography and Environmental Sciences at the Jan Kochanowski University in Kielce. She is the head of the Department of Socio-Economic Geography at this Institute. She is also the chairwoman of the Task Force for Rural Areas of the Committee for Spatial Economy and Regional Planning at Polish Academy of Sciences. Professor Kamińska conducts research on the development of rural areas and small towns in Poland, in particular on human and social capital in rural areas and small towns, depopulation of these areas and their multifunctional development. She has published over 150 scientific works.

Pavol Korec is full professor in the Department of Economic and Social Geography, Demography and Territorial Development at the Comenius University in Bratislava (Slovakia). He is the author of several monographs including *Regional Development of Slovakia in Years 1989–2004* (2005), *Modern Human Geography of Bratislava – Territorial Structures, Networks and Processes* (2013), *Human Geography Approaches, Philosophy, Theory and Context* (2018), *Industry in Nitra, Global, National and Regional Context* (2019), *Theories of Regional Development and Research Regions* (2020). His latest works became oriented towards two problems: changes in urban economy and spatial structure of Bratislava and less developed regions in global scale and in Slovakia.

Robert Krzysztofik is Director of Institute of Social and Economic Geography and Spatial Management, University of Silesia in Katowice (Sosnowiec, Poland). Robert Krzysztofik's research lies at the intersections of spatial analysis of densely populated areas, the theory of urban geography, dynamics of urban development and economic and social drivers of changes in the urban perspective. He is also interested in the socio-economic transformation of postindustrial and polycentric urban regions. Robert Krzysztofik is an author of more than 100 publications. He has also ten books among his scientific achievements (author, co-author, co-editor).

Josef Kunc, Ph.D., is associate professor at Masaryk University, Faculty of Economics and Administration, Brno, Czech Republic. He is an economic geographer and regionalist specializing in the transformation of the current socio-economic and spatial environment of cities and regions. Other areas of his scientific interest are issues of consumer preferences, shopping behavior, and retail management as well as special interest tourism. Josef Kunc is the author of over 80 research publications including articles in peer-reviewed journals, books, and book chapters. He has experience from being a leader of national research projects and a team member of national and international projects.

Gábor Lux, Ph.D., is senior research fellow at Hungarian Academy of Sciences, CERS Institute of Regional Studies. His main areas of research are industrial restructuring and industrial competitiveness in Central and Eastern Europe; industrial policy, urban economic governance, and evolutionary economic geography. He is the author, co-author, or editor of 75 publications in Hungarian, English, and Russian, and co-editor of *The Routledge Handbook to Regional Development in Central and Eastern Europe*.

Mirosław Mularczyk works as an associate professor (professor extraordinarius) in the Institute of Geography and Environmental Sciences of Jan Kochanowski University in Kielce (Poland). He conducts research mostly in settlement geography. His publications in this field concern the development of regional settlement systems and small cities issues. Moreover, his interests include issues related to geography didactics and tourism geography. There are more than 100 scientific, popular science and didactic publications in his academic achievements.

Slavomír Ondoš is associate professor in the Department of Economic and Social Geography, Demography and Territorial Development at the Comenius University in Bratislava (Slovakia). He held research position at the Vienna University of Economics and Business, Austria (2008–2010) and teaching assistant position at the University of Nebraska Omaha, USA (2005). He participates in project linking academic community, non-profit and commercial sector. His writings examine emergent cooperative dynamics in socio-economic processes. His latest work becomes oriented towards the understanding of the innovation ecosystem and how the creation and commercial value of knowledge can be shaped by public leadership.

Dieter Rink holds a Ph.D. in philosophy and works as a senior scientist at the Helmholtz Centre for Environmental Research – UFZ in Leipzig, Germany, at the Department of Urban and Environmental Sociology. His research fields include sustainable urban development, urban governance, social movements, urban ecology and urban nature, urban shrinkage and regrowth, housing and housing policy. He is teaching urban sociology at the University of Leipzig and has been a member and coordinator of diverse national and international research projects, numerous publications on the aforementioned issues. His research is strongly interdisciplinary oriented and includes international comparative urban studies.

Irena Smolová is associate professor at the Department of Geography of the Faculty of Science of Palacký University in Olomouc. She is concerned with issues of physical geography and geography for education. Her publication portfolio consists of dozens of original articles also registered in the Web of Science or Scopus.

Zdeněk Szczyrba is associate professor at the Department of Geography of the Faculty of Science of Palacký University in Olomouc. He is concerned with issues of geography of services and urban geography. His publication portfolio consists of dozens of original articles also registered in the Web of Science or Scopus.

Petr Tonev, Ph.D., is assistant professor at Masaryk University, Faculty of Economics and Administration, Brno, Czech Republic. His research interest encompasses issues of geographical organization of space and society (regional taxonomy, spatial interaction modeling), urban and spatial planning, population geography, and regional development as well as cartographic visualization. He was a co-applicant of national research projects and a team member of several national projects. Research results of these projects were a basis for publications in respected national and international journals. He has also experience in the creation of strategic documents on different hierarchical levels outside academia.

Manuel Wolff is a research fellow at the Humboldt-Universität zu Berlin in the Department of Geography and guest researcher at the Department of Urban and Environmental Sociology Helmholtz Centre for Environmental Research in Leipzig, Germany. His main research interest is settled in urban and regional development within the context of urban (re)growth and shrinkage and the associated impacts on land use changes and ecosystem services. Especially, he works on quantitative comparative analysis of human–environmental interactions and trends in European cities including aspects of accessibility, resource efficiency, and quality of life.

1 Conceptualising urban transition in Central Europe

Waldemar Cudny, Josef Kunc and Irena Dybska-Jakóbkiewicz

Research problem and aims

This book is an edited volume presenting broadly understood socio-economic, functional and spatial transformation taking place in the post-socialist cities situated in selected countries of the Central European region. The region of Central Europe is located between the areas of Western and Eastern Europe. It is distinguished on the basis of several factors, including geographic location, history and common socio-economic elements. In the scientific literature, the region of Central Europe is presented differently. It is often identified with the countries of the Visegrad Group (V4 countries) (see: Gorzelak 1996; Tiersky 2004). This organisation was formed on the basis of a political agreement, signed in 1991, between countries with common history and culture. Since the dissolution of Czechoslovakia, the Visegrad Group includes Poland, Hungary, the Czech Republic, and Slovakia (http://www.visegradgroup.eu/). According to Lewis (2014), the Central European region encompasses Poland, the former German Democratic Republic (GDR), Czechoslovakia (later the Czech Republic and Slovakia), and Hungary. Johnson and Johnson (1996) have a wider view of Central Europe. According to them, the region encompasses Germany, Poland, the Czech Republic, Slovakia, Austria, Hungary, Slovenia, and Croatia.

Central Europe also occurs in divisions made by various international institutions. In European Union documents, the region includes countries such as Austria, Croatia, the Czech Republic, Hungary, Poland, Slovakia and Slovenia, eight Länder from Germany, and nine regions from Italy (Interreg Central Europe, Citizens' summary, Annex D, 2015). The World Bank distinguish in Europe a region consisting of similar European countries called Central Europe and the Baltics. According to the World Bank, this region includes Bulgaria, Croatia, the Czech Republic, Estonia, Hungary, Latvia, Lithuania, Poland, Romania, the Slovak Republic, and Slovenia.

In this book, cities of different size categories (from small-scale cities through to medium, large and extra-large cities to urban networks) and status are included in the analysis. They are located in the area of five selected post-socialist countries belonging to the Central European region (Figure 1.1). In reference to the division presented by Lewis (2014), urban areas from Eastern Germany (Chemnitz, Leipzig), Poland (Łódź, Kielce, Katowice), the Czech Republic (Olomouc, Brno), Slovakia (Bratislava, Nitra), and Hungary (Pécs) were accepted for research. In the case of

DOI: 10.4324/9781003039792-1

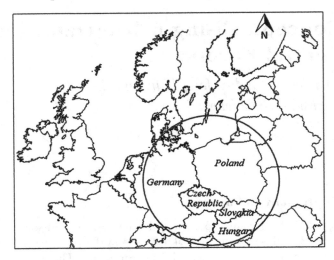

Figure 1.1 Location of Central European countries.
Source: Authors' elaboration.

Germany, the research area was limited to the cities belonging to former Germany Democratic Republic (DGR).

The research problem examined in the book is the in-depth analysis of the structural socio-economic changes ongoing in post-socialist cities of Central Europe. The main research aim is to present a comprehensive summary of the processes of socio-economic and demographic changes taking place in the studied cities after the fall of communism. It is about identifying the factors that have initiated the transformation of economy and society, giving a comprehensive presentation of transformation processes and their effects on the contemporary socio-economic, demographic, functional and spatial structure of post-socialist cities in the region of Central Europe.

The analyzed urban areas have undergone far-reaching political, socio-economic, functional and spatial changes in the last 30 years. These changes began with the collapse of communism and the centrally planned economy system in the region of Central Europe. The beginning of this period, often referred to as post-socialist transformation, dates back to 1989. Then in Poland, the round table talks began as a result of which the first partly democratic elections took place in this country (June 1989). The next important events were the Velvet Revolution in Czechoslovakia (November 1989), the fall of the Berlin Wall (November 1989) and the reunification of Germany (1990).

The consequence of the aforementioned political processes were the multifaceted socio-economic, demographic and spatial changes taking place in the countries discussed in the book. They have significantly affected urban areas in Central Europe (Smith et al., 2008; Hoff, 2011; Kavaliauskas, 2012; Cudny, 2012; Schweiger and Visvizi, 2018). This book is an attempt to summarise the main long-term processes of changes taking place in these urban areas and to identify contemporary and future trends in their socio-economic and spatial development.

The thirty years of post-socialist transition

It would be useful to broaden the discussion and introduce a comprehensive study in the field of urban environment focusing on the interconnecting geographical, social and economical aspects of urban areas located in the Central European region. The discourse on the transformation of cities most often presents examples from the broadly understood Western world. Some publications concern Asian countries, including China (Lin, 2004; Wang and Zhang, 2005). At present, however, there are only a few comprehensive books in English concerning the area of Central Europe (Stanilov, 2007; Hoff, 2011; Schweiger and Visvizi, 2018). Most often the books regarding this area take the regional (Gorzelak, 1996; Lux and Horváth, 2017) or historical approach (Gutkind, 1964; Kavaliauskas, 2012; Lewis, 2014). Therefore, it seems that our book will fill the gap in the international publishing market especially in relation to human geography, regional economy and comprehensive urban studies and management based on diversified examples.

The edited volume characterising cities in Central Europe is important because it presents cities from different countries, and of different types (especially in terms of size and function). It includes cities in typically post-socialist countries, but also presents a wide range of case studies from former East Germany (former GDR). This area was absorbed in 1990 by West Germany and was later transformed under the auspices of Western government. Therefore, this work makes it possible to compare the policies of urban development used in post-socialist countries (Poland, the Czech Republic, Slovakia, Hungary) with the policies applied in Western countries (Germany after reunification). In addition, the book combines two issues. First of all, it is an attempt to comprehensively summarise 30 years of socio-economic, spatial, functional and structural changes in the area of post-socialist cities. Secondly, it describes the most current contemporary direction of their development. We consider these elements to be an additional value added of this publication.

As mentioned before, 30 years have passed since the beginning of the transformation period. The areas presented through the book became part of the European Union and currently, their political and economic significance in Europe is systematically growing. Therefore, it seems interesting from a research point of view to present a comprehensive analysis of the transformation of cities in this region.

The post-socialist cities presented in this book are well-defined in the scientific literature (Ott, 2001; Stanilov, 2007; Sýkora, 2009; Hirt, 2013; Kovács et al., 2019). Sýkora (2009, p. 394) defined it as

> Cities in the transition stage. They are characterized by dynamic processes of change rather than static patterns. The urban environment formed under the previous system is being adapted and remodelled to match the new conditions of the political, economic, and cultural transition towards the capitalist society. Many features of a socialist city suddenly stood in opposition to the capitalist principles. The contradictions between the market rules and the socialist urban environment led to the restructuring of the existing urban areas. With time, new capitalist urban developments are having more and more influence on the overall urban organization. The post-socialist developments bring the

re-emergence of some pre-socialist patterns, transformations in some areas from the socialist times, and creation of new post-socialist urban landscapes.

Liszewski (2001, p. 304) defined the post-socialist city as

> Any city which has been functioning in new political and economic conditions for over 10 years and earlier (before 1990) it was functioning in the conditions of real socialism for 45 years, regardless of the fact whether it was built in this system, or much earlier, subject to its ideology and laws (centralization of power, Lack of market economy, social and spatial egalitarianism, ideologization of life etc.).

Ondoš and Korec (2008) identified a number of stages in the development of the post-socialist cities. They identified early, mature and late post-socialist cities followed by the emergence of the early capitalist city that could be later transformer into the capitalist city.

The complex transformations of Polish cities after 1989 were described by Węcławowicz (2016). Interesting conclusions about the post-socialist cities of Central and Eastern Europe were described, among others, on the background of implementation of the international research program ReNewTown (Węcławowicz and Wątorska-Dec 2012, 2013). On this basis, a concept and strategy for the further development of post-socialist cities were developed (Węcławowicz, 2013). The issue of urban identity during the post-socialist transformation was discussed by Young and Kaczmarek (2008). Cudny (2011, 2012) developed a model of changes in a post-socialist city.

The transformations that took place in post-socialist cities after 1989 included the social, economic, functional and spatial changes (Kovács, 1999; Cudny, 2012). In the social structure, the transformation often resulted in processes of depopulation and societal ageing resulting from a decline in the natural increase in cities and the massive emigration of young people (Steinführer and Haase, 2007; Stenning and Dawley, 2009; Steinführer et al., 2010; Ott, 2001; Haase et al., 2012). The phenomenon of the second demographic transition has appeared in the cities of Central Europe in the last decades (Haase and Steinfuhrer, 2005).

The socio-demographic changes and changes in the spatial and functional structure of the post-socialist city were described by Szafrańska. She identified the changes taking place in large post-socialist housing estates and explained the main factors behind them (Szafrańska, 2014, 2015). Using the example of large prefabricated housing estates, she also discussed the processes of demographic aging of post-socialist cities (Szafrańska, 2017). Research on socio-economic segregation and environmental justice was conducted, among others, by Marcińczak and Rufat. Their results provided information on social, economic and environmental inequalities in post-socialist cities (Marcińczak, 2013; Rufat and Marcińczak, 2020). Research on the gentrification in cities is gaining growing attention (see: Grzeszczak, 2010; Liszewski and Marcińczak, 2012; Górczyńska, 2012, 2015; Holm et al., 2015).

In addition, especially in the 1990s, post-socialist cities, in particular, those based on traditional industries, have been affected by the phenomenon of de-industrialisation followed by mass unemployment (Jürgens, 1996; Mulíček and Toušek, 2004; Osman et al., 2015). Changes in industrial spaces were discussed, among others, by Płaziak (2014) and Sikorski (2019). There have been processes of gentrification, social segregation and socio-economic revitalisation of some areas (Kovács et al., 2013; Kaczmarek and Marcińczak, 2013; Kaczmarek, 2015; Holm et al., 2015). Some of the post-socialist cities as a result of heavy de-industrialisation suffered from a very bad socio-demographic situation and began to fall into the category of shrinking cities (Buček and Bleha, 2013; Rink et al., 2014).

Changes in post-socialist cities included the transition of their functional structure. Old functions decreased e.g. due to the collapse of heavy industry and new functions developed, including a plethora of services, creative industries and new branches of production (Cudny, 2012; Chapain and Stryjakiewicz, 2017; Chaloupková et al., 2018). The ownership structure of enterprises has also changed. There has been a massive commercialisation and privatisation process and a wide inflow of foreign investments. Due to this transition, Central European cities isolated so far were included in the world's internationalising and globalising economy (Sýkora, 2009; Jacobs, 2013).

Within the spatial structure, a number of morphological changes took place. The phenomenon of suburbanisation developed, and ghettoisation and space fragmentation processes appeared. In addition, some urban areas have been transformed into degraded spaces, and some have undergone redevelopment and revitalisation (Krisjane and Berzins, 2012).

The phenomenon of suburbanisation is one of the most extensively studied processes of post-socialist urban change. The analysis of this phenomenon was carried out among others by Lisowski et al. (2014). Bański (2017) also analyzed land-use changes in five countries of the former Eastern Bloc (the Czech Republic, Hungary, Poland, Romania, and Slovakia). He noted that the greatest loss of agricultural land occurred in the suburban zone of large agglomerations (i.e. Warsaw, Prague, Budapest), which was the result of increasing suburbanisation.

The multidimensional transition processes taking place in the post-communist cities of the Czech Republic were described by Sýkora and Ouředníček (2007), Stanilov and Sýkora (2012), and Kubeš (2015). Many case studies were presenting the capital cities of post-socialist European states. Stanilov and Hirt (2014) researched Sofia (Bulgaria), Grigorescu et al. (2012) Bucharest (Romania), Kovács et al. (2019) Budapest (Hungary), Pichler-Milanović (2014) Ljubljana (Slovenia), Leetmaa et al. (2014) Tallinn (Estonia), and Šveda et al. (2016) Bratislava (Slovakia).

The presented book is to fill the gap in the scientific literature, which concerns the area of Central Europe. While there is a rich literature presenting the changes of cities taking place in recent decades in other areas of the world like Western Europe and North America (Watson and Gibson, 1995; Eade and Mele, 2002; Kazepov, 2005; Hutton, 2008; Mollenkopf and Crul, 2012) or Asia (Champion and Hugo, 2004; Prakash and Kruse, 2008; Heitzman, 2008; Misra, 2013), Central European examples were much less frequently presented in geographical and socio-economic literature. Moreover, most studies focus only on one or two cities or metropolitan areas, while our book presents a wider view of diverse cities from different countries in the region.

The methods and the structure of the book

The research method is a way of scientific thinking and creation of new solutions and ideas. According to Apanowicz (2002, p. 59) the scientific method is "a set of theoretically justified procedures conceptual and instrumental, covering the entirety of the proceedings research aimed at solving a specific scientific problem". Runge (2006) also sees a scientific method as a way of thinking and discovering the solutions to scientific problems.

Methods are divided into qualitative and quantitative, according to the research procedure and type of materials undergoing analysis. The qualitative methods include the studies of perception based on the opinions of other people, interviews, and own observations of the researcher. The group of quantitative methods includes the use of numerical data obtained through empirical studies or from statistical sources and their presentation and analysis among others with the use of statistical indicators.

The book presents a variety of case studies. The chapters refer to individual cities such as Chemnitz, Leipzig, Nitra, Bratislava, Brno, Olomouc, Pécs, Łódź and Kielce. There are also chapters describing groups of cities, like the Katowice conurbation or the group of peripheral cities and towns in eastern Poland.

Various research methods were applied for the analysis presented in this edited volume. They included qualitative methods such as document analysis, literature review and Internet searches. One important group of research methods was the analysis of statistical data and the use of statistical indicators showing changes in the socio-economic, functional and spatial structure of the examined cities, like Webb's method used for the analysis of urban populations types in the towns and cities from eastern Poland (Chapter 12). Such methods can be included in the group of quantitative research methods. Questionnaire surveys were also used, which turned out to be useful in analyzing the life plans of young people living in the Polish city of Kielce. In this case, it was about examining family and procreation plans and linking them with demographic and social changes in the studied city.

The book consists of 13 chapters, including Introduction, Conclusions and chapters presenting case studies on various types of cities from Eastern Germany, Poland, the Czech Republic, Slovakia and Hungary. The chapters presenting individual case studies are preceded by a comprehensive introduction (Chapter 1), in which the subject of research, the scientific aims, research area, methods used in the book and its structure are presented. The edited volume closes with the Conclusions, in which the most important conclusions resulting from the analysis carried out in individual chapters were presented.

Case studies were carefully selected in order to present not only cities from different countries from Central Europe but also cities of various scales and functions. The idea was to check whether similar transformation processes take place in post-socialist cities with different scales and functions.

Chapter 2, written by Vladimír Ira and Martin Boltižiar, analysed the case of Nitra, a city located in the western part of the Slovak Republic. Nitra is the sixth-largest city in Slovakia according to the number of inhabitants. This chapter presented the multidimensional, i.e. demographic, social, functional and spatial transformations that took place in the city after 1990. The authors emphasised the analysis of the land-cover changes during the transformation from communism to democracy. The most important directions of urban space transformation were identified. The transition included changes in urban forms, architecture, the spatial-functional structure of land use. Moreover, socio-demographic changes and functional transitions, including the transition of the structure of the industrial sector and services, were presented.

Chapter 3 concerning Olomouc (ca. 100,000 inhabitants), a city in the Czech Republic located in the region of Central Moravia, was written by a team of scientists consisting of Zdeněk Szczyrba, Irena Smolová, Martin Jurek and David Fiedor. This chapter presented the innovative role of universities as part of the creative environment and specifically with the analysis and description of the structure of the creative centre at Palacký University in Olomouc and its system of functioning. Strategic planning documents at the level of the university, city and region were studied and statistical data of Palacký University were analyzed. During the three decades of transformation, the creative environment at Czech universities changed significantly, which often became the driving force of creative regional economies. The authors identified more than twenty research centres, laboratories, and other forms of creative centres in Olomouc. Universities cooperation was started with employers from the region and from all over the Czech Republic, top domestic and foreign scientists were employed. There has also been noticed an increase of granted patents. The gradual commercialisation of research results is obvious, however, its intensity has remained low so far. This is a certain handicap for the future operation of the university's prestigious scientific infrastructure.

The next, Chapter 4 written by Birgit Glorius, presented the case study of Chemnitz (formerly Karl-Marx-Stadt). The author looked into the complex changes which took place after the fall of communism in one of the biggest cities located in East Germany. After the reunification of Germany, Chemnitz entered a difficult period of socio-economic transition. The changes included the fall of the traditional (Fordist) textile industry and urban shrinkage. Traditional industries were replaced by modern branches of production and more medium-sized firms were located in the city. Later on services developed and, large changes in the housing areas occurred. Moreover, after the initial demographical decline at the beginning of the socio-economic transformation, the city attracted new immigrants, including refugees. These socio-economic changes were accompanied by the introduction of neoliberal and diverse policies of urban development, social change, and the creation of the modern urban brand and its perception.

Wioletta Kamińska and Mirosław Mularczyk described Kielce—a medium-sized city. The city is located in south-eastern Poland and is the capital of the Świętokrzyskie Voivodship. Kielce has ca. 196,000 residents, and currently represents a shrinking city type. Kielce developed as a manufacturing centre based on metallurgical, electromechanical, and mineral production. After the fall of communism, most of the traditional branches of industry collapsed or underwent in-depth restructuring. The post-socialist transformation brought diverse socio-economic impacts, including negative consequences for the natural increase and migration in the city. Since the end of the 20th century, Kielce has suffered from a negative migration balance and a natural decrease, which were the results of economic crisis and social modernisation processes undergoing in Poland after

1989. The chapter by Wioletta Kamińska and Mirosław Mularczyk aims to present the life plans of young people living in Kielce and to determine the impact of these plans on the socio-economic future of Kielce. The chapter presented the socio-economic situation of the city as well as the restructuring process that took place after the fall of communism. On this background, the authors characterised the results of a survey conducted among academic youth in Kielce and related to their future life plans and procreative behaviours. The research results were then analysed against the concept of the second demographic transition.

Chapter 6, written by Gábor Lux, analysed the evolution of manufacturing industries of Pécs, a shrinking city in Southern Hungary. The city's growth was a result of rapid industrialisation during the 19th and 20th centuries. However, after the fall of communism, Pécs faced de-industrialisation processes and abandoned its mining and industrial heritage in favour of becoming a cultural city. The chapter by Gábor Lux questions the possibility of the successful realisation of these plans. It draws attention to a simplistic understanding of restructuring plans, leading from industrial specialisation to the growth of service oriented cultural economy. The author describes the appearance of "hidden sectors", important secondary and tertiary activities that can get caught in vicious circles of decline, policy neglect and network disintegration. The chapter examines the transformation of the city's industries and the patterns of industrial decline. The structure of foreign direct investment, endogenous and, knowledge-based industries underwent examination, along with the institutional background that impacted their development. Gábor Lux highlights the importance of diversified industrial structures, the relevance of local structures, and the role of institutions in the management of the future of manufacturing in the post-transition era.

Another chapter is devoted to Bratislava – the capital city of Slovakia (Chapter 7). Pavol Korec and Slavomír Ondoš presented the socio-economic development of the city after the system change in 1989. The role of the city grew significantly after the division of Czechoslovakia and the establishment of the Slovak Republic as an independent state in 1993. The social structure of Bratislava changed and the role of jobs requiring advanced qualifications rose. The city attracted international investments and creative specialists from other regions of Slovakia and from abroad. The internationalisation of capital and labour resources progressed in the next decades. In the subsequent years after the establishment of Slovakia, its capital recorded the rise of quantitative (i.e. number of firms, GDP per inhabitant) and qualitative growth indicators (i.e. commercial activities, social structure and, spatial development). The rise of the city's economy and the transition of its society rushed when Slovakia entered the European Union. New opportunities arose in the barrier-free, market-friendly European Union. In line with its socio-economic change, Bratislava experience a transition of urban space, including changes in the urban landscape and its functional-spatial structure.

Chapter 8 was written by Josef Kunc and Petr Tonev and concerned the city of Brno. This is the second-largest city of the Czech Republic (ca. 400,000 inhabitants), located in the south-eastern part of the country. This chapter characterised the transformation of the economic base ongoing in the city in the last 30 years. The authors carefully explained the transition of traditional industry and its impacts on the city's economic, political, and spatial structures. Brno remained a strong industrial centre for more than 200 years. Its economic base was formed by such branches of industrial production as machinery, armaments, and textile production. The industrial specialisation of Brno was the major factor influencing the economic and the functional-spatial structure of the city. However, after the democratisation of life and the introduction of the free market economy, the city entered the period of de-industrialisation. Traditional industries lost their economic dominance, and new branches of production appeared as well as services rose. The transition from communism to the capitalist system also brought massive changes in the urban environment. However, the industrial heritage left a highly significant trace on the urban space of Brno where the remains of industrial traditions are still visible.

Chapter 9 regarded Leipzig, a large city from East Germany was written by a team of German scientists which included Dieter Rink, Marco Bontje, Annegret Haase, Sigrun Kabisch and, Manuel Wolff. Leipzig is an example of a successful redevelopment of a former East German city. German reunification was followed by deep de-industrialisation that affected cities from the eastern part of the country. This process resulted among others in the dynamic shrinkage of many of them. These processes were also visible in Leipzig. The city suffered from a de-industrialisation and shrinkage during the 1990s. However, in the 2000s, the city experienced re-growth and, in the 2010s, dynamic re-growth. The redevelopment was so intense that in the 2010s Leipzig became the fastest-growing city in Germany. Massive public investments, subsidies and support programmes from central and regional governments were the basis for the revitalisation of the urban economy and space in Leipzig. These programmes also mobilised private capital, which joined the investment processes. However, despite the advantages of re-growth, some unsuspected disadvantages appeared. They included, among others, a shortage of affordable housing and schools or growing traffic congestion. Chapter 9 presented and explained the challenges and problems of "growth after shrinkage" (i.e. re-growth) at the example of Leipzig. The authors focused on three municipal policies regarding the housing market, public schools and public transportation. The analysis showed that re-growth affected the aforementioned three municipal policies and created pressure for a reaction. Urban policy-makers responded to this pressure with various programmes and measures which attempt to overcome the disadvantages of re-growth in the city.

Chapter 10 was written by Jolanta Jakóbczyk-Gryszkiewicz and presented the case study of Łódź, Poland's third-largest city. The city developed in the 19th century as a large industrial city and continued to be the leading Polish centre of textile production until 1989. After the fall of communism, Łódź underwent intensive de-industrialisation and urban shrinkage due to a deep economic crisis. Traditional industries collapsed and most old-fashioned textile production plants went bankrupt. Industrial decline, combined with depopulation, made Łódź the fastest-shrinking Polish big city. After 2000, the city received new investments comprising modern branches of production (e.g. household appliances, computer production) and new services. The inclusion of Poland in the European Union in 2004 created another boost for the economic development of the city. Łódź profited from its favourable location, industrial traditions, low prices of urban land and commercial spaces (i.e. office spaces). Despite the revival noticed in the last two decades, the economic breakdown of the 1990s still foreshadows the socio-economic situation of the city. The values of such socio-economic indicators as the level of salaries, prices of land, the level of natural increase and migration balance are still among the lowest of all the Polish big cities. Chapter 10 presented the history of the city, the results of the system transformation of the 1990s, and the current opportunities and threats for the city of Łódź.

Chapter 11 discusses the socio-economic and spatial changes of the Katowice conurbation located in southern Poland. A conurbation is a polycentric agglomeration of towns and cities, developed on the basis of mining and industrial production. The Katowice conurbation, characterised in Chapter 11, developed in the 19th and 20th centuries into one of the leading coal mining and industrial regions in Europe. The Katowice conurbation is currently the biggest urban space in Poland, encompassing 54 communes, and a population of ca. 2.4 million. Robert Krzysztofik analysed the advantages and disadvantages of urban transformation that occurred in Katowice conurbation after the fall of communism in 1989. The analysis included the presentation of reduction of the coal-mining, metallurgy and textile industries. The rising role of the region's services, in-depth socio-economic restructuring and, modernisation was also presented through the chapter. The socio-economic results of system change affected the demographic and social situation of the Katowice conurbation (i.e. depopulation, population ageing, urban shrinkage). Moreover, important spatial and environmental problems occurred as the result of urban transformation. They include functionally degraded areas, spatial conflicts, environmental pollution and the polycentric structure of the urban space undergoing analysis.

Chapter 12 was written by Wioletta Kamińska and Mirosław Mularczyk. This chapter differs from the previously presented case studies because it is devoted to the group of peripheral cities and towns in eastern Poland. The authors presented the relationship between population growth and the migration balance. In addition, they evaluated the influence of the aforementioned elements on the process of population ageing in the cities and towns under study. The chapter also presented correlations between the population size, the location of urban areas, population ageing and population dynamics. The analysis proved that the ageing processes in the urban spaces located in the peripheral locations of Eastern Poland are more intensive than the processes ongoing in cities and towns located in other parts of the country. This is mostly the result of the negative net migration rate noted mostly among women and young people in the urban spaces undergoing investigation.

The edited volume ends with conclusions where the most important results from all individual chapters are drawn together. Moreover, the conclusions summarise the main features of a transition process from socialist to post-socialist (neo-liberal) city.

References

Apanowicz, J. (2002) *Metodologia ogólna*. Gdynia, Bernardinum.
Bański, J. (2017) The consequences of changes of ownership for agricultural land use in Central European countries following the collapse of the Eastern Bloc. *Land Use Policy*, 66, 120–130.

Buček, J., Bleha, B. (2013) Urban shrinkage as a challenge to local development planning in Slovakia, *Moravian Geographical Reports*, 21(1), 2–15.

Chaloupková, M., Kunc, J., Dvořák, Z. (2018) The creativity index growth rate in the czech republic: A spatial approach. *Geographia Technica*, 13(1), 30–40.

Champion, T., Hugo, C. (2004) *New forms of urbanization. Beyond the urban–rural dichtomy*. Ashgate, Aldershot.

Chapain, C., Stryjakiewicz, T. (Eds.) (2017) *Creative industries in Europe: Drivers of new sectoral and spatial dynamics*. Cham, Springer.

Cudny, W. (2011) Model przemian miasta postsocjalistycznego – przykład Łodzi, *Studia Miejskie*, 4, 153–159.

Cudny, W. (2012) Socio-Economic Changes in Lodz – Results of Twenty Years of System Transformation, *Geografický časopis*, 64(1), 3–27.

Eade, J., Mele, Ch. (2002) *Understanding the city: Contemporary and future perspectives*. Oxford, Blackwell.

Gorzelak, G. (1996) *The regional dimension of transformation in Central Europe*. London-New York, Routledge.

Grigorescu, I., Mitrică, B., Kucsicsa, G., Popovici, E. A., Dumitraş,cu, M. and Cuculici, R. (2012) Post-communist land use changes related to urban sprawl in the Romanian metropolitan areas, *Human Geographies–Journal of Studies and Research in Human Geography*, 6(1), 35–46.

Grzeszczak J. (2010) Gentryfikacja osadnictwa. Charakterystyka, rozwój koncepcji badawczej i przegląd wyjaśnień. Instytut Geografii i Przestrzennego Zagospodarowania im. Stanisława Leszczyckiego PAN, Warszawa.

Górczyńska, M. (2012) Procesy zmian społecznych w przestrzeni Warszawy – gentryfikacja, embourgeoisement czy redevelopment?, [in]: J. Jakóbczyk-Gryszkiewicz (ed.), *Procesy gentryfikacji w mieście Cz. I. XXV Konwersatorium Wiedzy o Mieście*, pp. 245–255, Wydawnictwo UŁ, Łódź.

Górczyńska, M. (2015) Gentryfikacja w polskim kontekście: krytyczny przegląd koncepcji wyjaśniających, *Przegląd Geograficzny*, 87(4), 589–611.

Gutkind, E.A. (1964) *Urban development in Central Europe*. Collier-Macmillan, London.

Haase, A., Grossmann, K., Steinführer, A. (2012) Transitory urbanites: New actors of residential change in Polish and Czech inner cities, *Cities*, 29(5), 318–326.

Haase, A., Steinfuhrer, A. (2005) Cities in East-Central Europe in the Aftermath of Post-Socialist Transition, *Europa*, XXI, 97.

Heitzman, J. (2008). *The city in South Asia*. Routledge, London-New York.

Hirt, S. (2013) Whatever happened to the (post)socialist city? *Cities*, 32(Suppl. 1), 29–38.

Hoff, A. (ed.) (2011) *Population ageing in central and eastern Europe: Societal and policy implications*. Ashgate, London.

Holm, A., Marcińczak, S., Ogrodowczyk, A. (2015) New-build gentrification in the postsocialist city: Łódź and Leipzig two decades after socialism, *Geografie*, 120(2), 164–187.

Hutton, J. (2008) *The new economy of the inner city*. Routledge, London-New York.

Interreg central Europe, Citizens' summary, Annex D. (2015) Interreg CENTRAL EUROPE, Vienna, DoA 01. 06.2019, document available at https://www.interreg-central.eu/Content.Node/documents/Citizen-s-summary-published.pdf

Jacobs, A. (2013) The Bratislava metropolitan region. *Cities*, 31, 507–514.

Johnson, L., Johnson, L. R. (1996) *Central Europe: enemies, neighbors, friends*, Oxford University Press, USA.

Jürgens, U. (1996) City profile Leipzig. *Cities*, 13(1), 37–43.

Kaczmarek, S. (2015). Skuteczność procesu rewitalizacji. Uwarunkowania, mierniki, perspektywy. *Studia Miejskie*, 17, 27–35.

Kavaliauskas, T. (2012) *Transformations in Central Europe between 1989 and 2012. Geopolitical, Cultural, and Socioeconomic Shifts*. Lexington Books, Lanham (Mar.).

Kazepov, Y. (2005) *Cities of Europe: Changing Contexts, Local Arrangement and the Challenge to Urban Cohesion*. Blackwell, Oxford.

Kovács, Z. (1999). Cities from state-socialism to global capitalism: an introduction. *GeoJournal*, 49(1), 1–6.

Kovács, Z., Farkas, J.Z., Egedy, T., Kondor, A.C., Szabó, B., Lennert, J., Baka, D., Kohán, B. (2019) Urban sprawl and land conversion in post-socialist cities: The case of metropolitan Budapest. *Cities*, 92, 71–81.

Kovács, Z., Wiessner, R., Zischner, R. (2013) Urban renewal in the inner city of Budapest: Gentrification from a post-socialist perspective, *Urban Studies*, 50(1), 22–38.

Krisjane, Z., Berzins, M. (2012) Post-socialist urban trends: new patterns and motivations for migration in the suburban areas of Rīga, Latvia, *Urban Studies*, 49(2), 289–306.

Kubeš, J. (2015) Analysis of regulation of residential suburbanisation in hinterland of post-socialist 'one hundred thousands' city of České Budějovice, Bulletin of Geography, *Socio-economic Series*, 27, 109–131.

Leetmaa, K., Kährik, A., Nuga, M., Tammaru, T. (2014) Suburbanization in the Tallinn Metropolitan Area, [in:] K. Stanilova and L. Sýkora (ed.), *Confronting suburbanization: Urban decentralization in postsocialist Central and Eastern Europe*, pp. 192–224, Oxford: Wiley-Blackwell.

Lewis, P. G. (2014) *Central Europe since 1945*, Routledge, London; New York.

Lin, G. C. (2004). Toward a post-socialist city? Economic tertiarization and urban reformation in the Guangzhou metropolis, China. *Eurasian Geography and Economics*, 45(1), 18–44.

Lisowski, A., Mantey, D., Wilk, W. (2014) Lessons from Warsaw: The lack of co-ordinated planning and its impacts on urban sprawl, [in:] K. Stanilov and L. Sýkora (eds.), *Confronting suburbanization: Urban decentralization in postsocialist Central and Eastern Europe*, pp. 225–255, Oxford: Wiley-Blackwell.

Liszewski, S. (2001) Model przemian przestrzeni miejskiej miasta postsocjalistycznego. In Jażdżewska, I., ed. *Miasto postsocjalistyczne – organizacja przestrzeni miejskiej i jej przemiany: XIV Konwersatorium wiedzy o mieście*, pp. 303–310. Łódź, Wydawnictwo UŁ.

Liszewski S., Marcińczak S. (2012) Geografia gentryfikacji Łodzi: studium dużego miasta przemysłowego w okresie posocjalistycznym. [in:] J. Jakóbczyk-Gryszkiewicz (ed.), *Procesy gentryfikacji w mieście. Część I. XXV Konwersatorium Wiedzy o Mieście*, pp. 71–87, Wyd. Uniwersytetu Łódzkiego, Łódź.

Lux, G., Horváth, G. (2017) *The Routledge handbook to regional development in Central and Eastern Europe*, Routledge, London-New York.

Marcińczak S. (2013) *Segregacja społeczna w mieście postsocjalistycznym*. Bukareszt,

Kaczmarek S., Marcińczak S. (2013) The blessing in disguise: Urban regeneration in Poland in a neo-liberal milieu, [in:] M.E. Leary and J. McCarthy (eds.), *The Routledge companion to urban regeneration*, pp. 98–106, Routledge, London and New York.

Misra, R.P. (2013) *Urbanisation in South Asia: Focus on Mega Cities*. New Delhi, Cambridge Univerity Press India.

Mollenkopf, J., Crul, M. (eds.) (2012) *The Changing Face of World Cities. Young Adult Children of Immigrants in Europe and the United States*. New York, Russell Sage Foundation.

Mulíček, O., Toušek, V. (2004) Changes of Brno industry and their urban consequences. *Bulletin of Geography* (socio-economic series), 3(1), 61–70.

Ondoš, S., Korec, P. (2008) The rediscovered city: a case study of post-socialist Bratislava, *Geografický časopis*, 60, 199–213.

Osman, R., Frantál, B., Klusáček, P., Kunc, J., Martinát, S. (2015) Factors affecting brownfield regeneration in post-socialist space: The case of the Czech Republic, *Land Use Policy*, 48, 309–316.

Ott, T. (2001). From Concentration to De-concentration – Migration Patterns in the Post-socialist City, *Cities*, 18(6), 403–412.

Płaziak, M. (2014) Przemiany funkcji handlowo-usługowych w mieście postsocjalistycznym na przykładzie Nowej Huty, [in:] E. Kaczmarska and P. Raźniak (eds.), *Społeczno-ekonomiczne i przestrzenne przemiany struktur regionalnych*, pp. 85–100. Kraków: Oficyna Wydawnicza AFM.

Pichler-Milanović, N. (2014) Confronting suburbanization in Ljubljana: From "urbanization of the countryside" to urban sprawl. [in:] K. Stanilov and L. Sýkora (eds.). *Confronting suburbanization: Urban decentralization in postsocialist Central and Eastern Europe*, pp. 65–96. Oxford: Wiley-Blackwell.

Prakash, G., Kruse, K.M. (2008) *The Space of Modern City. Imaginaries, Politics and Everyday Life*. Princeton, Princeton University Press.

Rink, D., Couch, C., Haase, A., Krzysztofik, R., Nadolu, B., Rumpel, P. (2014) The governance of urban shrinkage in cities of post-socialist Europe: policies, strategies and actors, *Urban Research & Practice*, 7(3), 258–277.

Rufat, S., Marcińczak, S. (2020) The equalising mirage? Socioeconomic segregation and environmental justice in post-socialist Bucharest, *Journal of Housing and the Built Environment*, 35, 917–938.

Runge, J. (2006) *Metody badań w geografii społeczno-ekonomicznej–element metodologii, wybrane narzędzia badawcze, Wyd*. Uniwersytetu Śląskiego, Katowice.

Schweiger, Ch, Visvizi, A. (2018) *Central and Eastern Europe in the EU: Challenges and Perspectives Under Crisis Conditions*. New York, Routledge.

Smith, A., Stenning, A., Rochovská, A., Świątek, D. (2008) The emergence of a working poor: labour markets, neoliberalisation and diverse economies in post-socialist cities, *Antipode*, 40(2), 283–311.

Stanilov, K. (ed.) (2007) *The Post-Socialist City. Urban Form and Space Transformations in Central and Eastern Europe after Socialism*. Dordrecht, Springer.

Stanilov, K., Hirt, S. (2014) Sprawling Sofia: Postsocialist suburban growth in the Bulgarian capital, [in:] K. Stanilov and L. Sýkora (eds.), *Confronting suburbanization: Urban decentralization in postsocialist Central and Eastern Europe*, pp. 163–191. Blackwell, Malden.

Stanilov, K., Sýkora, L. (2012) Planning markets, and patterns of residential growth inmetropolitan Prague, *Journal of Architectural and Planning Research*, 29(4), 278–291.

Steinführer, A., Bierzynski, A., Großmann, K., Haase, A., Kabisch, S., Klusáček, P. (2010) Population decline in Polish and Czech cities during post-socialism? Looking behind the official statistics, *Urban Studies*, 47(11), 2325–2346.

Steinführer, A., Haase, A. (2007) Demographic change as a future challenge in East Central Europe, *Geografiska Annaler B*, 89(2), 183–195.

Stenning, A., Dawley, S. (2009) Poles to Newcastle: grounding new migrant flows in peripheral regions, *European Urban and Regional Studies*, 16(3), 273–294.

Sikorski, D. (2019) Wybrane kierunki i aspekty przemian funkcjonalnych terenów przemysłowych we Wrocławiu w latach 1989–2016, *Prace Komisji Geografii Przemysłu Polskiego Towarzystwa Geograficznego*, 33(4), 227–240.

Sýkora, L. (2009) Post-socialist cities. In Kitchin, R., Thrift, N., eds. *International Encyclopedia of Human Geography*, pp. 387–395. Elsevier, Oxford.

Sýkora L., Ouředníček, M. (2007) Sprawling post-communist metropolis: Commercial and residential suburbanization in Prague and Brno, the Czech Republic, [in:] E. Razin, M. Dijst and C. Vazquez (eds.), *Employment Deconcentration in European Metropolitan Areas*, pp. 209–233. Dordrecht: Springer.

Szafrańska, E. (2014) Transformation of large housing estates in post-socialist Łódź (Poland), *Geographia Polonica*, 87(1), 15–24.

Szafrańska, E. (2015) Transformations of large housing estates in Central and Eastern Europe after the collapse of communism, *Geographia Polonica*, 88(4), 621–648.

Szafrańska E. (2017) Starzenie się mieszkańców wielkich osiedli mieszkaniowych w mieście postsocjalistycznym – przykład Łodzi, *Space-Society-Economy*, 20, 43–64.

Šveda, M., Madajová, M., Podolák, P. (2016) Behind the differentiation of suburbandevelopment in the hinterland of Bratislava, Slovakia, *Czech Sociological Review*, 52(6), 893–925.

Tiersky, R. (2004) *Europe Today*. Rowman & Littlefield Publishers, Lanham.

Wang, S., Zhang, Y. (2005) The new retail economy of Shanghai. *Growth and Change*, 36(1), 41–73.

Watson, S., Gibson, K. (1995) *Postmodern Cities & Spaces*. Oxford, Blackwell.

Węcławowicz G. (2013) *Transnational Development Strategy for the Post-socialist Cities of Central Europe*. Warszawa, Instytut Geografii i Przestrzennego Zagospodarowania PAN.

Węcławowicz, G. (2016) Urban development in Poland, from the socialist city to the post-socialist and neoliberal city, [in:] V. Szirmai (ed.), *"Artificial towns" in the 21st Century. Social polarisation in the new town regions of East-Central Europe*, Institute for Sociology. pp. 65–82, Centre for Social Sciences Hungarian Academy of Sciences, Budapest.

Węcławowicz, G., Wątorska-Dec, M. (2012) Kształtowanie nowego oblicza postsocjalistycznych miast Europy Środkowo-Wschodniej, *Przegląd Geograficzny*, 84(4), 639–647.

Węcławowicz, G. and Wątorska-Dec, M. (2013) Polepszanie warunków życia w miastach Polski i Europy Środkowej i w ich postsocjalistycznych dzielnicach według koncepcji projektu ReNewTown, *Studia Miejskie*, 12, 35–43.

Young, C., Kaczmarek, S. (2008) The Socialist Past and Postsocialist Urban Identity in Central and Eastern Europe: The Case of Łódź, Poland, *European Urban and Regional Studies*, 15(1), 53–70.

2 Post-socialist urban change and its spatial patterns

The case of Nitra

Vladimír Ira and Martin Boltižiar

Introduction

Post-socialist cities are cities under transformation and their urban landscapes formed under decades of socialism are being adapted to new conditions shaped by the political, economic, social and cultural transition to capitalism (Sýkora 2009). In recent decades, post-socialist cities and their societies have experienced a dramatic political, economic, social and cultural change. Trends and patterns of change embedded in the overall process of transformation have profoundly influenced the spatial adaptation and repositioning of post-socialist cities. The unique development and transformation challenges in post-socialist cities were linked to the changes in the external environment (national and global), which were much more dramatic and revolutionary, and to changes in the internal environment (both the urban system and the city itself). The impact of transitions to democracy (systemic political change), to markets (systemic economic change) and to decentralised systems of governance on cities triggered the rise or fall of certain cities, and led to specific responses to the processes of change in several domains (areas). Sýkora and Bouzarovski (2012) understand the post-socialist transition as a broad, complex and lengthy process of social change, which proceeds through a multitude of particular transformations. They identify three transitions: institutional reconfigurations; transition in the domain of social organisation; and the practice and reconfiguration of the urban landscape.

Post-socialist urban change has created a mosaic of diverse urban experiences. Tsenkova (2008) claims that despite the diversity, the spatial transformation has three principal dimensions: (1) new spaces of production/consumption; (2) social differentiation in residential spaces associated with growing inequality; and (3) conflicts and selective urban development associated with new models of governance and institutional transformation.

Musil (2005) stresses that the observable urban changes in post-socialist cities can be linked to the synergy of several causal factors, including privatisation and the restitution of property, the reintroduction of a market for property and land, the de-industrialisation of cities, a growing presence of foreign investors in industry as well as in services, successive commodification of housing, including deregulation of rents, growing income differences, the weakening of the welfare state system and of urban public transport systems, the decentralisation of city

DOI: 10.4324/9781003039792-2

governments, increased stress on environmental quality, some liberalisation of immigration policies, the individualisation of values and the increasing plurality of lifestyles.

The reshaping of the townscape, the physical and functional transformation of the urban space, the comeback of the importance of land rent, and an increasing number of actors competing for space, renaissance of self-government, the increase of social and spatial differentiation, the transformation of employment structure, are from the perspective of Matlovič (2004), the most important general trends in the transition cities.

An important publication, *The Post-Socialist City* edited by Stanilov (2007), analysed fifteen years after the beginning of the transition period and pointed out that the numerous changes that have taken place in the way the urban space is produced and restructured in the post-socialist CEE cities have had both a positive and a negative impact on the built environment and the quality of life of its residents.

The main undergoing changes in the former socialist cities of Central and Eastern Europe and relatively reliable account of the transformation processes can be found in several handbooks, monographies and articles. There exist several contributions which try to analyse and explain in a complex way the general features of post-socialist cities and various aspects of their transformation. Several studies (e.g. Andrusz et al. 1996, Enyedi 1998, Kovács 1999, Musil 2005, Tsenkova 2008, Sýkora 2009, Sýkora and Bouzarovski 2012) explore the main features of the transition to democracy, markets and decentralised governance, and on the characteristics of multilayered processes of spatial transformation in post-socialist cities. Many authors (among others, for example Węcławowicz 1997, Matlovič et al. 2001, Cudny 2006, 2012, Stanilov 2007, Ondoš and Korec 2008, Young and Kaczmarek 2008, Matlovič and Nestorová-Dická 2009, Kabisch et al. 2010, Nae and Turnock 2011, Haladová and Petrovič 2017, Kovács et al. 2019) tried to define the most important demographic, social, economic, functional, morphological and land-cover changes which took place in post-socialist cities as a result of the political and economic transformation after 1989. Hamilton et al. (2005), in their publication dedicated to transformation of cities in Central and Eastern Europe, analysed inter- and intra-urban transformation of a number of capital cities: Berlin, Warsaw, Budapest, Prague, Ljubljana, Sofia, Riga, Tallinn, Vilnius, and Moscow.

Thanks to the rich literature on post-socialist urban transformation we can have quite an interesting picture of new inter-urban as well as intra-urban processes. A relatively complex picture is provided by an analysis of more than 180 articles concerning the urban geography aspects of European post-socialist cities and their near hinterland, published between 1990 and 2012 in the international journals (Kubeš 2013). According to this analysis, the most frequent article topics include the socio-spatial structure of the city and its transformation, followed by urban planning and management in the city and suburbanisation and urban sprawl. A smaller number refers to physical spatial structure, housing structure and functional spatial structure of the city and its transformation.

Whereas traditional approaches to the analysis of urban change are mostly concerned with measuring, mapping and classifying objective characteristics, the behavioural approach applied in a few studies is in contrast. It is subjective, with an emphasis on studying the urban change as it seems to be rather than as it is (Young and Kaczmarek 1999, Ira 2003, Kunc et al. 2014). These studies portray the urban change as it is seen through the prism of personal experience.

In order to assess the post-socialist urban change in Nitra more effectively, some of the studied elements were compared to those in some Central European cities (Ira 2003, Cudny 2006, 2012, Stanilov 2007, Tsenkova 2008, Matlovič and Nestorová-Dická 2009, Kubeš 2013, Kovács et al. 2019). The analysis of urban change was based on the quantitative statistical data from the Statistical Office of the Slovak Republic and data published by authors dealing with various aspects of transformation in post-socialist Nitra. It was supplemented with qualitative data from several case studies used for the analysis of changes in the physical, functional and socio-demographic structure of the city. The basic source of information was the literature regarding the city, its transformations and urban land-cover change after 1989 (Bugár et al. 2008, Trembošová 2009, Repaská and Bedrichová 2013, Haladová and Petrovič 2015, 2017, Korec and Popjaková 2019a, 2019b). The information necessary to write this chapter was also obtained by means of field research carried out in Nitra.

The basis for the land-cover maps for the years 1990 and 2019 were coloured orthophotos from the companies Eurosense and Geodis, s.r.o. Bratislava, with 1 m pixel resolution. Their interpretation was realised by analogue vector digitisation in computer environment using ArcGIS software. The map legend was based on the CORINE Land Cover classification level 3 (Feranec and Oťaheľ 2001). A supplementary source of information was a field reconnaissance survey aimed at verifying the current state of the land-cover areas obtained by the interpretation of current orthophotos.

In our study, we focused on the interpretation of land-cover changes and its spatial structure on the example of the cadastral area of Nitra City (the study area was delimited by cadastral boundaries before 1990 when the city had the largest area in history—146 km²), especially in relation to anthropogenic pressures on the landscape. The identified changes are mainly the result of human-driven changes in land use. In this sense, it is possible to classify individual types of changes and interpret them according to the nature and intensity of their impact as the aforementioned anthropogenic pressures. These are then understood as the main drivers of changes in the study area. This principle is used as a first step in DPSIR analyses (driving forces, pressures, state, impacts, responses) in ecological and also geographic modelling (Petit et al. 2001). The local pressures (the expansion of urban built-up areas, the intensification of agriculture, drainage, the abandonment of agricultural land, afforestation and deforestation) are related to changes in land use and the resulting fragmentation of individual classes of land cover.

The first aim of this study is to present a comprehensive summary of the processes of urban changes taking place in Nitra after the fall of communism. The study identifies the factors that have initiated the transformation of economy and society, giving a comprehensive presentation of transformation processes and their effects on the contemporary physical, functional, socio-demographic and spatial structure of post-socialist cities. The second aim is to analyse the land cover in 1990 and 2019 and to interpret land-cover change and its spatial differentiation on the example of the cadastral area of Nitra City, especially in relation to anthropogenic pressures on the urban landscape. It will enable to show how current demographic, social, economic and political processes shape the structure of post-socialist city and its landscape.

The study area

The city of Nitra is the administrative, industrial and cultural centre of the Nitra region in southwest Slovakia (Figure 2.1). It is also an important road junction, educational and scientific centre with two universities. Nitra lies along the Nitra River, 85 km northwest of the capital Bratislava. The cadastral area of the city currently covers 100.45 km². The altitude ranges from 138 m to 587 m. The city of Nitra consists of 13 urban neighbourhoods. The territory of the city is situated on the border of the Pannonian Basin (Danubian Plain) and the Carpathian System (Tribeč Mountain) with a share of different landscape types (Figure 2.2). Nitra has the earliest written evidence of its existence in all of Slovakia; it was referred to in 828 as Nitrava. At that time, it already was the seat of the ruler of the Nitra

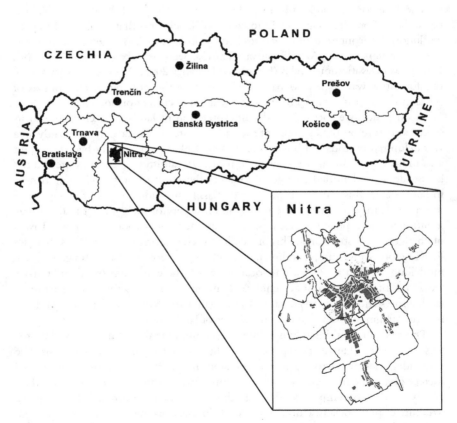

Figure 2.1 Study area of the Nitra City and its location within Slovakia.
Author: M. Boltižiar created by ArcGIS.

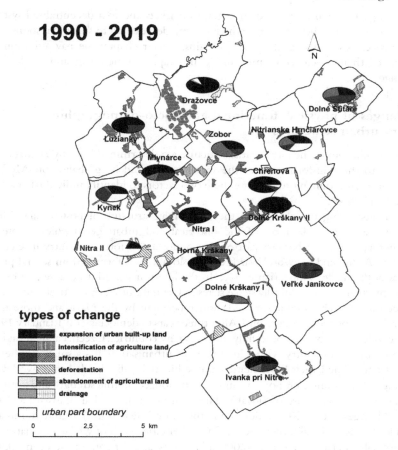

Figure 2.2 Changes in land cover between 1990 and 2019.
(Authors: M. Boltižiar, G. Bugár by ArcGIS)

Principality. Later, it became a stronghold and religious centre. In the 13th century it became a landlord's town of the bishop and thus remained an important centre of both education and the church. Town privileges were acquired in 1248. In the middle of the 19th century, the population approached 10,000 inhabitants, and in the second half of the 19th century industrial enterprises (distillery, soap production, steam mill, and machinery) were established. The development of the city accelerated after the foundation of the Czecho-Slovak Republic. In March 1945 the city was subjected to heavy bombing, which led to the destruction of a large proportion of the buildings in the historic centre. In the second half of the 20th century Nitra expanded to large housing estates, new industrial plants, school buildings and the exhibition area Agrokomplex. With the population of 78,353 inhabitants (as of 31 December 2019), the city of Nitra currently ranks as the sixth-largest city in Slovakia, with a population density of 770 inhabitants/km^2. In recent years, however, the population has stagnated or has even started to decline.

The impact of transitions to democracy, to markets and to a decentralised system of governance on the city during the last three decades led to specific responses to the processes of change in several domains. In our chapter we pay attention to three of them: socio-economic and demographic; functional; and land-cover change.

Changes of physical, functional and socio-demographic intra-urban structures

Current changes in the city's spatial structure are determined both by factors connected to the transformation of the social system and by globalisation. Methodologically, it is difficult to separate these two factors to identify individual changes (Matlovič 2004).

Understanding the changes in the intra-urban structure of post-socialist Nitra requires a brief explanation of the historical and urban geographical contexts. A century ago, Nitra became part of Czechoslovakia, a former country in Central Europe. At the end of World War I, Czechoslovakia was formed from several provinces of the Austro-Hungarian empire. In the interwar period it became one of the most prosperous and politically stable states in Central Europe. A multi-ethnic democratic state, Czechoslovakia survived dismemberment by the Nazis and more than four decades of communist regime. After three years of democracy, on 1 January 1993, Czechoslovakia separated peacefully into two new countries, Czechia and Slovakia.

Slovakia, as a country with a lower rate of urbanisation, has a small number of larger cities. The largest Slovak cities can be hierarchically divided into two levels, taking into account their population size, function, economic strength and position in the regional structure of Slovakia. The position of the capital Bratislava and the second-largest city of Košice (located in the east of the country) is quite dominant; at a lower hierarchical level there are four other cities, each of whose population in the socialist period exceeded 80,000 inhabitants, and for some time even 90,000 inhabitants. The processes of independence (administrative segregation) of peripheral rural municipalities and the process of suburbanisation have caused the population of these cities to decrease quite significantly after 1989, and at present only the city of Prešov is approaching a population of 90,000 inhabitants. Žilina, Banská Bystrica and Nitra all have less than 80,000 inhabitants. Demographic development in the largest Slovak cities, in which less than 1.0 million inhabitants live, i.e. approximately one-fifth of the population of the Slovak Republic. At the same time, the population of these cities is declining relatively rapidly, in contrast to the developments up to 1989, when socialist industrialisation and urbanisation caused a rapid increase in the population of (not only) this group of cities. In these cities, significant human capital has historically been, and still is, concentrated. Traditionally, these cities have a favourable educational structure, a significant share of non-productive and higher value-added sectors, commercial and non-commercial services, and a substantial part of the country's capacities of university education, research and development. The demographic development of these cities also deserves special attention in this respect. Equally important is the fact that cities, not only from a social, economic and cultural point of view, but also from a demographic point of view, are the initiators of change, the nuclei from which change and innovation are diffused in space (Šprocha et al. 2016).

Post-socialist change in Bratislava is most influenced by globalisation, a significant strengthening of quaternary functions, the position of the capital of an independent state and significant changes in the lifestyle of the population. The change in other major cities has been influenced by the strengthening of universities, research institutions and regional self-government functions. In Žilina and Nitra, the construction of large plants in the automotive industry also played an important role. A negative migration balance and a decline of fertility rate have led to the decrease of population in Nitra. The population has been redistributed from the city to the suburban hinterland. Residential suburbanisation has expanded in the majority of the hinterland (former rural settlements). A decrease in the city population, combined with an ageing population, can be considered as the most alarming issue. Looking at the shrinkage in Nitra, it becomes obvious that currently the city has been affected by population loss in a moderate but constant way (87,569— as of 31 December 1996, 83,692—as of 31 December 2009 and 78,353 as of 31 December 2019). Shrinkage represents a real challenge for a city, its inhabitants and decision-makers (Buček and Bleha 2013). If a city loses population, some of the dynamics of urban development begin to change. A new situation arises for a lot of urban functions (e.g. demand for public transport, demand for service infrastructures for the elderly and young generations, demand on the housing market, the investment decisions of companies, owners and developers, resources at hand of urban governance).

The changing social structure of the society and increasing inequality in Nitra were reflected in differentiated forms of housing. Selected segment of population moved to several locations within the city in order to accentuate their private property and social status. These are mainly enclosed and locked residential areas inhabited by residents who place an emphasis on their status and security. Thus a part of the public space in Nitra has, in recent years, been increasingly replaced by explicitly private or quasi-public spatial forms (gated communities, exclusive suburbs, new or gentrified urban districts, etc.) that are easily accessible only to certain segments of the population.

Despite the fact that industry has played a very important role in the modern history of Nitra, from the second half of the 19th century, the city was not an important industrial centre of Slovakia for a long time. Nitra was one of the cities that became important industrial centres in the period 1948–1989. Until then, Nitra paid more or less for an unfavourable location in the Slovak railway network. Prior to 1989, plants in the food industry, the chemical industry, and the engineering and electrical engineering industries were represented in Nitra. The building materials industry, the woodworking industry, the textile and clothing industry and the printing industry were also of great importance. The process of transformation of industry in Nitra after 1989 had all the features of the transformation of Slovak industry. There have been significant changes in the ownership, sector, size and, of course, in the spatial structure, its distribution within the city. The industrial structure changed significantly after 1989, which was closely related to the restructuring process in general, but also to the establishment of new companies. The development of industry in Nitra after 1989 brought significant changes in a number of localities in the city. Industrial development in the city was concentrated in three types of locations: in two industrial parks (larger integrated investment projects), in three production clusters and in ten equipment-production clusters. Paralleled by de-industrialisation and the shift to service-based urban economy, new industrial development has been directed to greenfield projects. The construction of industrial parks and the concentration of industrial plants in them, and especially the construction of the Jaguar Land Rover Slovakia car plant, gradually require a significant change in the road network and other infrastructure elements. The year 2018, when the Jaguar Land Rover (JLR) automobile plant started operations in Nitra, can be considered the end of the qualitative development of the industry in this city. Nitra, with the arrival of Jaguar Land Rover, is finally ranked among the key industrial centres in Slovakia (Korec and Popjaková 2019a). JLR claimed to be attracted to the area by a number of factors: tariff-free access to the EU; productive, but relatively lower-wage workers; a strong supply base; low-cost logistics; and upgraded infrastructure (Jacobs 2017).

Retailing and its activities, together with services, are in the long term the most significant vehicles for change, new trends and innovations in the tertiary sector of Nitra. The structure of the retail sector has been transformed considerably across city in the three most recent decades. Consumption patterns in the 1990s, especially in the urban environment of transition countries, were accepted with unusual dynamics. Rapid adaptation of Western norms by the markets was supported by the entry and dominance of Western capital (Kunc and Križan 2018).

Until 2002, the most important new shopping centres in the city built on a greenfield site were concentrated on the periphery (Hypermarket Tesco, Metro Cash & Carry and Billa), taking into account accessibility criteria, land rent, approaching to the consumer in the residential area, and parking requirements. After 2002, the shopping centres (OC Centro Nitra and ZOC Max) were concentrated in the housing estate Chrenová, where they densified the housing development. Centro Nitra grew on an open area used as a parking space, ZOC Max was built in a locality, which, according to the original zoning plan, was referred to as a sports zone (Trembošová 2009). The new shopping centre Mlyny Gallery, with its unique appearance, has become a new landmark in the city centre.

The adaptation of the Slovak higher education system to neoliberal principles resulted in qualitative and quantitative changes and has led the individual higher education institutions to make an effort to increase their performance and efficiency (Matlovič and Matlovičová 2017). After the initial phase of extensive development of higher education institutions and their faculties in Nitra (in the 1990s) followed the phase of consolidation caused by the demographic stagnation or decrease, increasing competition of higher education institutions in Slovakia and abroad and pressure on quality and the performance of research. Currently, there are two university-type institutions in Nitra. The older of the two, the University of Agriculture in Nitra, was established by the Czechoslovak government in 1952. In 1996, the University of Agriculture was renamed the Slovak University of Agriculture in Nitra (SUA). At present, SUA is a top Slovak research and educational institution of international importance (5,707 students as of 31 October

2019), operating in a wide range of scientific fields, including genetics, biotechnology, agriculture, environment, food sciences, horticulture, landscape engineering and design, engineering, economics, management, marketing with a focus on rural regions and regional development. Constantine the Philosopher University in Nitra (UKF) was established in 1996. UKF is the fourth-largest public university in Slovakia formed as a modern European general university. A total of 6 998 students (as of 31 October 2019) study at the Faculty of Arts, Faculty of Central European Studies, Faculty of Education, Faculty of Natural Sciences and Faculty of Social Sciences and Health Care. The uniqueness of UKF in education is represented by teacher training programmes in Hungarian and the training of teachers, social workers and public education for Roma ethnicity. Among other things, significant structural changes in secondary and tertiary education can be expected, coupled with significant changes in industry, especially in the automotive sector.

Apart from the development of higher education institution in Nitra, the advent of the 21st century brought a new wave of development processes based on creativity and innovation, improving the position of the city within creative industries in Slovakia.

The internationalisation of higher education, science and research, as well as foreign investment in trade, but especially in the automotive industry, has created a new phenomenon that the city will have to focus on. A small group of foreign managers and high-salaried employees of foreign companies formed a specific segment of new inhabitants of the city. By contrast, labour immigrants constitute the lower tier of the socio-economic hierarchy in some parts of the city which could lead to the formation of new ethnic enclaves. It is a case of the ever-increasing presence of foreign workers (mainly from Ukraine, Serbia, Romania, Albania and other countries). Nitra is gradually becoming a multi-ethnic and multicultural city, where, in addition to the historical ethnicities of the city, Slovak, Hungarian and Roma, these foreigners are increasingly visible (Korec and Popjaková 2019b).

Summary of post-socialist changes in Nitra's urban structures over the last three decades is provided in the overview Table 2.1.

Urban land-cover change analysis

Land cover in the year 1990

In 1990, at the beginning of transformational changes, the largest area was covered by the class of non-irrigated arable land, which, in the form of large-block fields, accounted for more than half of the study area (54%). Mosaics of fields, meadows and permanent crops were located along the edges of settlements (4%). Vineyards as a part of agricultural land accounted for 2% of the area. They were found especially at the foothills of Tribeč mountain on the south-southeastern slopes. The grasslands, mainly meadows, covered forest-free parts of Tribeč mountain, as well as smaller areas in a flat territory. Up to 15% of the area was occupied by forests, mostly deciduous, less mixed and coniferous. They mainly covered the Zobor massif. Woodcuttings were represented by a very small proportion (below 1%). Up to 14% of the area was covered by urban fabric, mainly areas of concentrated

Table 2.1 Three areas of post-socialist urban change in Nitra

Areas of urban change	Characteristics of post-socialist urban change
Changes of physical (morphological) structure	– increasing morphological diversity of built-up areas, new architecture – physical decline of large-scale housing estates in 1990s – revitalization of prefabricated panel houses in last two decades – construction of gated communities – new constructions on the territory of inner city (commercial, business and residential) – extension of built-up area induced by construction of roads, industrial plants, newly built single-family houses and multi-apartment houses (especially on city edges)
Changes of functional structure	– significant development of residential function especially in suburbs – deindustrialszation (traditional industries) and later dynamic development of new industrial sites – commercialisation of city centre and city sub-centres (arrival of department stores, banks, business services, transformation of retailing) – conversion of apartments into offices and business in the city centre – new functional utilisation of brownfields – tertiarisation and quarterisation, increasing importance of higher education and research sectors – modernisation of technical and transport infrastructure – weakening of agricultural function in suburban hinterland
Changes of socio-demographic structure	– changing daily life of residents as an impact of urban restructuring – differentiations of quality of life in the city neighbourhoods – new character of daily mobility especially in suburbs – declining number of residents in the city ('urban shrinkage') – gradual ageing of population in the city (especially in large-scale housing estates) – increasing number of residents in suburbs during last two decades and its influence on the composition of residents – significant transformation of employment structure by economic sectors (especially significantly increased proportion of services) – increasing number of persons employed in creative industry – arrival of new residents to some inner parts and displacement of original residents – gradually formed small gated communities – partial deterioration of social structure in originally socialist housing estates – increasing social inequalities and social segregation of the population – higher incidence of socially marginalised population groups – increasing number of foreigners (employment in automotive industry)

settlement fabric (13%), which was represented mainly by built-up areas in in the outskirts of the city (where the substantial part consists of single-family houses), but also by typical housing estates with 3 to 14 storeyed buildings with shops and services in Chrenová, Klokočina, Čermáň and Diely (1%). Concentrated built-up areas of prevailingly residential or multifunctional character, and historical core are typical for the city centre (1%). Industrial, commercial and transport units were situated mainly on the outskirts of the city but also within and made up 7% of the area. Settlement vegetation (1%) was located mainly around the city centre (city park). Watercourses (mainly the regulated flow of the Nitra River) and small water bodies had a minimum proportion (less than 1%) in terms of their share.

Land cover in the year 2019

The general process of urban transformation (after 1989) concerns all major cities in Slovakia. A specific phenomenon is the construction of shopping and logistics centres in the hinterland of larger cities. There was also an increase in individual house construction within their hinterland. Similar to the Nitra City (Jaguar Land Rover) also in the hinterland of other cities, the house construction was associated with the construction of large industrial areas, especially, of the automotive industry: Bratislava—Volkswagen, Trnava—PSA Peugeot-Citroen and Žilina—KIA Motors (Feranec et al. 2018).

Compared to 1990, urban development has stabilised at 20% of the area (built-up areas and settlement vegetation together). The R1 expressway was built and also, as a result of suburbanisation processes, the area of residential development in the surrounding municipalities forming the urban parts (Štitáre, Janíkovce, etc.) has increased. The most significant change, however, is the significant increase in the industrial site located on the northwest outskirts of the Dražovce urban area. This is related to the newly built large site of the British Jaguar Land Rover automobile plant, which was put into operation in 2017. Its construction significantly influenced the landscape of the city, as can be viewed from the Zobor hill.

The area of arable land decreased to 49%, mainly due to housing construction, while the share of forests remained unchanged (15%). Map legends from 1990 to 2019 were reclassified to CORINE Land Cover classes for the purpose of interpreting each type of change according to the methodology used.

The relatively short, almost 30-year period from 1990 to 2019 within the Nitra cadastral area shows 12% of classified changes, with the largest type being the expansion of urban built-up areas (7%). Overall, 88 % of the area remained unchanged (Figure 2.2). The most significant changes in the built-up and developed areas occurred after 1990. On the one hand, it was the result of spatial development of industry and, on the other hand, it was caused by housing development, mostly in the suburban areas. The Old Town Centre (Nitra I) has undergone this rapid development earlier, especially at the beginning of the 20th century. Together with the above-mentioned housing estates and industrial areas, the share of built-up area in the urban neighbourhood Nitra I is up to 80%. Arable land is the most widespread land use in almost all cadastral areas, with the exception of urban neighbourhoods lying on the boundary of Zobor hills and lowland (Zobor, Dolné

Štitáre, Nitrianske Hrnčiarovce and Dražovce). Forests cover most of the slopes of the Zobor hills (Dražovce, Zobor, Nitrianske Hrnčiarovce, and Dolné Štitáre) and their area did not change significantly. In the lowland part, originally large forest areas were deforested already in the earlier historical period.

This analysis of land use changes is an essential element of the ecological model based on the DPSIR assessment principle. This methodology analyses the causal relationships between the state of the land use, pressures that caused it, drivers of these pressures (socio-economic activities), impact of these changes on the landscape or selected landscape feature (e.g. biodiversity) and potential response (in decision-making and direct activities).

Figure 2.2 illustrates the changes in land cover of Nitra City in the studied years 1990 and 2019. Map shows spatial distribution of individual types of changes. Circle diagrams are showing the mutual ratio of classified types of changes (in terms of anthropogenic processes) in the cadastral areas of fifteen urban neighbourhoods.

Summary

The "past" continues to survive within the post-socialist cities. In spite of the spontaneous development, various functional, social, economic, cultural and land use transformations this past survives in the city's appearance, in its principal spatial structures, in the main features of urban development, and in the everyday practices.

Three decades of post-socialist urban geography show that it is possible to approach the post-socialist change, or "transition", through various perspectives. In our chapter devoted to post-socialist Nitra, we tried to point out, on the one hand, the basic transformation changes of physical, functional and socio-demographic character, on the other hand we analysed the basic characteristics of land cover changes within the city territory.

The decades of post-socialist transformation significantly shaped the urban spatial organisation. Urban development, sometimes decline, and restructuring were conditioned by new principles and mechanisms based on democratic policy-making. The spatial structure of post-socialist Nitra contains new elements of urban landscapes, however significant and relatively large urban sections still resemble the socialist-era city. In accordance with Sýkora (2008), we state that an important part of the core transformations of political and economic systems were accomplished within few years, but changes in the city structure have been going on for several decades and will continue for years to come. Post-socialist Nitra is a city in transition. Despite the shrinking process it is characterised by the dynamic developmental changes. The urban environment formed by four decades of the previous system is being adapted and modified to new political, social, economic, and cultural conditions. The post-socialist urban development is gaining more important impacts on the overall urban organisation, although there are still remaining some socialist patterns in the new urban landscape.

Land-cover change in Nitra due to urban restructuring and urban sprawl have clearly changed the image of the city (Haladová and Petrovič 2017). The second stage of transformation, when spatial development became characterised by mass movements of people and jobs from the core city to the suburbs generated urban sprawl. The land-cover change data showed that a gradual shrinkage of natural and agricultural surfaces took place in the period between 1990 and 2019. This was mainly the result of processes affecting the location of infrastructure, industry and housing. It was also shown by our data that during the last three decades the tempo of land-cover conversion has been significant, especially in the wider peri-urban zone.

Transforming and changing socialist past is visible within the city. Nitra as regional centre has attracted human resources, changed the employment structure, mainly in the tertiary and quaternary sectors, reorganised production and non-production activities and directed functional and spatial development into new model of post-socialist of urban environment. This chapter provided brief evidence on trends and processes of change in the post-socialist city. The complexity of that change is equally important for the future research agenda, as are the major problems of the multiplicity of interrelated economic, social, institutional and spatial processes of the current phase of transformation (Tchenkova 2008). The study of the post-socialist city today is hardly possible without an interdisciplinary framework. Applying different approaches and critical reflections on the post-socialist city (including geographical ones) is likely to bring new views and concepts of changing urban environment in the future.

The results of our analysis show many similarities and some differences between Nitra and other Slovak cities (or smaller Central Eastern European cities) during the process of their intensive intra-urban transformation in the 1990s from "socialist" to "post-socialist" cities. To some extent the impact of globalisation, and to a large extent European integration, and the internationalisation of the Slovak economy and society, together with developmental policies and regulations, have all had profound effects on the inner urban structure. Major policy changes and initiatives are needed in Nitra, which should improve its competitiveness in Slovak and Central Eastern European city networks, while preserving sustainability and the quality of life for its local citizens.

Funding

This paper was supported by Scientific Grant Agency of the Ministry of Education, science, research and sport of the Slovak Republic and the Slovak Academy of Sciences (projects VEGA No 2/0024/21 and No 1/0880/21) and project APVV-18-0185.

References

Andrusz, G., Harloe, M., & Szelenyi, I., eds. (1996). *Cities After Socialism: Urban and Regional Change and Conflict in Post-Socialist Societies.* Oxford: Wiley-Blackwell.

Bugár, G., Petrovič, F., Hreško, J., & Boltižiar, M. (2008). Krajinnoekologická interpretácia zmien druhotnej krajinnej štruktúry mesta Nitra. In V. Herber (ed.). *Fyzickogeografický sborník 6: Fyzická geografie a trvalá udržitelnost* (pp. 104–110). Brno: Přírodovědecká fakulta MU v Brně a Česká geografická společnost.

Buček, J., & Bleha, B. (2013). Urban shrinkage as a challenge to local development planning in Slovakia. *Moravian Geographical Reports*, 21(1), 2–15.

Cudny, W. (2012). Socio–economic changes in Lodz– the results of twenty years of system transformation. *Geografický časopis*, 64 (1), 3–27.

Cudny, W. (2006). Socio-economic changes in medium-sized towns in Poland during the transformation period. *Folia geographica* 10, 53–58.

Drgoňa, V. (2004). Assessment of thé landscape use changes in the city of Nitra. *Ekológia (Bratislava)*, 23 (4), 385–392.

Enyedi, G., ed. (1998). *Social change and urban restructuring in Central Europe*. Budapest: Akadémiai Kiadó.

Feranec, J., & Oťaheľ, J., (2001). *Krajinná pokrývka Slovenska*. Bratislava: Veda.

Feranec, J. Oťaheľ, J., Kopecká, M., Nováček, J., & Pazúr, R. (2018). *Krajinná pokrývka Slovenska a jej zmeny v období 1990–2012*. Bratislava: Veda.

Haladová, I., & Petrovič F. (2015). Classification of land use changes (model area: Nitra town). *Ekologia (Bratislava)*, 34(3), 249–259.

Haladová, I., & Petrovič, F. (2017). Predicted development of the city of Nitra in southwestern Slovakia based on land cover-land use changes and socio-economic conditions. *Applied Ecology and Environmental Research*, 15(4), 987–1008.

Hamilton, F. E. I., Dimitrovska-Andrews, K., & Pishler-Milanović, N., eds. (2005). *Transformation of cities in central and eastern Europe*. Tokyo: United Nations University Press.

Ira, V. (2003). The changing intra-urban structure of the Bratislava city and its perception. *Geografický časopis*, 55(2), 91–107.

Jacobs, A. J. (2017). *Automotive FDI in emerging Europe: Shifting locales in the Motor Vehicle Industry*. London: Palgrave Macmillan.

Kabisch, N., Haase, D., & Haase, A. (2010). Evolving reurbanisation? Spatio-temporal dynamics as exemplified by East German City of Leipzig. *Urban Studies*, 47(5), 967–990.

Korec, P., & Popjaková, D. (2019a). Priemysel v Nitre, od parného mlynu Arpád, cez Plastiku, n. p., k Jaguar Land Rover Slovakia. *Acta Geographica Universitatis Comenianae*, 63(1), 103–134.

Korec, P., & Popjaková, D. (2019b). *Priemysel v Nitre: globálny, národný a regionálny kontext*. Bratislava: Univerzita Komenského v Bratislave.

Kovács, Z. (1999). Cities from state-socialism to global capitalism: an introduction. *GeoJournal*, 49(1), 1–6.

Kovács, Z., Farkas, Z. J., Egedy, T., Kondor, A. C., Szabó, B., Lennert, J., Baka, D., & Kohán, B. (2019). Urban sprawl and land conversion in post-socialist cities: The case of metropolitan Budapest. *Cities*, 92, 71–81.

Kubeš, J. (2013). European post-socialist cities and their near hinterland in intra-urban geography literature. *Bulletin of Geography. Socio-economic Series*, 19, 19–43.

Kunc, J., & Križan F. (2018). Changing European retail landscapes: New trends and challenges. *Moravian Geographical Reports*, 26(3), 150–159.

Kunc, J., Navrátil, J., Tonev, P., Frantál, B., Klusáček, P., Martinát, S., Havlíček, M., & Černík, J. (2014). Perception of urban renewal: reflexions and coherences of socio-spatial patterns (Brno, Czech Republic). *Geographia Technica*, 9(1), 66–77.

Matlovič, R. (2004). Tranzitívna podoba mesta a jeho intraurbánnych štruktúr v ére postkomunistickej transformácie a globalizácie. *Sociológia*, 36(2), 137–148.

Matlovič, R., Ira, V., Sýkora, L., & Szczyrba, Z. (2001). Procesy transformacyjne struktury przestrzennej miast postkomunistycznych (na przykładzie Pragi, Bratysławy, Olomuńca oraz Preszowa). In Jażdżewska, I., ed. *Miasto postkomunistyczne – organizacja przestrzeni miejskiej i jej przemiany. Konwersatorium wiedzy o mieście, 14* (pp. 9–21). Lódź: Uniwersytet Łódźski.

Matlovič, R., & Nestorová-Dická, J. (2009). The city of Košice in the context of post-socialist transformation. *Questiones Geographicae*, 28B(2), 45–56.

Matlovič, R., & Matlovičová, K. (2017). Neoliberalization of the higher education in Slovakia: a geographical perspective. *Geografický časopis*, 69(4), 313–337.

Musil, J. (2005). Why socialist and post-socialist cities are important for forward looking urban studies. Paper presented at *conference "Forward Look on Urban Science"*, Helsinki, 26–28 May, 2005.

Nae, M., & Turnock, D. (2011). City profile: The new Bucharest: Two decades of restructuring. *Cities*, 28(2), 206–219.

Ondoš, S., & Korec, P. (2008). The rediscovered city: a case study of post-socialist Bratislava. *Geografický časopis*, 60(2), 199–2013.

Petit, S., Firbank, L., Wyatt, B., & Howard, D. (2001). MIRABEL: Models for Integrated Review and Assessment of Biodiversity in European Landscapes. *AMBIO: A Journal of the Human Environment*, 30(2), 81–88.

Repaská, G., & Bedrichová, K. (2013). Prejavy rezidenčnej suburbanizácie v mestských častiach Nitry – Klokočina a Kynek. *Geografické informácie*, 17(1), 75–92.

Stanilov, K., ed. (2007). *The post-socialist city: Urban form and space transformations in Central and Eastern Europe after socialism*. Dordrecht: Springer.

Sýkora, L (2009). Post-socialist cities. In R. Kitchin & N. Thrift (eds). *International encyclopaedia of human geography*, Vol. 8. (pp. 387–395). Oxford: Elsevier.

Sýkora, L., & Bouzarovski, S. (2012). Multiple transformations: Conceptualising the post-communist urban transition. *Urban Studies*, 49(1), 43–60.

Šprocha, B., Vaňo, B., Jurčová, D., Pilinská, V., Mészáros, J., & Bleha, B. (2016). *Demografický obraz najväčších miest Slovenska*. Bratislava: Infostat.

Trembošová, M. (2009). Nitra – mesto obchodných centier. *Geografické štúdie*, 1, 69–79.

Tsenkova, S. (2008) Managing change: the comeback of post-socialist cities. *Urban Research & Practice*, 1(3), 291–310.

Węcławowicz, G. (1997). The changing socio-spatial patterns in Polish cities. In Kovács, Z., and Wiessner, R., eds. Prozesse und Perspektiven der Stadtentwicklung in Ostmitteleuropa. *Münchener Geographische Hefte*, 76. (pp. 75–82). Passau: L.I.S. Verlag.

Young, C., & Kaczmarek, S. (1999). Changing the Perception of the Post-Socialist City: Place Promotion and Imagery in Łódź, Poland. *The Geographical Journal*, 165(2), 183–191.

Young, C., & Kaczmarek, S. (2008). The socialist past and postsocialist urban identity in Central and Eastern Europe. The case of Łódź, Poland. *European Urban and Regional Studies*, 15(1), 53–70.

3 The university as the creative hub

The case of the city of Olomouc after 1989

Zdeněk Szczyrba, Irena Smolová, Martin Jurek and David Fiedor

Introduction

Universities have played an important role in serving as centres of innovative research as early as their first period of accelerated growth during the Early Modern period. The number of European universities rose from 29 in AD1400 to 73 in AD1625 and these institutions had a strong influence on both European religion and society in the periods of the Renaissance and the Reformation (Grendler 2004). Subsequent development of the modern society was shaped by industrial revolution, leading to urban growth and changes in the structure of workforce, with an increasing demand for highly skilled and educated workers and professionals not only for industry, but also for the emerging services sector. As places of systematically exercised critical thinking, universities also played their role in shaping political ideas, with students or academics entering public debate with city administration or even with the government.

In this context, the innovative role of universities is closely linked to creative ideas and the capability to offer creative solutions to end users. Universities are considered an integral part of the economy, which they also help form (Banks 2018, Schlesinger 2016). According to Howkins (2001) or Florida (2002), creative economy is based on the new creative workforce and creative industries in close relation to cities. So-called "creative cities" play an irreplaceable role in the creative economy (Landry 2000, Egedy 2016). Florida (2002) claims that the cultural sector is essential for the area of creativity and the economic development of cities and regions. Evans (2009) notes that cities and regions often use culture and creativity as a tool for economic development and he links the economic development of cities and regions to local participants and "hubs" at universities. Universities as creative hubs are developed with the aim of supporting creativity in the academic environment and generate creative activities in support of the growth of the local creative economy (Evans 2009, Freeman 2004).

There are many examples of creative hubs at universities across Europe. Schlesinger (2016) mentions an illustrative example of the foundation of five big university consortia in the United Kingdom, four of which were labelled "knowledge exchange hubs for the creative economy" while the fifth one (Research and Enterprise in Arts Creative Technology) was founded as a research centre for author rights and new business models in the creative economy. Ashton and Comunian

DOI: 10.4324/9781003039792-3

(2019) carried out desktop research at British universities with the aim of finding out whether they have infrastructure and activities that could be seen as a "creative hub". They found out that a large number of hubs (more than one hundred) were working at British universities in the year of their study, distributed unevenly among dozens of institutions, with most institutions having at least one hub, but some up to four. Creative hubs commonly function at other universities in the world, usually on the basis of cultural and creative centres or as a research and development infrastructure (labs).

The establishment of creative hubs at Czech and other central European universities follows the mainstream of the "creative movement" in the world (Egedy 2016, Rembeza 2018), albeit with a certain delay. Egedy (2016: 91) states that cities in Central and Eastern Europe suffer from a lack of co-operation between participants, and not only between economic subjects but also universities, policy-makers and local administration.

The aim of this contribution is to discern the structure of the creative hub at Palacký University Olomouc and to describe how it functions. In our study we search for answers to the following questions:

1. Universities in Europe and elsewhere have recently undergone a restructuring of their activities towards an increase in their creative potential and involvement in the creative economy (Evans 2009, Ashton and Comunian 2019). What creative activities are typical for Palacký University Olomouc and which of them may be regarded as important for the development of the city and region? We assume that the creative activities of the university are in keeping with the economic goals of the development for the city of Olomouc, or for the Olomouc Region, speaking about the framework of their strategic documents.

2. Research at foreign universities proves that universities create their own creative spaces through which they help produce creative ideas and through which they provide services to end-user companies (Delgado et al. 2020) or facilitate the involvement of students (Jankovska 2008). What creative spaces were built by Palacký University Olomouc in support of the creative hub, and how do these spaces serve their users? Based on our academic experience at Palacký University Olomouc, we know that the university has certain creative spaces, yet we lack detailed information on their use for non-academic sphere.

3. University hubs are created to provide concrete creative services to small and medium-sized businesses in the creative sector (Comunian and Gilmore 2015, Virani 2015). What creative services are provided in this regard by Palacký University Olomouc, to what types of customers and from which parts of the region? This question is legitimate in regional science with its aim towards the development of the concept of "triple helix" (Etzkowitz and Leydesdorff 2000, Etzkowitz 2011).

The main methods used in this contribution were the review of literature on university creative hubs, and the analysis of university documents (annual reports, strategic plans etc.) and planning documentation at the city and regional levels (strategic plans of the city, regional development strategy, regional innovation strategy etc.). Long-term observation of the academic environment at the university (since 1995) was used as a supplementary source of information. In addition to data derived from the analysed documents, we used a data base of research centres in the Czech Republic, compiled by the main author. All the data are publicly accessible.

The changing role of universities

In a knowledge-based economy, higher education institutions are vital to raising competitiveness of business. The presence of a university broadens the city functions not only by attracting sharp minds from a wider region, but also by facilitating research with innovations that may be applied in business. Lambooy (2002) argues that innovation systems today are based on a dynamic interaction of firms, scientific research organisations (like universities) and regional or national governments. Etzkowitz and Leydesdorff (2000) have described this interaction as the "Triple Helix", a non-linear model of innovation in which university, industry and government are not synchronized into a stable relationship but rather influence each other by communicating intentions, devising strategies, and creating projects while continuously harmonising their infrastructure in order to facilitate co-operation.

The demand for innovation is not solely on the side of business companies. Within the knowledge management strategies and knowledge-based urban development approach (Ardito et al. 2019), local governments enable the creation of smart city projects, in which entrepreneurs often collaborate with universities or which even give rise to university-related start-up companies. This whole process leads to the hybridisation of the roles of individual stakeholders (Ferraris et al. 2018), supporting a more prominent role of the university.

In Central Europe, the role of universities in relation to their city and region has changed with the socio-political shift and economic transition of the decades following the collapse of socialism after 1989. The demand for tertiary education has been rising in most of the OECD countries since the 1990s, yet while the percentage of 25–34-year-olds with tertiary education has almost doubled in the OECD average between 1998 (23.8%) and 2018 (44.5%), in the countries of Central Europe it has rather tripled or quadrupled, starting from a lower base: in Poland from 11.8% to 43.5%, in Slovakia from 11.3% to 37.2%, in the Czech Republic from 10.5% to 33.3% and in Hungary from 13.9% to 30.6% (OECD 2020). Broadening the offer of university study programmes, establishing new faculties at the existing universities and opening new universities and colleges (both public and private) was supported by the governments in those countries in an effort to raise the attainment in tertiary education towards Western European standards and thus prepare their population for the changing labour market demands (Pachura 2017, Matlovič and Matlovičová 2017).

This transition led to the acquisition and construction of buildings that would serve the growing numbers of university students and academic personnel. Properties abandoned in the 1990s due to political and economic changes were often acquired and restored by universities, or brownfield and greenfield sites within the city limits were used for the construction of new university campuses. The growing number of university students also influenced the housing market of the cities, as it reacted to the higher demand for student accommodation by expanding the segment of flats purchased as investment and rented to students.

According to Lorber (2017), knowledge institutions are key to the successful implementation of projects that provide regional development. By establishing business incubators and partnership networks they facilitate the transfer of new methods and good practices towards local industry and government. They can even initiate the regeneration of old industrial zones after decades of their decline, which happened e.g. in the city of Maribor in Slovenia. The University of Maribor brought the city institutions to an interdisciplinary project for sustainable regeneration of the traditional but degraded industrial zones of the city in the period from 2017 to 2028. A reindustrialisation process is underway with the goal of creating innovative urban areas vital for a sustainable urban development of the city.

In the context of the aforementioned, universities are perceived as a source of the knowledge economy, which is founded upon a wide spectrum of innovation, especially for the industrial sector. However, universities cannot be denied their contribution to the long-term development of art and culture in cities and regions (Goddard and Chatterton 1999). As Ashton and Comunian (2019) claim, the involvement of universities within the creative economy has three basic levels: (i) the universities' activity in their creative and cultural context, i.e. on the university campus; (ii) creative knowledge generated by universities; and (iii) linking creative human potential (scholars, graduates, experts) onsite as well as outside of university campuses. With respect to the current needs of the knowledge economy, it is most desirable to establish "creative platforms" based on the third level.

In recent years, we have seen universities establishing so-called "creative hubs" as a logical outcome of the efforts of universities to transfer creative opportunities outward (Comunian and Gilmor 2015). Their establishment in different parts of the world responds to the goal of connecting universities with the employment sector, which is the so-called third role of universities (Ptáček and Szczyrba, 2017). The basic characteristics of creative hubs are the following: (i) they provide services to small and medium-sized enterprises as well as micro-enterprises in the creative sector; (ii) they focus on the early stage of creative small and medium-sized enterprises and micro-enterprises; (iii) individuals (brokers) help them provide and maintain relations within as well as outside the hub; (iv) they have become critically needed for the existence and overall sustainability of the local creative ecology (Virani 2015: 5).

Creative hubs take various forms. Ashton and Comunian (2019) defined a total of seven types of creative hubs, ranging from centres as temporary infrastructure, incubators, research and creative centres or workspaces shared by students, to talent competitions and business support networks. Every university chooses its own path for establishing a creative hub, which is based on the knowledge of the academic environment, creative disciplines and the university's research profile.

In the typology of creative hubs (Virani 2015), universities are labelled as "training institutions" providing both "hard creative services" (studio space, labs, etc.) and "soft creative services" (collaborative opportunities or business support). Other types are incubators or service centres for companies, which are often established by universities and which help in transferring technology and know-how from the academic environment into business.

Universities actively develop activities aimed at collaboration with the business sector. For this purpose, they construct modern innovation spaces of various forms (fab lab, maker space, co-work, etc.). For companies, these spaces are opportunity for the development of a wide spectrum of innovative projects that may integrate the users at an early stage of the process of design of a new product. These spaces represent an opportunity to create relations with industry and at the same time they support education and research. As teachers and students from various study programmes meet in the innovative space, it helps carry out more complex projects (Delgado et al. 2020).

City of Olomouc and its university—University and the city of Olomouc

Olomouc is one of the oldest cities in the Czech Republic, with the original settlement originating during the 10th century. With a population of 100,000, it is the sixth-largest city in the country. Due to its geographical location in the centre of the historical territory of Morava, Olomouc has always been a significant centre along trade routes right from the beginning. The symbol of continuity and bearer of the city's specific features is mainly the historical centre, which was an intact formation right up until the end of the 1860s. This formation separated the open space in the countryside and the rural suburbs by fortification zones.[1] The beginnings of this urban complex date back to the 11th century. However, the historical centre did not experience a major boom in construction until later, during the Renaissance and especially the Baroque eras (Ptáček et al., 2007). This period also includes the establishment of the local university.

The presence and interaction of today's Palacký University Olomouc with the city of Olomouc spans more than 440 years. In the 16th century, Olomouc was the capital of the Margraviate of Moravia and served as a multifunctional settlement with certain features of a metropolis (Miller 2002) thanks to its regional political, economic and episcopal function. Most of its population was German, craftsmen and teachers were coming from German cities and Protestant ideas were soon taking root in Olomouc. In reaction to this, the Jesuits were invited to Olomouc by the bishop in 1566 to run a college which was granted the right (in 1573) to provide university education (Jakubec 2009). In the course of history, the role of the university in Olomouc evolved. In the Thirty Years' War, the city lost its status of the capital of Moravia (relocated to Brno in 1641 as a consequence of moving land administration and judicial institutions) and Olomouc was subsequently turned into a fortress. Accordingly, its administrative role was diminished and emphasis was placed on its military function while still serving as a seat of diocese and newly as an academic centre (after Prague, the university in Olomouc is the second oldest in the Czech Republic). Jesuits were a vital component of the baroque reconstruction and urban development of the inner city, purchasing houses and plots for the construction of new buildings (Pojsl 2002). The end of the Jesuit order in 1773 led to a decrease in the importance of the university in Olomouc and most of its faculties were gradually closed; only theology continued to provide higher education. Traditional fields of higher education here still include philosophy, law and medicine, but all of them have a shorter education period than theology (Šantavý 1980). For example, Gregor Johann Mendel, the world-famous founder of genetics, completed his studies in philosophy at the local university.

The next stage in developing higher education in Olomouc began in 1946. At that time, the Czechoslovak parliament passed a law renewing the university under the new name of Palacký University Olomouc (Law No. 35/1946 Coll.). Over the next four decades, the university in Olomouc, in common with all of the other universities in the country, was under the control of the state and the Communist Party. This resulted in the abolition of the Faculty of Theology in 1950 (although it was briefly restored around 1968). The need to educate students in natural sciences led to the establishment of the Faculty of Science at the end of the 1950s (1958). Together with the faculties of philosophy, medicine and education, which were institutionalised at the time of restoring Palacký University in Olomouc, they formed its basis until 1989.

In retrospect, the year 1989 and those following were a major milestone in the development of Palacký University Olomouc. Democratic conditions were re-established and dynamic changes were instituted—measures and counts of number of faculty members, number of study programmes, number of students, internationalisation, scientific performance, etc. Gradually, Palacký University Olomouc expanded by four more faculties as opposed to its size at the end of the 1980s: Faculty of Theology (re-established in 1990), Faculty of Law (re-established 1991), Faculty of Physical Culture (1991) and the Faculty of Health Sciences (2008). Today, more than 20,000 students attend Palacký University Olomouc (as of 20 January 2020). For the sake of comparison, about 45,000 students attend the largest university—Charles University in Prague, and almost 30,000 student frequent Masaryk University in Brno. This makes Palacký University Olomouc the third-largest tertiary education institution in the Czech Republic in terms of student numbers. According to this criterion, the three largest universities have a one-third share of the total number of the country's university students. Palacký University Olomouc has a share of 7% (see Table 3.1).

Table 3.1 shows the development in the ranking of the largest universities in the Czech Republic in the past 20 years. Charles University in Prague maintains its position as number 1, according to the total count of students. The second largest is Masaryk University in Brno, even though there has been a sharper decline in the number of total students between 2010 and 2020. Palacký University Olomouc is third in the latest ranking, and it is worth noting that the total count of students has remained roughly the same between 2010 and 2020 despite the overall decrease (by 24%) in the total number of students at Czech universities in the past decade.

In terms of the spatial distribution of the buildings, Palacký University Olomouc has an extensive infrastructure, which has changed dramatically from its conditions in the late 1980s. There are dozens of buildings throughout the city, more or less functionally connected. Campuses are located in the city centre as well as in the city's east and west ends. Part of the university's facilities are historically linked with

Table 3.1 The largest universities in the Czech Republic (according to number of students to 31 December)

University / Ranking in 2000 / 2010 / 2020	2000	%	2010	%	2020	%
Charles University (Prague) / 1 / 1 / 1	35,400	18.6	49,747	12.6	48,434	16.2
Masaryk University (Brno) / 2 / 2 / 2	18,083	9.5	39,454	10.0	31,521	10.5
Palacký University Olomouc / 7 / 5 / 3	12,287	6.4	21,935	5.5	21,824	7.3
Czech University of Life Sciences Prague / 10 / 6 / 4	7,874	4.1	20,765	5.2	20,683	6.9
Brno University of Technology / 3 / 4 / 5	15,934	8.4	22,206	5.6	18,643	6.2
Czech Technical University in Prague / 3 / 3 / 6	18,334	9.6	22,517	5.7	17,388	5.8
Czech universities in total	190,204	100.0	395,982	100.0	299,396	100.0

Source: Ministry of Education, Youth and Sports 2020.

the city centre (rectorate, library, Faculty of Theology). Another part consists of the science centre buildings, which were built a few years ago in order to strengthen the university's innovation potential. This study pays great attention to this part of the university as it is here that significant generation of creative ideas and opportunities takes place. This piece of research infrastructure was specifically built for this purpose (see below). In terms of the facility's amenities, it is the newest state-of-the-art building in the Palacký University Olomouc. At the same time, there is a high concentration of creative human potential also originating from abroad.

Palacký University Olomouc is a key facility for Olomouc, whether it is student activities within the city (for study or leisure purposes), or as a source of employment. In the first case, studentification (Smith 2005) is a fundamental influence in Olomouc. It is mainly about offering accommodation to students off-campus (private accommodation) and then the students' leisure activities (frequenting shops, restaurants, cafés, cinemas, theatres, etc.). For these facilities located in the city, students are often the basis for their activities. Mutual social and economic interlinking is beneficial for both the city and the university. Universities and the private sector contribute both to the development of human capital and to the overall prosperity of the region (Zvara et al. 2013, Benneworth 2010). There is no doubt that Olomouc is a university city and that its students create a distinctive social and cultural environment. On the other hand, the university in Olomouc is one of the largest employers in the city and region with more than 4,000 employees. Approximately half of them are academics.

Involvement of universities in regional structures has been subject to long-term research (Elliott et al. 1988, Goddard and Chatterton 1999). The nature of the co-operation corresponds with the size and significance of the university for the development of the city and region. A common form of co-operation between the university and the city is working on projects of strategic planning, when attractiveness of the city for its dwellers, investors, tourists, and other users is being strengthened (van den Berg and Russo 2003). On the other hand, universities by means of their activities (education, R&D) engage in wider co-operation within the framework of regional development (Arbo and Benneworth 2007).

The relations of Palacký University Olomouc and the city of Olomouc have been forming over the course of several centuries and can only be seen as close and important on or both sides. The city needs the university, and vice versa, which is a case for other European universities as well (Ingrová 2019). Analysis of strategic documents of the city of Olomouc, namely of the strategic plan for the period 2017–2023, confirmed long-term co-operation between the university and the city. In addition to a wide range of smaller common activities, the city collaborated on creating a business incubator at the Science and Technology Park of Palacký University Olomouc and in 2016 it also participated in the project "Cultural and Creative Industries" (Box 3.1). This project can be regarded as the inception of systemic support of the development of creative economy of the city and region. Other institutions in the region participated in the project (regional administrative authority, economic chamber etc.). In the design and implementation part of its strategic plan, the city of Olomouc declares systemic support of the university in R&D. The city intends to use the local potential and increase the intensity of co-operation between firms, universities, and other research institutions. An upcoming project, in which the city of Olomouc participates by preparing the development area, is the National Biomedical and Biotechnological Park.

In the aspect of regional innovation strategy of the Olomouc Region (as of its 2020 update), Palacký University Olomouc is a key player in the region's innovation system. Regional support is primarily aimed at assistance in implementing innovation projects of the firms, using selected tools of innovation vouchers, knowledge transfer partnership, regional proof-of-concept fund, seed fund etc. Further on, the regional administration decided to support firms also using creative vouchers or incubation programmes for knowledge firms. The range of support to innovative and creative firms in the region is rather broad, however, it is in its early stage. Only the upcoming years will show whether this support is sufficient and effective.

University as a creative hub

For centuries, the university in Olomouc has been perceived as an important educational centre in the Moravian region. Its activities focused mainly on the areas of culture and art, with regard to the fields of study offered at the university (see above). In accordance with Ashton and Comunian (2019), the university was intertwined with the city's economy, especially at the level of creative and cultural context. In the second half of the 20th century, the university's creative activities began to develop further in connection with the state's policy on research, which

---

BOX 3.1   The Cultural and Creative Industries in the Olomouc Region

A project accomplished by Palacký University Olomouc in co-operation with the city of Olomouc, Olomouc Regional Authority and other regional institutions several years ago (2016). The aim of the project was to situate companies in the region that operate in sectors classified as cultural and creative, and bring a specific economic benefit to the region, especially in terms of employment. Some of the mapped sectors are closely related to subjects taught at the university, thereby representing useful synergy. Among the most frequent in the survey were companies in the following sectors: books and press, software development or architecture.

Other activities within the project were seminars, which presented the outputs of the project to the public, and further development of individual sectors was discussed. All of these and other activities linked to cultural and artistic industries in the region were covered under the "Creative Olomouc" brand (https://kreativniolomouc.cz). In this respect, it is a continuation of the project, which directed the university to the next stage of developing creative activities for cultural and artistic sectors in the region.

Source: Palaščák and Bilík 2017

---

followed five-year plans at the time, just like other sectors of the economy. Universities thus participated in research that was part of a state-run economy, in broadly defined areas of the economy (i.e. the Higher Education Act 1980, § 67). The Czechoslovak Academy of Sciences of the time played a key role in defining research tasks, especially in relation to universities (Meeting of the Federal Assembly of the CSSR, 1973). In this period, universities were subject to directive management of the Communist Party.

A key change for universities in former Czechoslovakia was the Velvet Revolution in November 1989, which initiated a broad emancipation process in tertiary education. Under the Higher Education Act (1990), universities became self-governing entities that were allowed to freely teach, create and research (§ 2). As previously mentioned, new faculties were established at Palacký University Olomouc. There were new fields of study and the number of students and staff increased overall. First of all, it was all about cultivating creative activities within the university. Only later did the university become more deeply connected with employers.

The next section of this study presents findings about the current forms of creative hubs at Palacký University Olomouc. By creative hub, we mean the part of university infrastructure where creative ideas are concentrated and "creative platforms" are constructed, i.e. the creative connection between the university and its environment. In this respect, we can single out two creative hubs with different focus at Palacký University Olomouc, each of which has its creative spaces. These spaces correspond to some types of creative hubs listed in the Ashton and Comunian (2019) typology. These creative spaces were also identified on the basis of data analysis at Palacký University Olomouc. Only research centres and laboratories functioning on faculty level (not departmental level) were considered in the analysis. A total of 20 research centres or laboratories were identified. Each of the eight faculties of the university has at least one research centre or laboratory: only one at the Faculty of Health Sciences and at the Faculty of Education, while a total of five centres at the Faculty of Science. The situation at the Faculty of Science is quite distinct in comparison with other faculties, because two of the four centres are large science centres with hundreds of employees. A similar situation is at the Faculty of Medicine and Dentistry with one large science centre (see below). The following text presents selected types of creative hubs at Palacký University Olomouc. Their selection is subjective, aiming to illustrate the achieved level of development of creative hubs at Palacký University Olomouc.

A creative cultural and artistic hub

The foundation of the cultural and artistic centre of Palacký University Olomouc is made up of the numerous activities in the humanities fields at the university. There are many of them and, therefore, it is necessary to make a purposeful selection, which in its nature corresponds to the generation of creative activities overstepping the boundaries of the city or region. In this sense, *Academia film Olomouc (AFO)* is definitely prestigious for the university. It is an international festival of popular science films with a history spanning more than 50 years (founded in 1966). Simultaneously to the screening of the films with a scientific theme, lectures are given by invited guests (in the past, e.g., evolutionary biologist Richard Dawkins). At present, the festival is one of the largest educational events in Europe, visited by thousands of visitors each year (Palacký University Olomouc).

The activities of the *Palacký University Olomouc Art Centre* are very extensive and consist of creative artwork and teaching. The centre was opened in 2002 in the building of the "old university" in the city centre. Before renovation, the building was used for military purposes. Today, it is a creative space at the university and an important spot for cultural events in the city. There are more than just auditoriums; there are studios as well as film and theatre halls. The AFO festival takes place here, along with a number of other cultural events (e.g. Olomouc Baroque Festival). The centre is a base for teaching history of art, film and theatre production, television studies, etc. (Palacký University Olomouc).

A creative scientific research hub

This hub at Palacký University Olomouc consists of several creative types, each of which has its own creative focus and mission, and are also different in nature. First of all, it should be mentioned that the university in Olomouc, just like most universities around the world, addresses the connection of research and business through its own incubator. This is specifically the *Science and Technology Park of Palacký University Olomouc*, which was put into operation in 2000. Construction

Table 3.2 Palacký University Olomouc cooperation with the applied sector in the years 2015–2019

	2015	2016	2017	2018	2019
Proof-of-concept projects	5	10	5	7	3
Supported spin-off	6	6	6	0	0
Prepared industrial rights	2	9	3	28	23
Licencing contracts	0	11	8	6	8
Revenue from contractual research (in 000s CZK)	24,139	24,379	23,983	29,238	28,057
Share of revenues in the budget (%)	0.6	0.7	0.7	0.7	0.6

Source: Annual reports of Palacký University Olomouc 2015–2019.

took place in three phases, with the last block of buildings (block 3) being opened in 2015. This timeline illustrates the demand of creative companies for contemporary spaces (offices, laboratories, etc.) serving as places of business in connection to the creative environment of the university. For the needs of the university, it addresses the questions of its intellectual property and the commercialisation of scientific discoveries. Table 3.2 gives a closer look at the co-operation between Palacký University Olomouc and the applied sector.

Another type of creative centres are research centres. They are diverse in their subjects and reflect the creative ability to research and present selected phenomena. There are several of them at Palacký University Olomouc, and they are engaged in research in various fields. One of them is the *Joint Laboratory of Optics of Palacký University Olomouc and the Institute of Physics of the Czech Academy of Sciences*. This research centre is historically one of the "oldest" at the university (founded in 1984). The basis of the laboratory's activities is basic and applied research in defined areas of optics. One part of the laboratory deals with advanced approaches focused on laser and optical technologies. This specialisation is used by the laboratory for entering into large international collaborations researching cosmic rays (e.g. the Pierre Auger Observatory or the CTA—Cherenkov Telescope Array). The second part participates in the CERN-ATLAS international experiment, which is implemented on a particle accelerator at CERN in Switzerland (http://jointlab.upol.cz).

In the next part of the text, we focus on science centres at Palacký University Olomouc that were built about ten years ago as part of the construction of a relatively large network of science centres in the Czech Republic. The construction also included the acquisition of state-of-the-art instrumentation. More information about the programme for the construction of science centres in the Czech Republic is provided in Box 3.2.

Figure 3.1 shows the spatial location of centres in the country. It shows that the university in Olomouc received financial support for three research centres, all implemented in the period 2007–2013. We can see that some university cities in the Czech Republic built five or more science centres (Plzeň, Brno, Ostrava). Let

BOX 3.2   Operational programme for Science and Research for Innovation—building "large infrastructure"

The aim of this programme was to strengthen the position of universities and scientific institutions in all regions through NUTS 2 (except for the capital city of Prague). The exclusion of Prague was influenced by its above-average economic position (given by GDP) within the NUTS 2 regions. On the other hand, many universities and research institutions based in Prague developed their scientific research infrastructure by building new research centres just outside Prague's administrative border (in the Central Bohemian Region). The operational programme supported 48 research centres (eight of which were categorised as excellent), including the necessary state-of-the-art technical equipment.

The programme was supposed to support a limited number of science centres; however, the reality was quite different. There is no use speculating why almost 50 science centres were established in a small country such as the Czech Republic. A few years after the expiration of the operational programme, the state—via national and European resources—subsidised operations within the so-called National Sustainability Programmes I and II. It was a commitment of the state to the European Commission to finance the sustainability of the established science centres. In addition, the centres are financed through grants and applied research.

Source: Ptáček and Szczyrba 2017, Ministry of Education, Youth and Sports.

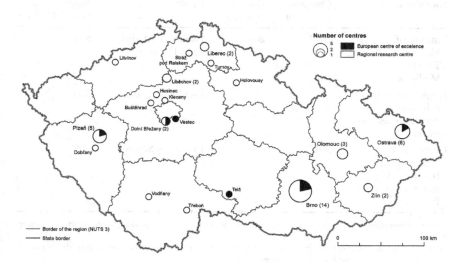

Figure 3.1 Science centres in the Czech Republic supported by the Operational Programme for Science and Research for Innovation (implemented in 2007–2013).

Source: Author's elaboration

us add that there are more universities and other scientific institutes that have built science centres. There was only one university in Olomouc.

In the university's structure, the centres are a part of the faculties. Each centre has a different subject focus (see below). Each science centre has a team of experts, including foreign contributors. There are also university staff who work at the centre, which is outside of their own faculty. From the beginning, the centres functioned as multidisciplinary and remain as such up to today. Certain specialized professions were in shortage and thus demanded from other faculties (e.g. mathematicians of chemists) as part-time job positions. Data on the number of employees in Table 3.3 apply to all employees as stated above, i.e. some of the employees appear twice as both university and science centre employees.

The *Regional Centre of Advanced Technologies and Materials (RCPTM)* deals with the development and transfer of high-tech technologies into medicine, industrial and environmental applications. It is a centre focused on research in the field of new metal structures, nanostructures, water treatment technologies, etc. Employees also include experts from abroad. Part of the centre's staff consists of teams working in another part of the university. This is the Joint Laboratory of Optics, which conducts some research work in cooperation with the RCPTM (https://www.rcptm.com/).

The *Centre of the Region Haná[2] for Biotechnological and Agricultural Research (CRH)* is a joint centre of Palacký University Olomouc, the Institute of Experimental Botany of the Czech Academy of Science and the Crop Research Institute. The centre focuses on research in the field of advanced biotechnology and agricultural

Table 3.3 Science centres at Palacký University Olomouc supported by the Operational Programme Science and Research for Innovation

Science centre	Operations commenced	Employees (approx.)	Faculty	Support (in mil. CZK)
Institute of Molecular and Translational Medicine	2013	200	Faculty of Medicine and Dentistry	861
Regional Centre of Advanced Technologies and Materials	2013	130	Faculty of Science	500
Centre of the Region Haná for Biotechnological and Agricultural Research	2012	200	Faculty of Science	808

Source: Operational programme for Science and Research for Innovation, documents of science centres.

Note: This is input support for the construction of the science centre. Other funds from the National Sustainability Programme I and II are not included here.

research. With its activities, the centre continues the agricultural tradition in the region. The centre employs two hundred scientists. Some of them are experts from abroad. Another group of experts works simultaneously in other areas of the university's structure, for example, the Laboratory of Growth Regulators operating at the Faculty of Science. The CRH is a Regional Branch Office of the European Biotechnology Federation (http://cr-hana.eu/).

The *Institute of Molecular and Translational Medicine* is a technological structure and platform for molecular-oriented founding and translational biomedical research. Although its founding was initiated by Palacký University Olomouc, other entities also participated, including the University Hospital Olomouc, the University of Chemistry and Technology Prague and the Institute of Organic Chemistry and Biochemistry of the Czech Academy of Science. At the same time, other experts from the university work there, i.e. from fields close to the centre's focus (information technology, mathematical modelling, optics, etc.). The centre's main areas of scientific interest are: the molecular basis of diseases, experimental therapeutics, translational medicine, etc. (https://www.imtm.cz/).

The research centres at Palacký University Olomouc focus primarily on basic research and their scientific results are published or applied for patent protection. This is demonstrated in Table 3.4. Increase in numbers of registered patents or utility models reflects the creative activity of the research centres; the increase coincides with the period of existence of those centres. Specific information on registered patents can be found on the websites of the research centres.

In accordance with Comunian and Gilmore (2015), research centres at Palacký University Olomouc provide creative services for business partners in the form of contractual research and as a catalogue of services listed on the websites of the research centres. The data available show that the research centres in Olomouc have been stagnating up to now, with their commercial activities and their economic output from commercial services showing lower intensity. This is namely caused by primary orientation on scientific results that yield in primary financial gains in the framework of national support of R&D in the Czech Republic. Thus, the centres are not currently forced to focus their activities more into the application sphere.

Table 3.4 Numbers of registered patents and utility models

	2009	2010	2011	2012	2013	2014	2015	2016	2017	2018	2019
Patents	0	14	12	11	16	13	13	12	19	25	26
Utility Models	5	3	9	11	14	5	19	10	7	11	4
Total	5	17	21	22	30	18	32	22	26	36	30

Source: Palacký University Olomouc.

Conclusion

This study answered its research questions. There are various forms and types of creative hub at Palacký University Olomouc. Research centres were identified at all eight faculties of the university. The analysis of strategic documents showed that the creative activities are in close relation to the needs of the city and region, which is a good signal. The university works with a wide range of creative spaces, most of them established after the year 1989.

The development of the creative environment in Czech universities after 1989 has changed significantly. It is evident, for example in Figure 3.1. The example of Palacký University Olomouc shows the influence of state policy in the field of science and research in the planning period of the EU 2007–2013. This period offered dynamic development of the university with respect to the creative centre. This is evident in the case of science centres, but also in other presented examples. During the past three decades, the university in Olomouc gained a vibrant mosaic of its creativity for the regional creative economy, which is the ultimate goal.

Co-operation with employers has been initiated and they find their business solutions at the university (within the framework of contractual research, granting licences, consultations, etc.). This does not only concern organisations from the region. We also have noted an increase of granted patents (Annual Report of Palacký University Olomouc 2019). We see gradual commercialisation of research results; however, its intensity has remained low to date. This is a certain handicap for the future operation of a prestige scientific university infrastructure.

With respect to the future, it will be interesting to observe the development trajectory of research centres at Palacký University Olomouc. Until recently, it existed independently within individual faculties. The situation has begun to change with the initiative of the leaders of the university to integrate the research centres into one "super-centre". These efforts for changing the organisation of the fundamental pillars of R&D at Palacký University Olomouc are underway since 2018. One of the concerned faculties (Faculty of Science) disagreed with such a profound change in the organisational structure of the university, which has led to a postponement in the foundation process of the research "super-centre". In the meantime, a case dealing with scientific ethics of several members of one of three research centres emerged (Hořejší 2020). Questions were also raised about the ambiguous nature of the partition of property and employee affiliation within the new organisational structure. The academic senate of the university approved the establishment of the "super-centre" called CATRIN (Czech Advanced Technology and Research Institute) and this new institute has formally commenced its operation at the beginning of 2021, yet issues remain with property and staff affiliation. After all, the new research "super-centre" may prove largely beneficial for Palacký University Olomouc as well as for the creative economy of the region. Time will tell whether this venture is a success.

Notes

1 Olomouc served as a city-fortress in the Austrian Empire. The Olomouc Fortress was established in 1742 and was abolished in 1866.
2 Haná is an ethnographic region located in Central Moravia. Its centre is Olomouc.

References

Arbo, P., Benneworth, P. (2007). Understanding the regional contribution of higher education institutions: A literature review. OECD Education Working Paper No. 9.

Ardito, L., Ferraris, A., Petruzzelli, A. M., Bresciani, S., Del Giudice, M. (2019). The role of universities in the knowledge management of smart city projects. *Technological Forecasting and Social Change*, 142: 312–321.

Ashton, D., Comunian, R. (2019). Universities as creative hubs: Modes and practices in the UK context. In: Gill, R., Pratt, A. C., Virani, T. E. (eds.) *Creative Hubs in Question: Place, Space and Work in the Creative Economy* (Dynamics of Virtual Work), Cham: Palgrave Macmillan, 359–379.

Benneworth, P. (2010). *University engagement and regional innovation*. Brussels: European Centre for Strategic Management of Universities.

Banks, M. (2018). Creative economies of tomorrow? Limits to growth and the uncertain future. *Cultural Trends*, 27(5): 367–380.

Comunian R., Gilmore A. (2015). *Beyond the creative campus: Reflections on the evolving relationship between higher education and the creative economy*. London: King's College London.

Delgado, L. Galvez, D., Hassan, A., Palominos, P., Morel, L. (2020). Innovation spaces in universities: support for collaborative learning. *Journal of Innovation Economics & Management*, 31(1): 123–153.

Egedy, T. (2016). Old and new residential neighbourhoods as creative hubs in Budapest. *Mitteilungen der Osterreichischen Geographischen Gesellschaft*, 158: 85–108.

Elliott, D. S., Stanford L., Meisel, J. B. (1988). Measuring the economic impact of institutions of higher education. *Research in Higher Education*, 28(1): 17–33.

Etzkowitz, H. (2011). The Triple Helix: University–industry–government innovation in action. *Papers in Regional Science*, 90(2): 441–442.

Etzkowitz, H., Leydesdorff, L. (2000). The dynamics of innovation: From National Systems and "Mode 2" to a Triple Helix of university-industry-government relations. *Research Policy*, 29(2): 109–123.

Evans, G. (2009). Creative cities, creative spaces and urban policy. *Urban Studies*, 46(5–6): 1003–1040.

Ferraris, A., Santoro, G., Papa, A. (2018). The cities of the future: Hybrid alliances for open innovation projects. *Futures*, 103: 51–60.

Florida, R. (2002). *The rise of the creative class: and how it's transforming work, leisure, community and everyday life*. New York: Basic Books.

Freeman, A. (2004). *London's Creative Sector: 2004 Update*. Greater London Authority Working Paper No. Report (April 2004). Available at: https://mpra.ub.uni-muenchen.de/52626/1/MPRA_paper_52626.pdf.

Goddard, J. B., Chatterton, P. (1999). Regional development agencies and the knowledge economy: Harnessing the potential of universities. *Environment and Planning C: Government and Policy*, 17(6): 685–699.

Grendler, P. F. (2004). The universities of the Renaissance and Reformation. *Renaissance Quarterly*, 57(1): 1–42.

Hořejší, V. (2020). Aféra v Olomouci a principy vědecké etiky. Vědavýzkum.cz

Howkins, J. (2001). *The creative economy: how people make money from ideas*. New York: Penguin Press.

Ingrová, P. (2019). *Univerzitní město* [Diplomová práce]. Olomouc: Univerzita Palackého v Olomouci.

Jakubec, O. (2009). *Město a biskupové. In Dějiny Olomouce, 1. svazek*. Olomouc: Univerzita Palackého v Olomouci.

Jankovska, M. (2008). Use of creative space in enhancing students' engagement. *Innovations in Education and Teaching International*, 45(3): 271–279.

Lambooy, J. G. (2002). Knowledge and Urban Economic Development: An Evolutionary Perspective. *Urban Studies*, 39(5–6): 1019–1035.

Landry C. (2000). *The Creative city: A toolkit for urban innovators*. London, Comedia – Earthscan.

Lorber (2017). Universities, knowledge networks and local environment for innovation-based regional development: case study of the University of Maribor. *Geografický časopis*, 69(4): 361–383.

Matlovič, R., Matlovičová, K. (2017). Neoliberalization of the higher education in Slovakia: a geographical perspective. *Geografický časopis*, 69(4): 313–337.

Miller, J. (2002). Raně novověká městská migrace – české země, královské Uhry, polsko-litevský stát: pokus o základní komparaci. *Český časopis historický*, 100(3): 522–554.

OECD (2020). Population with tertiary education (indicator). doi:10.1787/0b8f90e9-en (Accessed on 14 May 2020).

Pachura, P. (2017). University and space – in search of knowledge production places. *Geografický časopis*, 69(4): 301–311.

Pojsl, M. (2002). Restaurace katolicismu a státního absolutismu. In *Olomouc: malé dějiny města*. Olomouc: Univerzita Palackého v Olomouci.

Palaščák, R., Bilík, P. (2017). *Kulturní a kreativní průmysly na Olomoucku*. Olomouc: Univerzita Palackého.

Ptáček, P., Szczyrba, Z., Fňukal, M. (2007). Prostorové proměny města Olomouce s důrazem na rezidenční funkce. *Urbanismus a územní rozvoj*, 10(2): 19–26.

Ptáček, P., Szczyrba, Z. (2017). The role of universities in strengthening innovation: the case of EU structural funds in the Czech Republic. *Geografický časopis*, 69(4): 339–360.

Rembeza, M. (2018). Creative placemaking in Poland. Can art become an effective tool of urban regeneration? *5th International Multidisciplinary Scientific Conference on Social Sciences and Arts SGEM2018*.

Smith, D. (2005). 'Studentification' the gentrification factory? In: Atkinson, R. and Bridge, G. eds. *Gentrification in a global context: the new urban colonialism*. Housing and Society Series. Routledge, 72–89.

Schlesinger, P. (2016). The creative economy: Invention of a global orthodoxy. *Innovation: The European Journal of Social Science Research*, 30(1): 73–90.

Šantavý, F. (1980). *Organizace/pečeti a insignie Olomoucké Univerzity v letech 1573/1973*. Olomouc: Univerzita Palackého v Olomouci.

Van den Berg, L., Russo, A. P. (2003). *The student city. Strategic planning for student communities in EU cities*. Vienna: European Regional Science Association.

Virani, T. E. (2015). Re-articulating the creative hub concept as a model for business support in the local creative economy: the case of Mare Street in Hackney. Creativeworks London Working Paper Series No. 12

Zvara, J., Šašinka, P., Rybář, J., Hudeček, P. (2013). Vysoké školy jako aktér regionálního rozvoje. In *16. mezinárodní kolokvium o regionálních vědách. Sborník příspěvků* (16th International Colloquium on Regional Sciences. Conference Proceedings.), Masaryk University Press, 127–139.

Internet sources

Annual report Palacký University Olomouc 2009–2019

Law No. 35/1946, Coll., Zákon o obnovení university v Olomouci

Law No. 39/1980, Coll. Zákon o vysokých školách

Law No. 172/1990, Coll. Zákon o vysokých školách
Ministry of Education, Youth and Sports
Proceedings of the Federal Assembly of the Czechoslovak Socialist Republic 1973, available:
  https://efektivniuspory.cz/zprava-o-zamereni-vedeckotechnickeho-rozvoje-v-cssr/;
  https://kreativniolomouc.cz

4 In the shadow of Karl Marx

The case of Chemnitz and its multiple transitions

Birgit Glorius

Introduction

The system change from socialism to capitalism, and its consequences in terms of socio-economic conditions and urban structures, have been addressed by a broad range of academic disciplines, mostly applying the terms "transition" or "transformation". While the term "transition" implies a perspective that the post-communist countries underwent a continuous process of going "back to normal", resulting in a complete adaptation of the Western, capitalist, neoliberal ideas, the term "transformation" brings the contingency of social and spatial contexts to the fore, and suggests that alternative developments might be possible (Fassmann 2013; Meeus 2016).

Regarding the transformation of urban settings, research has developed the notion of "post-socialism" to highlight the "rupture between the modern and the postmodern" (Hirt 2008) that those cities faced after the end of state socialism. In economic terms, the most notable rupture was the sudden termination of Fordist production schemes and the move to a post-Fordist era, combined with the economic effects of globalisation. Regarding population development, a major rupture was the huge demographic decline at the beginning of the 1990s with massive emigration and dramatically dropping birth rates. Further relevant urban transformations were the capitalisation of land and property markets which led to a redefinition of urban geographies, the liberalisation of labour and consumer markets which resulted in growing social stratifications, and the decentralisation of public administrations, which allocated governance structures to the local level and thus opened up avenues for diversified processes of local governance and development goals (Milanovic 1997; Sykora 1998).

This chapter describes the main causes, processes and effects of the multidimensional socio-economic changes in Chemnitz since the fall of the Berlin Wall. The main focus lies on the economic and societal transformations since the 1990s and how these have been framed by various stakeholders. Conceptually, the contribution is based on geographical transformation theory focusing on socio-spatial change, and sociological theories on post-modernism, focusing on societal changes and how they are contextualised within different parts of society. Data for this chapter stem from statistical sources, academic papers, and from the author's own observations and data collections (notably a series of focused interviews with shop owners in Chemnitz), from several research and teaching projects carried out since 2013.

DOI: 10.4324/9781003039792-4

Chemnitz, a city of approximately 246,000 inhabitants (2019), is located in the south of the former German Democratic Republic (GDR), close to the Czech border. It is the third-largest city in the state of Saxony. As one of the forerunners of early industrialisation, Chemnitz became a hotspot of textile production and gained the nickname "the Saxon Manchester". During socialist times, Chemnitz was one of the most important centres of industrial production in the German Democratic Republic and was renamed "Karl-Marx-Stadt".

From Chemnitz to Karl-Marx-Stadt

The city of Chemnitz was first mentioned in 1143 and developed as an early industrial centre of textile production. At the end of the 19th century, local businesses turned to the production of machines such as weaving machines, and Chemnitz gradually developed into a vivid industrial city with supra-regional outreach (Viertel and Weingart 2002). By the end of the 19th century, Chemnitz was a renowned centre of textile and machine production. Further important sectors were iron foundries, metal goods manufacturing, electrical engineering, bicycle manufacturing, dyeing and the chemical industry. The population quickly increased from 10,000 in 1800 to 360,000 in 1930, accompanied by settlement expansions, which were typical for the late 19th and early 20th century. During World War II, the city centre of Chemnitz suffered devastating destruction. Nearly all historic buildings in the inner city were completely destroyed and, in total, the city lost almost two-thirds of its built fabric (Richter 1998).

After World War II, Germany was divided in two parts: While the territory that was occupied by French, British and American troops in the west of Germany and the western part of Berlin formed the Federal Republic of Germany (FRG), the territory occupied by Soviet troops in the east of Germany—where Chemnitz is located—became the German Democratic Republic (GDR).

The reconstruction process after World War II was oriented around socialist principles of urban planning; it comprised the erection of representative buildings for state and political institutions and the reshaping of the historical street grid with a focus on large street axes and public spaces used for public manifestations (Lindner 2011). For this purpose, some of the few remaining historic buildings in the inner city were demolished (Viertel and Weingart 2002). In 1952, Chemnitz became a county capital. In 1953, to mark Karl Marx Year, Chemnitz was given a new name: Karl-Marx-Stadt (Lindner 2011). Karl Marx was a German philosopher who, together with Friedrich Engels, became the most influential theorist of socialism and communism. He was born in the city of Trier and had no connections to Chemnitz. The denomination of the city merely followed the decision of the GDR regime, which argued it was appropriate given the strong traditions of the labour movement in Chemnitz and the achievements of the city in the reconstruction process. In 1971, a huge bronze bust of Karl Marx was erected in the centre of the city, which serves as a main meeting point for political demonstrations (Figure 4.1).

The vast demolitions at the end of World War II left roughly 100,000 people homeless in Chemnitz. As the city's population constantly grew in the years

Figure 4.1 Karl Marx Monument in the centre of Chemnitz.
Source: Author.

thereafter (from less than 250,000 in 1945 to 290,000 in 1950, 300,000 in 1975 and up to a peak of 320,000 in 1985), the fight against housing shortages was another important focus of urban planning. From the 1960s, several large new blocks of flats were erected, both in the inner city and in the suburbs, using the technology of industrialised building with pre-fabricated units. Contrary to the 19th century buildings, which were perceived as bourgeois, those new and uniform housing units underlined the egalitarian character of the socialist society (Rietdorf 1975). The largest was the "*Fritz-Heckert-Gebiet*", which housed roughly 80,000 people until 1990 (Viertel and Weingart 2002). The remaining buildings from the late 19th century—among them a considerable portion of Art Nouveau buildings—were neglected and entered a process of decay, sometimes leading to the complete decline of those neighbourhoods (Viertel and Weingart 2002; Beuchel 2006).

The economic strategy of the centralist GDR regime re-assigned Chemnitz the role of an industrial core with a focus on textile industry and machine construction (*Aufbaugesetz* 1950, §2). Chemnitz delivered 20% of the total industrial production, almost half of all textile machines and around one-third of machine tools in the GDR (Viertel and Weingart 2002).

Throughout the time of the GDR, the position of the three largest urban settlements in Saxony remained the same: Chemnitz/Karl-Marx-Stadt as the core of industrial production; Leipzig as the hub for international trade; and Dresden as the political capital of the state. Those ascriptions and the resulting image of the cities showed a strong persistence in the years after the fall of the Berlin Wall.

Political change and early transition period

After the system change in 1989 and German reunification in 1990, Chemnitz suffered an economic breakdown, followed by mass unemployment and huge population losses (18% between 1990 and 2011), due to internal migration and decreasing birth rates. The general decline was reflected upon negatively in and outside of Chemnitz, and Chemnitz earned the image of a stagnating, ageing and backward-looking town.

Since the 2000s, the economic and population situation has been stabilised again. Unemployment decreased and even the population development stabilised, mainly due to a positive migration balance, notably thanks to international migrants. However, the economic success of mainly small and medium-sized companies, infrastructure investments and public image campaigns have not changed the rather negative image of the city.

The early years of transition during the 1990s were characterised by the almost total breakdown of the centrally planned economy, leading to huge unemployment and considerable emigration, mainly to the Western part of Germany. A rapidly declining birth rate, combined with selective emigration, caused an accelerated demographic ageing. In 1990, the population of Saxony was around 4.7 million, and by 2008 it had dropped to around 4.1 million; 70% of this decrease was due to declining birth rates (Lowe and Nagl 2011).

Further fields of struggle were the restoration of the neglected public infrastructure and the imposition of new systems for almost all parts of everyday life, be it public administration, finances, social security or education. In large parts, the restoration meant a transfer of West German strategies and technologies to East Germany, often combined with a personnel transfer of West Germans into leading positions in public administration in East Germany.

The crisis of the centrally planned GDR economy resulted in massive supply shortages which worsened throughout the 1980s. The economic crisis, combined with the ruling system's inability to openly address existing problems, resulted in growing public and political unrest and in the formation of civic resistance. The public protest quickly gained ground after the municipal elections of 7 May 1989, which were marred by accusations of mass fraud, after it was claimed that voter participation was nearly 100% and an overwhelming majority of candidates on the official lists were elected (Geißler 1991). A further provocation was the celebration of the 40th anniversary of the GDR on 7 October 1989. The huge public demonstrations, combined with the refusal to publicly admit political and economic difficulties or start a reform process, triggered public protest throughout the country, which was answered by massive state force. In September 1989, the NGO "New Forum" was founded and rallied for civic engagement in order to obtain societal change. Further democratic groups developed in the wake of the "New Forum", mainly on religious or ecological grounds (Reum and Geißler 1991). In Karl-Marx-Stadt, those oppositional groups merged into the *Democratic Oppositional Platform* (DOP). Throughout November and December 1989, opposition groups and ruling governments formed *Round Tables* to find a collective pathway to overcome the political, economic and societal crisis of the country. The first *Round*

Table in Karl-Marx-Stadt met on 3 January 1990. It established a regular working routine throughout the year and thus managed to uphold the political void until the election of a new municipal government on 1 June (Reum and Geißler 1991). The results of the municipal elections in 1990 clearly displayed the new political diversity: while the strongest party, the Christian Democratic Union, gained 29 of 80 seats, the new city parliament united representatives of 13 different political groups altogether (Reum and Geißler 1991).

In the wake of the formation of non-governmental organisations (NGOs) during 1989, there was also an initiative calling for the restitution of the city's original name, Chemnitz, as the decision to rename the city Karl-Marx-Stadt on 10 May 1953 had been implemented without public participation. In April 1990, a public vote was organised, which resulted in a large majority who opted for the restitution of "Chemnitz". The municipal government followed the public vote and renamed the city (Reum and Geißler 1991).

During the first years of transition, Chemnitz suffered from an economic breakdown, as the formerly dominant state companies in the manufacturing industry lost state support, former trade relations to the other Central and East European countries were dissolved, and there was a massive price increase of products in East Germany due to the introduction of the West German currency D-Mark (Bachmann et al. 1995; Viertel and Weingart 2002). The main reform strategy in East Germany was to support the privatisation of state industries. This was achieved by the dissolution of state companies into smaller segments, of which the more productive parts were privatised, while the less productive parts were dissolved. In Chemnitz, more than 1,050 state companies were privatised, and 300 companies were restituted to their former owners (Bachmann et al. 1995). This resulted in massive job losses. According to documents from the trade union *IG Metall*, the number of employees in the 21 largest production companies in Chemnitz decreased from 42,675 to less than 7,408 within a period of only 27 months (January 1990–March 1993) (Feuerbach 2014). Between 1990 and 1995, the number of employees dropped from 165,081 to less than 128,858, and continued shrinking down to roughly 101,702 in 2005 (Feuerbach 2014).

While there was full employment during the socialist regime, guaranteeing a job for every citizen—notwithstanding that many of those jobs were rather unproductive—unemployment and all kinds of compensatory labour market positions became the new normal during the transition period. In 1991 in the labour market region of Chemnitz, one-third of the working population were without a regular job (Table 4.1): they were either unemployed (8.1%), had a contract but with little if any work to do, and thus received compensatory salaries ("*Kurzarbeit*", 21.7%), were enrolled in retraining measures or took part in so-called publicly sponsored jobs ("*Arbeitsbeschaffungsmaßnahmen/ABM*"), mainly in the service sector.

Regarding demographic development, there was a profound population loss which had already started before reunification: during the 1980s, a rising number of Chemnitz's citizens applied for emigration to West Germany, especially in 1989, when a total of 4,500 people emigrated. In the first years of transition, the population decreased from 294,244 (1990) to 249,757 (1998)—a loss of 15%. One major reason for this rapid decline was a sudden drop in birth numbers in 1990. While

Table 4.1 Labour market position of working population, Labour market district
Chemnitz, 1991

Position in the Labour market	Total number	Share of Workforce
Total workforce	294,141	100
Regularly employed	191,925	65.2%
Unemployed	23,950	8.1%
Short-time work ("*Kurzarbeit*")	61,735	21.0%
Publicly sponsored job ("*ABM*")	7,831	2.7%
Retraining measures	8,700	3.0%

Source: BA 1992, p. 799, own calculations.

during the 1980s, birth numbers ranged between 3,000 and 4,000 per year, numbers started dropping rapidly in 1990 (2,837) until they reached the lowest level of 1,213 in 1994. They did not recover significantly in the years thereafter, leading to a negative reproduction balance of 1,000–2,000 per year. Just as in other Central and Eastern European States during the transition period, the decline of birth numbers resulted from a postponement of fertility decisions, due to the unstable conditions in the context of the economic breakdown and growing unemployment, and as a response to new opportunities for individual life, such as higher education or travelling (Billingsley 2010; Frejka 2008). A further reason for population losses and declining birth rates was emigration, which was especially strong at the beginning of the 1990s, when thousands of Chemnitz's residents moved to West Germany to escape unemployment and find better living conditions. A final aspect of demographic decline was suburbanisation. During the most dynamic period of suburbanisation between 1994 and 1998, 17,500 people moved from Chemnitz to the adjacent municipalities, accounting for 44% of the total population loss during that period (Köppen 2005). The young age of the migrants, combined with the drop of birth rates, resulted in an accelerated demographic ageing and a profound gap in the composition of age cohorts. Since the end of the 1990s, population decrease has slowed down, among others due to administrative reforms that integrated some of the suburban municipalities into the municipality of Chemnitz, what brought additional inhabitants into the city. Between 1990 and 2011, the total population decline was 18%.

Urban restructuration in Chemnitz after 1990 had to pursue a number of difficult tasks: the largely dysfunctional city centre had to be revitalised, the historic housing from the late 18th and early 19th centuries needed support, while the rapidly increasing residential mobility of Chemnitz's inhabitants made any planning forecast more than difficult. Furthermore, planning and reconstruction processes were complicated by lengthy ownership clearances and restitution processes of historic buildings, and by the implementation of new regulations due to reunification.

The revitalisation of the city centre was one of the most pressing issues. Due to the destructions caused by World War II bombings and the incomplete reconstruction during the GDR period, there were huge blighted zones in the inner city, which thus lacked urban functionality and aesthetics. In 1992 an initial master-plan was developed and entered into a broad participatory process, which revealed major divergences regarding the focus for further development. While city planners and politicians favoured public functionality in the city centre, private investors were oriented towards the highest economic output (Richter 1998). A second masterplan in 1998 set the tone for inner-city reconstruction: it focussed on the restoration of the old street grid and the erection of buildings which re-interpreted the historical density of the city centre. Several warehouses and inner-city shopping centres were built towards the end of the 1990s, all of them with ambitious, yet mismatched, architecture (Richter 1998). The whole process of functional reconstruction was challenged by a parallel development of large shopping and trade centres on the outskirts of Chemnitz. This development responded to the growing demand for differentiated consumption after 1990, which could not immediately be met in the city centre. Between 1990 and 1995, four shopping centres were built on the edges of the city, with a total retail trade area of 120,000 m^3, while the city centre only provided about 50,000 m^3 at the end of the 1990s (Richter 1998).

A second pressing issue during the early transition period was the improvement of the housing situation. A large proportion of the historic buildings erected before 1918 needed extensive restoration, which was mainly carried out by private investors. Many of those buildings were in such a disastrous state that they had to be demolished. More and more brownfield sites appeared in the inner-city neighbourhoods due to demolitions, but there was also increasing public protest against the further destruction of Chemnitz's urban heritage. At the same time, considerable construction activities started, as investors assumed a huge demand from the population for larger, modern housing units. Public co-funding schemes and tax reduction programmes fuelled this process. Thus, not only individual homes, but also larger housing units were built in the suburban regions of Chemnitz, which added to the total housing stock during the 1990s and gradually provoked a shift from housing shortages to housing vacancies (Köppen 2005).

Chemnitz as a shrinking city

At the beginning of the 2000s, a new discourse entered debates on urban development: the discourse of shrinking cities. With reference to the termination of the industrialised Fordist age, a paradigmatic shift was hypothesised, which would bring the period of growth to an end (Rieniets 2009; Wolff and Wiechmann 2018). Urban shrinking was first described in cities with typical Fordist industries, such as the car industry in Detroit (Beauregard 2009; Hollander et al. 2009; Pallagst 2009). Those cities experienced unprecedented economic crises, followed by big waves of outmigration, which were paralleled by sub- and counter-urbanisation processes, usually led by middle-class white households.

In Europe, notably Central and Eastern Europe, urban shrinking was discussed in the context of political, economic, demographic and social restructuring in the context of post-socialism, combined with other "post-" phenomena such as post-Fordism, post-modernism and post-industrialism (Baron et al. 2010; Steinführer and Haase 2007; Turok and Mykhnenko 2007). Großmann (2007) stressed that in relation to urban settlements, the term "shrinkage" must be understood "as a process of socio-spatial restructuration in the context of a constantly decreasing population, leading to changes in and between social and material spaces, and a reduction of connectivity between those spaces" (Großmann 2007, p. 27, translated from German by the author). In this period, which extended from around the turn of the millennium to 2012, the increasing polarisation of society became evident, not only with regard to social status or attitudes, but also as specific urban quarters— under conditions of shrinkage—were in a competitive struggle against one another. This was true for Chemnitz, where the demographic and economic stabilisation during the 2000s could not prevent the growing social polarisation.

Demographic and economic stabilisation

After the turn of the millennium, the population numbers of Chemnitz have stabilised: While between 1990 and 1999, the population loss amounted to 33,395 people or 11.3%, the population decrease between 2000 and 2009 was just 13,833 people, or 5.4%. Suburbanisation processes have come to an end, and since the mid-2000s there have even been net migration gains regarding the mobility between Chemnitz and suburban municipalities (Stadt Chemnitz 2016). However, the natural growth has remained negative, and population ageing has continued. Between 2002 and 2010, the median population age increased from 44.2 to 46.8 years, and the proportion of people younger than 21 dropped from 17% to 14.5%. The share of the "parent generation", those between 30 and 49 years old, dropped from 27.3% to 24.7% (Stadt Chemnitz 2008b, 2016). In the years to come, ageing will be one of the major challenges of Chemnitz's demographic development.

Alongside the demographic situation, the economic situation also stabilised at the start of the 21st century. After the shockwave of industrial shutdowns in the early 1990s and privatisations, the Chemnitz region again became one of the most important industrial production sites of East Germany. Between 1990 and 2002, 620 manufacturing industries with 30,000 employees settled in Chemnitz (Viertel and Weingart 2002). As a reaction to the de-industrialisation processes of the 1990s, the economic strategy of the municipal administration focussed on attracting businesses. Image campaigns were launched, and new business areas in the city's periphery were developed, while the historic industrial buildings in the inner city lost their function. Many of these mostly attractive brick-stone sites were redeveloped as areas for cultural activities, leisure, or offices, but some were destroyed.

The economic development during the 2000s shows that Chemnitz regained its old strength as a hotspot of producing industries. In particular, the automobile industry resettled in the region (for example, Volkswagen Sachsen GmbH), and a large number of specialised small and medium-sized companies developed in the field of automobile and machine production. The Chemnitz University of Technology gained a role as a major knowledge hub, initiating knowledge networks among small and medium-sized companies and supporting start-ups in the field of micro-electronics, robotics and other technologies (Feuerbach 2014).

Between 2005 and 2015, the number of employees increased from 101,702 to 112,011, and the unemployment rate dropped from 19.4% to 9.1%. However, there were also growing social disparities, with the number of people receiving social welfare payments rising from 2,987 in 2005 to 4,656 in 2015 (Stadt Chemnitz 2007, 2016).

Dealing with housing vacancies

Due to the investments in (re)construction during the 1990s, combined with population decrease, massive housing vacancies appeared as a new problem towards the end of the 1990s. At the beginning of the 2000s, about 35–40% of all pre-war buildings were vacant in the urban agglomerations of Saxony. Vacancies in the pre-fabric housing neighbourhoods of the GDR amounted to 25%, and there were also vacancies in those newly constructed housing units of the 1990s (Köppen 2005). In the whole former GDR, there were approximately one million housing vacancies, which made up 13% of the total housing stock (Pfeiffer et al. 2001). In Chemnitz, housing vacancies were estimated at between 41,000 and 43,000 apartments in 2000, which amounted to a share of 23–25% of the total housing stock (Stadt Chemnitz 2002). The vacancies were not evenly distributed over the city, but concentrated in certain residential areas: the biggest settlement from GDR times with pre-fabricated houses, "*Fritz-Heckert-Gebiet*", lost up to 20% of its population between 2002 and 2006. Around 3,700 former inhabitants moved to other districts, while around 1,000 of them left Chemnitz (Stadt Chemnitz 2008b).

After 2000, a state programme ("*Stadtumbau Ost*") supported the demolition of houses in East German cities in order to stabilise the housing markets. Between 2000 and 2007, 8,660 apartments were demolished in the *Fritz-Heckert-Gebiet*, which lowered the vacancy rate from 31% to 24% (Stadt Chemnitz 2008b). Many inhabitants of the *Fritz-Heckert-Gebiet* objected to the demolition strategy. They complained that the strategy of the city administration, which foresaw that the shrinking process should start from the periphery to re-establish urban density in the city, ignored the citizens' interests and furthermore lowered the attractiveness of the neighbourhood, resulting in accelerated population loss (Grossmann 2007).

Meanwhile, the inner-city areas with pre-1918 housing regained attractiveness, as more and more pre-war buildings were restored. However, the pace of restoration varied, due to the varying perceived attractiveness of those neighbourhoods. The most telling difference in Chemnitz can be observed in the neighbourhoods of "*Kaßberg*" and "*Sonnenberg*". Both districts were developed during the city expansions of the late 19th and early 20th centuries. Historically, the *Kaßberg* area was inhabited by the upper classes. The neighbourhood has a large number of very representative Art Deco buildings, with spacious and green courtyards, and large alleys lined by trees. Restoration work started early during the 1990s, and in the early 2000s, the area had a positive migration balance, as more and more people moved into the newly restored buildings (Richter 1998). The *Sonnenberg*, on the other hand, was erected as a working-class neighbourhood. It is more densely built up than the *Kaßberg*, with fewer green spaces and smaller apartments. It is also located close to the former manufacturing sites. The pace of restoration was much slower in the *Sonnenberg*, which was due to the perception of investors that the area might be less profitable than, for example, the *Kaßberg* (Feuerbach 2014). Between 2002 and 2006, the area had a negative migration balance of 6.6%, mainly due to movements to other neighbourhoods (Stadt Chemnitz 2015). Even though numerous houses have been demolished since 2000, the area still had a vacancy rate of 34% in 2007 (Stadt Chemnitz 2008b).

As a consequence of growing housing vacancies and selective population movements into more attractive districts, the city of Chemnitz faced an increasing social stratification from the 2000s. In those residential areas where rental prices were low and vacancies high, the proportion of those on social welfare was highest. Thus, in 2006, the highest share of households who received social welfare payments was found in the *Sonnenberg* (31%). Also in the *Fritz-Heckert-Gebiet*, where many more affluent households moved out and mainly poorer families and elderly people stayed, the proportion of those receiving social welfare payments amounted to 16% (Stadt Chemnitz 2008b).

Alternative strategies of urban development

The urban developments outlined above were not free of conflict, as there were competing perspectives regarding the further development of Chemnitz. Großmann (2007), who analysed the major discourses of urban development in the city, found a domination of rationalistic, growth-oriented perspectives, arguing that economic development was the only viable development strategy. This attitude was accompanied by a strong resistance against bottom-up policies or experimental urban developments. From this perspective, vacant houses or brownfields were perceived as a danger to urban identity (Großmann 2007). On the other hand, there are also more integrative approaches towards urban development. Representatives of integrative approaches stress the interrelations of urban developments, regarded as spillover effects from segmented development strategies in the housing sector in some parts of the city, which can trigger developments of other parts of the city. Within the integrative approach, sustainable development strategies are favoured, highlighting the potential of Chemnitz's inhabitants and the necessity of participatory approaches of urban development and urban governance (Großmann 2007).

The effects of those competing perceptions can be demonstrated through the discourses around an experimental development project during the 2000s: the "*Brühl*" district. Located near the main station, this inner-city neighbourhood, erected around 1910, was a lively inner-city district until the end of the 1980s, with a small, but well-developed commercial area. Since then, many houses had been abandoned, and due to the new development of the inner-city retail trade area, almost all shops in the *Brühl* pedestrian zone (Figure 4.2) had closed. In 2006, the municipal housing company GGG—owner of a large number of houses in the *Brühl* district—initiated the concept "Working and Living in *Brühl*". People who moved to the *Brühl* were offered housing for a very moderate monthly rent of 3 euros/m². For shopowners who started a business in the *Brühl*, the monthly rent was only 1.50 euros/m². The GGG company did some light renovation works to modernise heating, electricity and sanitary facilities, but further renovations were the responsibility of the tenants (*371 Stadtmagazin* 2009).

Soon, various creative businesses and a café were opened, and more and more young people moved to the neighbourhood. It slowly regained its former liveliness and redeveloped a positive image. Another major impulse for the *Brühl* was the decision to renovate a large pre-1918 industrial building located at the edge of the district and redevelop it as the main university library. The university's engagement and the close proximity to the central station raised the interest of professional investors. The GGG sold parts of its housing stock to investors, and, soon afterwards, renovation work started all over the area. The rapid development, which can be interpreted as early gentrification, was criticised by tenants and shopowners. In a series of eight interviews carried out by the author in 2014 as part of a teaching project on the topic of urban development, many interviewees criticised the contradictions between the initially participatory and rather non-commercial approach towards the area's development, and the re-introduction of free market rules as soon as the participatory revitalisation strategy worked out:

> For a while it looked as if the *Brühl* would develop in an interesting direction. But this [the participatory approach] has never really been practiced. You never know how it will turn out in the end, but I think it won't develop how I imagined it. Probably, there will be no place for me any more. (…) In the end the decisions are made by those who establish the housing prices, not by those who want to create something here.
> (Interview 4, Shop owner, *Brühl*; translated from German by the author)

> Some of the houses are renovated in a way that the tenants won't be able to afford the rent any more. And this is not what we wanted here. This is not how I imagine the *Brühl*, as a neighbourhood for young people, creative people and students. This is the wrong politics.
> (Interview 2, Café owner, *Brühl*; translated from German by the author)

Another central point of criticism concerned the urban governance strategy, aiming to manage creative processes with the instruments of top-down urban planning. But without a certain freedom and serenity, according to a city councillor, creative potential cannot unfold:

> You can compare alternative culture in a city with wild growth in a garden: you cannot develop it with a plan, but you can obstruct it. (…) Chemnitz lacks the serenity to let something grow uncontrolled, the willingness to give room to creativity. You cannot develop an alternative culture in a targeted manner…
> (City councillor, in *371 Stadtmagazin* 2009; translated from German by the author)

Recent urban re-imaginations and societal developments

The most recent urban developments in Chemnitz comprise a major shift of urban image-building strategies, connected with the struggle to cope with the

Figure 4.2 Entry gate to the *Brühl* boulevard.
Source: Author.

consequences of demographic ageing, the effects of societal cleavages, and increasing diversity. These developments must be understood in the wider frame-work of demographic and political changes in Germany since 2014/15, notably the massive increase in the arrival of asylum seekers and the politicisation of migration and asylum thereafter. This led to the rise of the right-wing populist party *AfD* (*"Alternative für Deutschland"* / *"Alternative for Germany"*), which entered the Federal Parliament for the first time after the 2017 elections with 12.6% of the votes.

In demographic and economic terms, the city of Chemnitz developed in a posi-tive way after 2010. Unemployment decreased from 12.7% in 2010 to 7.6% in 2017, and the number of employees increased from 106,864 to 115,677 over the same period (Stadt Chemnitz 2011, 2018). From 2010, there was a moderate popu-lation growth, amounting to a positive balance of around 10,000 until 2017. However, the population structure continuously changed towards older age groups (60 and older), which made up more than one-third of the population in 2017 (Stadt Chemnitz 2011, 2018). There was also an increasing internationalisation during those recent years, with the proportion of foreigners rising from 3.0% in 2010 to 7.6% in 2017. This resulted from net immigration of international students at the Technical University of Chemnitz (numbers increased from 617 in 2008 to 3,001 in 2017), and—especially since 2015—the allocation of asylum seekers. In March 2018, around 6,000 people with an "asylum background" (asylum seekers or persons with refugee or subsidiary protection status) lived in the city (Stadt Chemnitz 2018).

Urban re-imaginations and socio-cultural modernisation

In recent years, the city administration made large efforts to improve the image of Chemnitz, which is mostly perceived as an industrial city with few attractions. Various image campaigns during the 2000s emphasised the past position of the manufacturing industry for the city and at the same time depicted Chemnitz as an innovative, future-oriented and competitive business location. In 2007, the slogan *"City of Modernism"* was introduced. This refers to significant developments in Chemnitz in the period of industrial and architectural modernism (1920s) and aims to maintain "modernism" as a basic principle for future development, meaning openness to new and experimental developments in the fields of industrial production, culture, art and society (Stadt Chemnitz 2008a).

Also the new cultural strategy, titled "Give Culture Space", stressed the intention to open up towards cultural innovations as an "engine of urban development" (Stadt Chemnitz 2019), whereas before its introduction, image and development campaigns solely focused on industrial traditions and technical innovations. The new strategy aimed to promote the settlement of creative people in the urban space. The existing vacancies are now perceived as potential for the creative class, who are expected to revive the city and raise the attractiveness of formerly neglected neighbourhoods such as *Sonnenberg*, where an area of 2,000 m^2 has been developed as an incubator centre for creative businesses and start-ups (Kreativhof "Die Stadtwirtschaft"). In 2018, the support program "KRACH" was established. Winners of the annual competition can stay rent-free in a municipal business centre for three years and are furthermore supported financially and in terms of knowledge transfer to enable the development of business innovations (Förderprogramm Kreativraum Chemnitz KRACH). In addition, the successfull application of Chemnitz for European Capital of Culture 2025 fits into the new strategy. The title of the bid-book *"Aufbrüche. Opening Minds. Creating Spaces"* sums up the leitmotif to develop a future-oriented self-image for Chemnitz, based on the ruptures and contradictions of the past (Stadt Chemnitz 2020).

Those campaigns refer to the conceptualisation of the creative class as a driver of urban regeneration processes, prominently put forward by Landry (2000) and Florida (2002). Florida (2002) conceptualises the connection of high concentrations of creative workers and their possible influences on urban economies. As a necessary condition for the establishment of the creative class in urban space, he argues, not only is hard infrastructure such as mobility hubs and attractive office spaces necessary, but also soft infrastructure such as the mindset of a city.

In Chemnitz, many creative stakeholders question the success of these new strategies. They stress that the selective emigration processes of the transition years and the lack of diversified immigration since then has resulted in the absence of an urban bourgeoisie which could support alternative approaches (Palarz 2019). Thus, large parts of the (ageing) resident population are backward-looking; they resist processes of renewal and innovation and tend to selectively highlight everything that got worse since the GDR era, but barely draw any positive impulse from recent developments (Feuerbach 2014).

Intelmann (2019) denotes the recent cultural modernisation of Chemnitz as a "passive revolution", as it is not supported by the majority of the local population. The minority positions of a weak civil society, he argues, is supported by the city administration for the sake of city branding against the resistance of the local majority. However, this strategy will only be continued as long as the alternative approaches of the creative class are compatible with the neoliberal logics of a post-Fordist urban economy.

Societal ruptures: the "Chemnitz Incident"

As already described above, the stabilisation of demographic and economic structures does not mean a societal harmonisation. Notably after the numerous arrivals of asylum seekers in 2015, debates on migration and asylum politics in Germany were increasingly intermingled with debates about the effects of the transition period and the role of East and West Germans during this time. Especially older East Germans who actively experienced life in the GDR and during transition complained about missing respect for their efforts and hardship during those times. Debates around the perceived subordination of East Germans were now paralleled by debates on asylum seekers' deservingness and raised the claim of many East Germans to first solve societal injustice resulting from transition before taking care of newly arriving migrants.

In the campaigns for the parliamentary elections in 2017, these arguments were especially picked up by the AfD, which gained a large number of votes in East Germany and has a specifically strong threshold in Chemnitz, where they received 24.3% of the city vote in 2017. The dynamic of this discourse can be highlighted by the example of the "Chemnitz incident" of August 2018, when a resident of Chemnitz was stabbed during a fight with three asylum seekers and died. On the day after the crime, the AfD and some local extremist groups from the far right mobilised via social media to join a demonstration. In the late afternoon, around 800 people gathered at the scene of the crime in Chemnitz city centre, close to the Karl Marx Monument. Later, groups of hooligans roamed through the streets, looking for foreigners and attacking them. Some days later, around 6,500 people joined a demonstration of the right-wing populist local initiative "Pro Chemnitz" at the Karl Marx Monument, while the counter-demonstration of major civil society organisations only mobilised about 1,000 people. In the days and weeks after the initial incident, there was a massive increase of violent attacks against foreigners, foreign restaurants and left-wing activists.

The incidents caused a strong echo in the media at the national and even international level, triggering debates about a re-establishment of Nazi ideology in Germany, but also debates about safety and security in the light of increasing immigration. In Chemnitz and Saxony more widely, meanwhile, not only was the "incident" politicised, but also its public representation via the media. Many Chemnitz citizens protested against the prejudice that all inhabitants were Nazis and claimed their right to join anti-asylum demonstrations. In a series of public debates, initiated and documented by the local newspaper *Freie Presse*, participants delivered explanations for the lack of distance between "ordinary citizens" and violent right-wing extremists during the demonstrations. As one elderly participant argued:

> People go out on the street because they are dissatisfied, whether left or right. Since the system change [of 1989/90], life has changed for the worse: Lack of infrastructure in the countryside, different wages in East and West, better childcare in the West. We accepted it [as] we thought it would get better. But the politicians are only concerned about themselves. What happens now is a call for help from the population. People have been dissatisfied for a long time. In GDR times, there was more care for poor people, and there was more neighbourly help.
>
> (Chemnitz resident, female, 73 years old, translated from German by the author; in Peters 2018, p. 11)

In this resident's statements, we can clearly see the subjective feeling of relative deprivation, not only in economic terms, but also regarding social recognition. The self-interpretation of collective loss and deterioration is paralleled by the perception of inferiority and of having no influence regarding political processes which are imposed from "above" (see Intelmann 2019).

The question why those xenophobic, pessimistic and backward-looking positions are found in Chemnitz, but to a lesser extent in other East German cities such as Leipzig, can be partly answered with respect to the demographic development. The selectivity of mobility since 1990, which has been characterised by a loss of mainly a younger and better educated population, has resulted in the dominance of older, rather conservative and anti-modernist milieus in the city. Unlike cities like Leipzig, for example, there never was a significant counter-mobility which would have helped to compensate the population losses in quantitative, but also in qualitative terms (see Weiske 2015). Thus, bottom-up modernisation processes in Chemnitz and efforts towards a more liberal society have special difficulty in gaining ground.

Conclusion

In this concluding part, I will wrap up the main results from the analysis and discuss them against the backdrop of transformation theory, emphasising the typical features of post-socialist transition and also highlighting the specifics of Chemnitz's urban development since 1989. The conceptual discussions on the terms *transition* and *transformation* have shown that there are diverging assumptions regarding the development paths of post-socialist cities. The example of Chemnitz is specifically suitable to exemplify the struggles of post-socialist cities to align their development to past strengths, deal with the multiple ruptures caused by the system change, and at the same time develop new avenues of development and find a modern urban identity.

Like many other post-socialist cities, Chemnitz had to cope with a strong demographic and economic decline after the political change and initially had difficulties finding its place within the new political and economic system. But there were also path dependencies that made Chemnitz's urban development special after the reunification. As Table 4.2 shows, the pre-transformation conditions, notably the city's development during the time of socialism, represented an important backdrop for all further developments.

As a city with industrial traditions, Chemnitz was predestined for the typical accumulation regimes and production methods of Fordism, which were characterised by hierarchical regulation regimes, a strong state and little societal differentiation. After the system change, Chemnitz had significant difficulties leaving this outdated development path and modernising its economic, urban and governance structures. This not only affected demographic and economic development, as has been shown, but also the transformation of urban structures and society. Thus, for many years during the transition period, Chemnitz remained oriented to top-down governance processes. In the terminology of transformation theory, the period of early transition followed the idea of "transition" as a development "back to normal", where the "normal" was constituted by the realities of West Germany in the 1990s.

Only very recently have new modes of governance been introduced, such as participatory approaches, the management of plurality and the support of bottom-up processes. This can be interpreted as the adoption of the "transformation" concept, stressing the contingency of social and spatial contexts. Transformation suggests that alternative developments might be possible, and they leave more space for individualised urban development strategies, based on the contingency of social and spatial contexts, and a sensitivity for bottom-up processes and a multiplicity of stakeholders. These new strategies are widely supported by urban political and administrative stakeholders, and also by parts of the creative class, and thus receive more space and attention. However, the "Chemnitz incident" highlights the fragility of those alternative and open developments and the strong persistence of anti-modern

Table 4.2 Main aspects of post-socialist transition in Chemnitz

	Before transition	Early transition	Late transition
Demography	Stable	Shrinking, ageing	Ageing, diversification
Economy	Industrial centre	Dismantling of industries, economic transformation	Diversification, cluster developments
Image/Identity	Working-class image	Shrinking city	Competing images and stakeholder groups (conservative working class, urban bourgeois, technocrats)
Governance	Centralist planning, top-down policies	Adoption of neoliberal principles, top-down policies prevail	Neoliberalism, introduction of participatory approaches

Source: Author's own compilation.

attitudes in the population. Large parts of society are afraid of social diversity and are therefore particularly susceptible to authoritarian ideas, combined with the call for a strong state. Thus, for the years to come, the reconciliation of the different parts of society and the negotiation of diversity will be a major task. This includes the activation of the passive parts of the population and the introduction of participative formats of urban and social development. The successful application for European Capital of Culture is currently giving major impulses for these development tasks. The winning of the title of European Capital of Culture means a great chance for Chemnitz's further societal and cultural development.

References

371 Stadtmagazin (2009). Brühl vs. Reitbahnviertel, 371 Stadtmagazin, 03, available at (http://www.371stadtmagazin.de/371magazin/items/bruehl-und-reitbahnviertel-in-chemnitz.html)

Bachmann, A., Dresler, A., & Tietze, W. (1995). Chemnitz in Sachsen entwickelt ein neues wirtschaftliches Profil. *GeoJournal*, 37(4), 539–556.

Baron, M., Cunningham-Sabot, E., Grasland, C., Riviere, D., & van Hamme, G. (eds.) (2010). *Villes et regions européennes en décroissance, maintenir la cohesion territoriale.* Paris: Hermès-Sciences.

Beauregard, R.A. (2009). Urban population loss in historical perspective: United States, 1820–2000. *Environment and Planning A*, 41(3), 514–528.

Beuchel, K.J. (2006). *Die Stadt mit dem Monument. Zur Baugeschichte 1945–1990.* Chemnitz: bd Dämmig.

Billingsley, S. (2010). The post-communist fertility puzzle. *Population Research Policy Review*, 29, 193–231.

Bundesanstalt für Arbeit (BA) (1992). *Arbeitsmarkt 1991: Arbeitsmarktanalyse für die alten und die neuen Bundesländer.* Nürnberg: BA.

Fassmann, H. (2013). Transformationsforschung und die wirtschaftliche und gesellschaftliche Transformation in den ehemaligen COMECON-Staaten. In H. Gebhardt, R. Glaser and S. Lentz (eds.). *Europa – eine Geographie* (pp. 227–233). Berlin, Heidelberg: Springer.

Feuerbach, F. (2014). Stadt- und Quartiersentwicklung unter Schrumpfungsbedingungen in Deutschland und den USA – Das Fallbeispiel der Partnerstädte Chemnitz, Sachsen und Akron, Ohio (PhD-Thesis). Leipzig: Universität Leipzig.

Florida, R. (2002). *The rise of the creative class—and how it is transforming leisure, community and everyday life.* New York: Basic Books.

Frejka, T. (2008). Overview chapter 5: Determinants of family formation and childbearing during the societal transition in Central and Eastern Europe. Demographic Research, Special Collection 7: Childbearing Trends and Policies in Europe, 139–170.

Gesetz über den Aufbau der Städte in der Deutschen Demokratischen Republik und der Hauptstadt Deutschlands, Berlin (Aufbaugesetz). Vom 6. September 1950. Gesetzblatt der DDR, 1950, Nr. 104, pp. 365–367.

Geißler, S. (1991). Vorgeschichte. In M. Reum & S. Geißler (eds). *Auferstanden aus Ruinen – und wie weiter? Chronik der Wende in Karl-Marx-Stadt/Chemnitz 1989/90* (pp. 9–26). Chemnitz: Verlag Heimatland Sachsen GmbH.

Großmann, K. (2007). *Am Ende des Wachstumsparadigmas? Zum Wandel von Deutungsmustern in der Stadtentwicklung. Der Fall Chemnitz.* Bielefeld: transcript.

Hirt, S. (2008). Landscapes of postmodernity: Changes in the built fabric of Belgrade and Sofia since the end of socialism. *Urban Geography*, 29(8), 785–810.

Hollander, J.B., Pallagst, K., Schwarz, T., & Popper, F. (2009). Planning shrinking cities. *Progress in Planning*, 72(4), 223–232.

Intelmann, D. (2019). Sieben Thesen zur urbanen Krise von Chemnitz. *Bemerkungen zu den Ereignissen seit dem 26. August 2018.* sub\urban zeitschrift für kritische stadtforschung, 7(1/2), 189–202.

Köppen, B. (2005). *Stadtentwicklung zwischen Schrumpfung und Sprawl. Auswirkungen der Stadt-Umland-Wanderungen im Verdichtungsraum Chemnitz-Zwickau.* Tönning, Lübeck and Marburg: Der Andere Verlag.

Landry, C. (2000). *The Creative City: A toolkit for urban innovators.* London: Earthscan.

Lindner, U. (ed.) (2011). *Chemnitz-Karl-Marx-Stadt und zurück.* Zwickau: Chemnitzer Verlag.

Lowe, D., & Nagl, W. (2011). Bevölkerungsentwicklung und Wanderungsströme von 1991 bis 2008 für Ostdeutschland und Sachsen. ifo Dresden berichtet 2/2011. Dresden: ifo.

Meeus, B. (2016). Migration and postsocialism: A relational-geography approach. In A. Amelina, K. Horvath and B. Meeus (eds.). *An anthology of migration and social transformation. European perspectives* (pp. 87–100). Cham: Springer.

Milanovic, B. (1997). *Income, Inequality and Poverty during the transition from planned to market economy.* Washington, DC: The World Bank.

Palarz, J. (2019). Die Rolle der Kunst in der Stadtentwicklung von postsozialistischen Städten - ein Vergleich von kunstorientierten Maßnahmen in den Partnerstädten Chemnitz und Łódź und deren Wirkung (Bachelor Thesis). Chemnitz: TU Chemnitz (unpublished).

Pallagst, K. (2009). Shrinking cities in the United States of America: Three cases, three planning stories. *The Future of Shrinking Cities*, 1, 81–88.

Peters, J. (2018). "Die Leute sind schon lange unzufrieden." Freie Presse 6 October 2018. Chemnitz.

Pfeiffer, U., Simons, A., & Porsch, L. (2001). Wohnungswirtschaftlicher Strukturwandel in den Neuen Bundesländern. Bericht der Kommission. *Kurzberichte aus der Bauforschung*, 42/4, 181–189.

Reum, M., & Geißler, S. (1991). *Auferstanden aus Ruinen – und wie weiter? Chronik der Wende in Karl-Marx-Stadt/Chemnitz 1889/1990.* Chemnitz: Verlang Heimatland Sachsen GmbH.

Richter, T. (1998). *Chemnitz – Gestern, Heute und Morgen.* Leipzig: Verlagsgesellschaft Quadrat Leipzig.

Rieniets, T. (2009). Shrinking cities: Causes and effects of urban population losses in the twentieth century. *Nature and Culture*, 4(3), 231–254.

Rietdorf, W. (1975). *Neue Wohngebiete sozialistischer Länder: Entwicklungstendenzen, progressive Beispiele, Planungsgrundsätze.* Berlin: Verlag für Bauwesen.

Stadt Chemnitz (2002). *Integriertes Stadtentwicklungsprogramm.* Chemnitz: Stadt Chemnitz.

Stadt Chemnitz (ed.) (2007). *Statistisches Jahrbuch 2006/2007, Stadt Chemnitz – Die Jahre 2005 und 2006 in Zahlen.* Chemnitz: Stadt Chemnitz.

Stadt Chemnitz (ed.) (2008a). *Entwurf Städtebauliches Entwicklungskonzept – Chemnitz 2020. Beratungsvorlage.* Chemnitz: Stadt Chemnitz.

Stadt Chemnitz (ed.) (2008b). *Statistisches Jahrbuch 2008, Stadt Chemnitz. Das Jahr 2007 in Zahlen.* Chemnitz: Stadt Chemnitz.

Stadt Chemnitz (ed.) (2011). *Statistisches Jahrbuch 2011, Stadt Chemnitz. Das Jahr 2010 in Zahlen.* Chemnitz: Stadt Chemnitz.

Stadt Chemnitz (ed.) (2015). *Zwei Namen, eine Stadt. Spurensuche zur Rückbenennung von Karl-Marx-Stadt in Chemnitz.* Chemnitz: Stadt Chemnitz.

Stadt Chemnitz (ed.) (2016). *Statistisches Jahrbuch 2015/16. Die Jahre 2014 und 2015 in Zahlen.* Chemnitz: Stadt Chemnitz.

Stadt Chemnitz (ed.) (2018). *Statistisches Jahrbuch 2017/18. Die Jahre 2016 und 2017 in Zahlen.* Chemnitz: Stadt Chemnitz.

Stadt Chemnitz (ed.) (2019). *Stadt Chemnitz: Kulturstrategie der Stadt Chemnitz für die Jahre 2018 bis 2030. Kultur Raum Geben.* Chemnitz: Stadt Chemnitz.

Stadt Chemnitz (2020). Aufbrüche. Opening Minds. Creating Spaces. Bidbook available at https://www.chemnitz2025.de/en/bid-book/

Steinführer A., & Haase, A. (2007). Demographic change as future challenge for cities in East Central Europe. *Journal Compilation, Swedish Society for Anthropology and Geography,* 82(2), 183–195.

Sykora, L. (1998). Commercial property development in Budapest, Prague and Warsaw. In G. Enyedi (ed.). *Social Change and Urban Restructuring in Central Europe* (pp. 109–136). Budapest, Hungary: Akademiai Kiado.

Turok I., & Mykhnenko, V. (2007). The trajectories of European cities, 1960–2005. *Cities,* 24(3), 165–182.

Viertel, G., & Weingart, S. (2002). *Geschichte der Stadt Chemnitz. Vom 'locus Kameniz' zur Industriestadt (Stadtgeschichte).* Gudensberg-Gleichen: Wartberg Verlag.

Weiske, C. (2015). Konflikte in einer alternden Stadt. In: Bundesinstitut für Bau-, Stadt-und Raumforschung (BBSR) (ed.). *Einheit und Differenz* (pp. 471–486). Stuttgart: Franz Steiner Verlag.

Wolff, M., & Wiechmann, T. (2018). Urban growth and decline: Europe's shrinking cities in a comparative perspective 1990–2010. *European Urban and Regional Studies,* 25(2), 122–139.

5 Young people's life plans and their impact on the demographic future of a shrinking city

Kielce case study

Mirosław Mularczyk and Wioletta Kamińska

Introduction

The processes of globalisation and European integration, which in the 1990s left a strong imprint on the countries of Central and Eastern Europe, including Poland, had an influence not only on the economic development of these countries, but also on the citizens' mentality and awareness level (see: Kotowska 1999, Szukalski 2004). The strong emphasis put on the needs of an individual, such as self-fulfilment and self-acceptance, was followed by a change of attitude towards the traditional model of family, maternity and the role of women in a modern society and family. That change was manifested by a rapidly decreasing fertility rate, which eventually led to population shrinkage and altered the demographic structures in individual regions. In many European countries, the continuous population decline made it necessary to reformulate the strategy of economic development and population policy. The state, regional and local authorities in those countries made every effort to modify the models of population procreation and to minimise the decline of the demographic potential due to migration outflow.

Young people have a strong influence on the demographic situation of individual countries, cities and villages. They take decisions about starting a family and having children. A number of studies (Solga 2013, Jończy et al. 2013, Rokitowska 2017) indicate that they also usually decide to change the place of living and working. Their level of education, financial situation, work opportunities, attitudes or moral patterns determine the size of the family and possible migrations.

Young people's behaviours as regards procreation, starting a family, attitudes and moral patterns vary, depending on the country and place of residence (Lesthaeghe 1991, Kurek and Lange 2013), as well as family and religious conditions (Lesthaeghe 2020). It is generally assumed that the young part of the society are usually more liberal and open as regards having a family and career in developed countries, large cities and central areas of individual regions around the world. The question that arises concerns the life plans, family preferences and career dreams of young people living in medium-sized cities, which have been losing their demographic potential for many years. Can young people's decisions concerning procreation and migration influence the future of medium-sized, shrinking cities?

DOI: 10.4324/9781003039792-5

In view of the above, the purpose of the chapter is to present the life plans of academic youth studying in Kielce and to define their potential influence on the demographic future of the city. Young people, especially students, are

> a barometer of changes and social moods. The situation of the young, the way they perceive the world, their aspirations and life pursuits are a measure of the transformations that have already taken place, as well as the distance that is still to be covered.
>
> (Szafraniec 2011, p. 11)

Students in Kielce form a relatively large social group. In 2018, it consisted of about 24,000 students, which placed the city at the 13th position among the university centres in Poland, with 121 students per 1000 inhabitants (Nowak and Wieteska, 2019). On the one hand, students are a very mobile social group, with high aspirations. They are young, unafraid of new challenges, ready to change the place of living in search of an interesting job and good living conditions. They usually do not have any family responsibilities (a spouse or children) and are not financially attached to a given place by having a house or an apartment. They may freely change the place where they live or work. On the other hand, due to their ability to search and process information, students adapt to new conditions and adopt new worldview models. That is why their decisions that regard starting a family and procreation are extremely important from the point of view of shaping the demographic future of urban centres.

The results will be interpreted with reference to the concept of the second demographic transition. The analysis conducted for the purposes of this chapter was based on a literature review, an analysis of statistical data provided by the Central Statistical Office (GUS), as well as on the results of a survey carried out among students of the largest university in Kielce—Jan Kochanowski University. The questionnaire used in the survey included questions regarding elements described in the second demographic transition concept (e.g. plans to enter into marriage, a consensual relationship, the age at starting the family, the number of children, preferred place of residence, etc.).

The second demographic transition theory vs. young Poles' procreative behaviours

Theory-wise, contemporary population changes, including young people's procreative behaviours, are explained through the conception of the second demographic transition (Van de Kaa 2003, Lesthaeghe 2010). It describes the process of population reproduction in contemporary developed societies, which is characterised by the lack of generational replacement on the one hand, and by non-traditional family patterns on the other (Szukalski 2005). According to the authors of this concept, the first demographic transition, which started in the late 18th century, did not end (as it was expected) with stabilised population development at the birth rate close to zero. In many countries of Northern and Western Europe, birth-rate values decreased to a level lower than the mortality rate. In the long run, it led to a population decline and threatened the process of generational replacement. It turned out that procreation models had changed, which was clearly visible in the falling fertility rate.

The predominant view presented in literature is that the main cause of changing procreation models was the broadly understood modernisation of socio-economic life (Kotowska 1999, Pruszyński and Putz 2016, Herudzińska 2012), for instance, in the growing individualisation. Societies accepted and propagated putting personal needs of self-acceptance and self-fulfilment before family interests. As indicated by Janiszewska (2013), never before, had emotions, feelings and individual needs determined the pace and degree of family development to such an extent. Changes in procreative behaviour were reinforced by Weberian rationalisation, which led to freeing whole societies from religious and family constraints, as well as traditional beliefs (Herudzińska 2012).

Apart from a significant decrease in the birth and fertility rates, contemporary population reproduction trends also include late marriages, the increasing number of separations, divorces and extra-marital births, the changing family model and household structure, the number of unwanted births, as well as voluntary childlessness becoming more and more common (Van de Kaa 1987, Kurek and Lange 2013).

Relating to the second demographic transition in North and West European countries, Lesthaeghe (1991; cited in Kotowska 1999) distinguished three stages of transformations. Stage I took place in 1955–1970. Its characteristic features included the falling number of first marriages, increasing number of divorces, decreasing fertility of women and a later age at first marriage. Stage II (1970–1980) was characterised by the growing popularity of informal relationships, such as cohabitation and LAT (living apart together), the growing number of extra-marital births, the later average age of women bearing their first child and the later average age at first marriage. Stage III began in the second half of the 1980s. It was marked by a stabilised number of divorces, falling number of second marriages, and further increase in the number of informal relationships, especially among divorced and widowed persons, a falling fertility rate in younger age groups (15–24 years) and a growing fertility rate in older age groups (30–39 years).

The demographic changes described above reached Central and Eastern Europe in the 1990s and were directly related to the economic and political system transformation (cf. Kurek and Lange 2013, Pruszyński and Putz 2016), which radically changed the labour market, as a result of education becoming increasingly significant and the decreasing sense of social security. The transformation also increased alternative costs, i.e. the loss of potential opportunities due to taking wrong decisions regarding marriage and/or motherhood. Another outcome of the transformation was the growing difficulty in combining the roles of a partner and a parent. These phenomena were intensified by, on the one hand, instability caused by the former system of values and principles, and on the other—by an increased sense of individual responsibility for one's career (Pruszyński and Putz 2016).

The conception of the second demographic transition has been repeatedly verified in empirical studies. Globally, evidence to confirm this theory have been presented by Lesthaeghe (2020), Coale (1973), Lesthaeghe, Van de Kaa (1986), and continentally—Lesthaeghe (1983), Philipov and Kohler (2001), Van de Kaa (1978),

Esteve et al. (2013), Frejka et al. (2010) Janiszewska (2013), Kurkiewicz (1998). There are also many publications presenting changes in procreative behaviour and family evolution in individual countries, e.g. in China (Feng and Quanhee 1996), the Philippines (Kuong et al. 2016), the USA (Lesthaeghe et al. 2016), Botswana (Mokomane 2005), etc.

In Poland, population changes referring to the second demographic transition have also been analysed. Research has been conducted on different spatial scales. The spatial aspect of the changes in procreative behaviour occurring in the whole country was explored by Kurek and Lange (2013). Their findings show the following changes in procreative patterns, which took place in Poland in the 1990s: a decrease in the number of births and fertility rate, later first pregnancies, an increase in the average age at having the first child, as well as a significantly higher percentage of extra-marital births. The authors quoted above noticed the spatial diversification of those changes. The decrease in the value of the total fertility rate was larger in the countryside than in cities. There were also differences between large cities, their surroundings and rural areas. The lowest fertility rate was identified in large urban centres, and it was higher in the suburban areas. The average age of women bearing children was higher in the eastern part of the country, while that of women bearing their first child was higher in large agglomerations and rural suburban areas than on the peripheries of the country.

Podogrodzka (2013) analysed fertility and marriage patterns in Poland during the system transformation period (1990–2010). Research demonstrated that in 1990–2004, the total fertility rate was systematically decreasing, both in urban and rural areas, but the decrease was larger in the countryside. After that period, the fertility rate slightly increased. The intensity of entering into marriage also slightly fluctuated in time. Before 2004, it had weakened and next increased again, both in urban and rural areas. Simultaneously, with time, the youngest persons were becoming clearly less inclined to get married (Podogrodzka 2013).

Family evolution and changes in the fertility rate among Polish women were also investigated by Szukalski (2004). He paid special attention to the problems of childlessness and having a small number of children. His research shows that in Poland, in the generations born in the first two decades of the 20th century, the percentage of childless women came up to several per cent, while among those born in the 1930s—only 2–8% did not have children. In comparison, the percentages of childless women born in the 1940s and 1950s were 4.5% and 3.5%, respectively (Szukalski 2004), while of those who were born in mid-1960s—15.5% (Beaujouan et al. 2016).

The problems of demographic changes in Poland were also studied in depth by Kotowska with her team (2014), and the results were presented in a book, entitled *Niska dzietność w Polsce w kontekście percepcji Polaków* (*Low fertility rate in Poland in the context of Poles' perception*). In this work, Kocot-Górecka (2014) notes that after 1989, the demographic dynamics in Poland rapidly slowed down, and the main characteristic of those changes was the decreasing number of births. According to this author, the total number of births in 1989–2003 decreased by nearly 40%. It happened at the time when women born in 1970–85 (the echo of the post-war baby boom) were at the peak of their procreative activity. After 2004, the number of births slightly rose. The author also noticed that the characteristic feature of the procreative changes was the increasing number of extra-marital births. The percentage of those births in the overall number of live births increased from 6% to 22%. The number of extra-marital births in the countryside increased by 189%, and in the city—by 151% (Kocot-Górecka 2014). The total fertility rate substantially decreased: in 1989, it was 2.07 and in 2012—1.3. At the same time, it must be emphasised that fertility rates which do not exceed 1.3 signify very low fertility.

Mynarska (2014) remarked that in Poland, after 1989, the number of new marriages fell and the age of getting married for the first time systematically increased. From the early 1990s, the average age of Polish women entering into marriage for the first time increased by three years. There was also a clear change in the age at which people most often got married, from 20–24 to 25–29 years (Kotowska and Giza-Poleszczuk 2010, after: Mynarska 2014). The authors of the cited publication referred to informal relationships as well. In the 1970s, only 9% of women started their relationship from co-habitation (90% of women got married without living together before). In the first decade of the 21st century, 60% of women chose cohabitation when developing their first relationship. In Poland, we could observe an increasing number of divorces, which, according to Mynarska (2014), resulted from the de-institutionalisation and destabilisation of family. In the 1970s and 1980s, the number of divorces per 1000 inhabitants oscillated around 0.15, and in the first decade of the 21st century, it was about 0.25–0.30.

Changes in the quality and structure of the family and informal relationships have been the subject of numerous studies (Szukalski 2004, Frątczak, Pęczkowski 2002, Slany 2002, Frątczak, Kozłowski 2005, Kotowska 2009, Kotowska and Giza-Poleszczuk 2010, Abramowska-Kmon 2014, Duszczyk et al. 2014). They show that in 1988–2011, a married couple with children represented the basic model of the family in Poland, but the percentage of this type of relationship decreased by 12%. The percentage of couples (informal and formally married) without children increased by 3%, and of single mothers—by 6%. More and more people decided to live without a formal or informal relationship. According to Abramowska-Kmon (2014), in 1988–2011, the percentage of bachelors increased by 5%, and the percentage of unmarried women—by 4%. At the same time, the percentage of unmarried young people (up to 29 years old) increased, which points to the weakening inclination to enter into matrimony.

The results of research on young people's family plans, conducted from the point of view of demographic changes, are very interesting. Janiszewska (2013) surveyed the opinions of young inhabitants of Łódź, regarding family and informal relationships. Research showed that one-quarter of all the respondents were staying in different forms of informal relationships. A positive and definitely positive attitude to cohabitation was declared by nearly half of the respondents. They claimed that the potential decision to get married may result from the desire to start a family and achieve stability, or from religious reasons. The motivation to start an informal relationship is often purely rational and instrumental (convenience, lack of commitments, the possibility to end the relationship easily) (Janiszewska 2013).

Similar research was conducted by Skibiński (2017). The researcher focused on the procreative and matrimonial attitudes among university students in the Silesian

Voivodeship. The study showed that nearly all respondents (90%) were interested in having children in the future, with a majority of this group thinking of having two. Planning one and many children was not very common. The respondents declared that they would carry out their procreative plans within three years or longer.

The research presented above also showed that young people, regardless of their marital status, were sexually active. The author found that it was clearly unrelated to procreation, which the respondents also confirmed by their answers to the next question, regarding contraception. The majority definitely accepted using different means to prevent unwanted pregnancy.

Kielce—a shrinking city

The problem of shrinking cities was broadly discussed in the literature on the subject for several decades (Downs 1997, Mykhnenko and Turok 2008, Cunningham-Sabot et al. 2013) and is still very popular with researchers. Despite this, adequate and accurate terminology has not been created yet. In the literature on the subject, we can find two approaches to defining the process of city shrinkage. In the first one, researchers consider the growing depopulation, related to the increasingly unfavourable economic situation of the city, while in the other one—city shrinking is treated as a process of social, economic and spatial transformations, which takes place in the conditions of a continuous population decline (Zborowski et al. 2012).

It is usually assumed that a shrinking city is a settlement unit characterised by a decreasing number of inhabitants, economic slowdown, falling employment rate and social problems resulting from the structural crisis (Cunningham-Sabot et al. 2009). Some researchers additionally define the minimum population of such a unit as 10,000 inhabitants (Wiechman 2007), the minimal population decrease rate as 0.15% annually (Stryjakiewicz et al. 2014), and the shortest period during which the demographic decline is recorded as five years (Pallagst 2010).

It is also difficult to identify and classify the factors responsible for the shrinking of cities and to choose the indicators describing this phenomenon. The most common causes of city shrinking include globalisation and de-industrialisation (Wu and Martinez-Fernandez 2009, Pallagst 2010, Cunningham-Sabot & Fol 2009). On the one hand, globalisation causes the development of global cities with a large concentration of services of the highest order, innovative enterprises and high technology. On the other hand, it generates negative phenomena, especially in industry, due to which the cities are not able to face the international competition. As a consequence, cities which have not managed to adjust to functioning in the conditions of the global, knowledge-based economy are becoming increasingly different from those which belong to the global network (Stryjakiewicz and Jaroszewska 2011). Accordingly, we can observe a population outflow from backward cities, which leads to a number of negative changes in the demographic and economic structure of individual settlement units. This means that the same factors which are responsible for the socio-economic development of some cities may trigger the shrinking process in others (Wu and Martinez-Fernandez 2009).

The literature presents a large number of different indicators of urban shrinking, including socio-demographic factors (the decreasing number of inhabitants, migration balance and population growth rate, demographic ageing), economic factors (the unemployment rate, local self-governments' incomes, the level of employment support), as well as spatial ones (the number of vacant houses, new apartments or issued house construction permits).

Shrinking cities have been identified all over the world, both in the developed countries (Cunningham-Sabot and Fol 2009) and in those which are at a lower level of economic development (Audirac and Arroyo-Alejandre 2010). Also, there are numerous descriptions of this type of cities in Eastern Europe (Mykhnenko and Turok 2008), including Poland, (Stryjakiewicz and Jaroszewska 2011, Musiał-Malago' 2017).

An example of a Polish shrinking city is Kielce,[1] which displays characteristic demographic and economic features (Musiał-Malago' 2018).

Kielce is a city in south-eastern Poland (Figure 5.1). Since 1999, it has been the capital of Świętokrzyskie Voivodeship, currently inhabited by ca. 195,000 people. In 1988–2019, the number of population decreased by over 16,000. The mean decrease rate was over 0.15% annually. A similar decrease rate was observed in 23 other cities with a comparable number of inhabitants (100–200,000) and of the same status (a city with *powiat* rights).

Since the second half of the 20th century, Kielce has been an academic city, with two state universities (Jan Kochanowski University and Świętokrzyska Technical University) as well as six private higher schools. In 2018, the largest of them, Jan Kochanowski University, educated about 11,000 students, which made up over 45% of all students in the voivodeship.

For many decades, the city was developing as a major centre of the Staropolski Industrial District, based on the metallurgic, electromechanical and mineral industry. In Kielce, there was a number of large industrial enterprises, concentrating a significant part of the regional workforce. After 1989, most of them did not adapt

Figure 5.1 The location of Kielce.
Source: Authors' elaboration.

to the rules of market economy, which resulted in their liquidation or restructuring, and, consequently, to employment reduction. The developing new labour market was unable to absorb the surplus of work force, which led to a growing unemployment. It was the primary cause of the population outflow, which could not be stopped by developing new functions in the city.

Negative demographic processes were noticeable in Kielce from the beginning of the transformation period. The economic impacts of the population decline were strengthened by the social modernisation processes described earlier. They could be seen in the changes of procreative patterns and the growing mobility of the population. In 1988–2019, the number of births drastically decreased from 2921 to 1681, i.e. by 43% (Table 5.1). It was a much larger drop in the number of births than the average value for Poland (31%). Analogically, the birth rate (per 1000 inhabitants) in Kielce, during the whole period in question, decreased by 40%, while in Poland—by 35%. In 2019, the birth rate for Kielce was about 8.61‰, while for the whole country—9.77%. Compared to cities with similar population and of the same status, the birth rate in Kielce was unfavourable. Out of the 23 cities inhabited by 100–200,000 people and holding the status of a city with *powiat* rights, only nine displayed a lower birth rate than Kielce.[2]

The natural population growth in Kielce was positive only at the beginning of the transformation period (1988–1995) and in the 21st century, it became negative (except 2010). At the end of the period in question, it was −2.6% for Kielce and −0.91% for the country. Among medium-sized cities with *powiat* rights, Kielce occupied the 12th position (in 11 cities the population growth rate was more favourable, and in another 11, the population natural decrease rate was higher than in Kielce).

During the period in question, in Kielce, we could also observe a worrying decrease in the total fertility rate. In 1995–2019, it dropped by 8%, from 1.319 to 1.210. In 2019, it was lower than the average for Poland: −1.410. Compared to other cities, the fertility rate in Kielce should also be seen as negative. According to Szukalski (2019), in Rzeszów, it was 1.416, in Olsztyn 1.570, in Zielona Góra 1.443, in Opole 1.255, Gorzow Wielkopolski 1.415 (data for 2018), while in Kielce −1.268.

The unfavourable demographic situation of the city, resulting from the changing procreative patterns, enhanced migrations. From the beginning of the transformation period, the migration balance was negative and towards the end of 2019 it was −2.4%. Numerous studies show that the most active in terms of migration are young, well-educated people. Thus, it can be assumed that the population changes do not refer just to the quantitative decline, but also to the quality of human and social capital of the city.

In Kielce, like in the whole country, people changed their attitude to getting formally married. Social activity as regards entering into matrimony definitely decreased, which was clear from the falling number of new marriages: it dropped from 1387 in 1988 to 839 in 2019, i.e. by over 60%. Calculating per 1000 inhabitants, the marriage rate in 2019 stood at 4.3, which was less than the national average (4.78). In comparison to other cities with *powiat* rights, populated by 100–200,000 people, this indicator should also be evaluated negatively. In 14 urban centres of this type, marriage rates were higher (from 5.55% in Gorzow Wielkopolski to 4.41% in Płock).

Table 5.1 Selected demographic data for Kielce in 1988–2019

Item	1988	1995	2000	2005	2010	2015	2019	1988–2019 rates in %
Number of population	211077	213777	213469	208193	202450	198046	194852	92.3
Live births	2921	1954	1686	1645	1954	1707	1681	57.5
Birth rate (per 1000 inhabitants)	14.2	9.2	7.9	7.9	9.62	8.60	8.61	60.6
Deaths	1607	1747	1719	1704	1863	20143	2178	135.5
Mortality rate (per 1000 inhabitants)	7.8	8.3	8.0	8.2	9.2	10.3	11.2	143.6
Population growth rate (per 1000 inhabitants)	6.4	1.0	−0.1	−0.3	0.45	−1.7	−2.6	x
Fertility rate	x	1.319	1.083	0.972	1.200	1.119	1.21	84.8
Migration balance of a person	739	−194	−563	−1026	−1122	0	−474	x
Migration balance (per 1000 inhabitants)		−0.9	−2.7	−4.9	−5.5	0	−2.4	x
Number of marriages	1387	1134	1215	1072	1235	859	839	60.5
Number of marriages (per 1000 inhabitants)	6.57	5.30	5.69	5.15	6.10	4.34	4.31	65.6
Number of divorces	x	x	193	200	355	355	289	149.7 (compared to 2000)
Number of divorces (per 1000 inhabitants)	x	x	0.7	1.0	1.7	1.8	1.5	214.3 (compared to 2000)

Source: Central Statistical Office data, stst.gov.pl

In 2000–2019, the divorce rate in Kielce doubled, from 0.7% to 1.5%. In absolute numbers, the number of divorces was not very large, but we should note the growing trend. Compared to the figures calculated for the whole country (in 2019—1.7%), the divorce rate in Kielce was low. This means that the inhabitants of Kielce are attached to the institution of marriage more than the citizens of other medium-sized urban centres.

In view of the facts presented above, it should be stated that the demographic changes that took place in Kielce during the transformation period can be explained through the concept of the second demographic transition. In 1988–2019, birth, marriage and fertility rates decreased and the number of divorces increased. The population outflow rate was negative, similar to the migration balance values. As a consequence, we can observe a substantial, systematic population decline and urban shrinkage.

It should be emphasised that Kielce is a particular example of a city that has been shrinking due to a number of reasons. Firstly, it is the capital of the Świętokrzyski region, where population problems had already been identified towards the end of the 20th century. The inhabitants' outflow at the turn of the 21st century had such an adverse impact on the economy that it was decided to create one of the two special demographic zones in Poland there. Population depression causes far-reaching social and economic consequences for Kielce and the whole region.

Secondly, the actual population decrease in Kielce is significantly larger than shown by official statistics, due to the incomplete record of external and internal migrations.

Thirdly, Kielce is the main centre of a voivodeship located in Eastern Poland. It is a peripheral area, in relation to the country and the European Union. After Poland joined the EU, those were the poorest areas in the whole European Union, with the GNP per inhabitant at 75% and below the average GNP for the EU. The economic situation of this region was worrying enough to create a special development program for Eastern Poland. Peripheral areas are threatened by the drainage of increase and development factors.

In addition to that, the forecasts prepared by the Central Statistical Office, regarding the changes in Kielce population potential by 2050 are very worrying. According to them, in 2014–2050, the number of Kielce inhabitants will have decreased by nearly one third.

Therefore, research on different aspects of the demographic future of cities like Kielce is very important. The trends that have been observed may be used to plan strategies for cities of similar demographic, economic and location conditions.

Survey results

Characteristics of survey respondents

The survey referred to in the chapter was conducted among 758 final year students of Jan Kochanowski University (UJK) in Kielce, the largest university in the region. The analysis included 750 verified questionnaires and took place in December 2019 and January 2020. The majority of the respondents were women (72%). The structure of the sample reflects the structure of students. Most respondents (nearly 65%) were Kielce residents, 21% came from the country and 14%—from a small town. All the respondents were attached to Kielce through the studies they had undertaken at the UJK. The majority of them (over 53%) rated their financial situation as good, about 33%—as average, 9%—as very good and 4%—as bad (Table 5.2).

Table 5.2 Respondents' characteristics

Characteristic	No. of respondents	Percentage
Sex		
women	546	72.8
men	204	27.2
Place of permanent residence		
village	156	20.9
small town	106	14.2
Kielce	486	64.9
Financial situation		
bad	32	4.3
average	248	33.1
good	400	53.3
very good	70	9.3

Source: Authors' elaboration.

Kielce students' plans regarding place of residence and starting a family

As it has been mentioned earlier, at the beginning of the 21st century, we could observe a clearly decreasing number of population in most Polish cities. A major cause of this situation was the growing emigration due to suburbanisation, as well as the outflow of young people to more attractive destinations (Solga 2013, Jończy et al. 2013, Rokitowska 2017). The same processes took place in Kielce. The survey showed that over 45% of the respondents were going to leave this city. Every fourth student planned to move to another Polish city, and about 15%—to live abroad. The smallest group (3%) declared settling down in the countryside. Nearly 44% of female students and nearly 50% of male students were going to leave Kielce. Considering the students' assessment of their own financial situation, the largest percentage of those who planned to migrate was observed in the group rating their financial status the lowest. The majority of this group were willing to move abroad. Among the respondents who rated their own financial situation higher, the percentage of those who were planning to leave their place of residence was smaller (Table 5.3).

Under 29% of the respondents, every third man and every fourth woman planned to stay at their current place of residence. Considering the financial situation of Kielce students, nearly half of those who rated it the highest were not going to leave their current place of residence. Among those declaring the lowest financial status, nearly 19% had such plans.

One fourth of the respondents did not have any specific plans for the future as regards place of residence. More women than men could not decide (ca. 29% and ca. 18%, respectively). The smallest group of respondents who could not decide were students who rated their financial situation as bad (Table 5.3).

Table 5.3 Respondents' plans regarding the future place of residence

Respondents		After completing education, I am planning to live:				
		At the current place of residence	In another Polish city	In another Polish village	Abroad	I don't know
				%		
	Total	28.5	26.9	3.2	15.2	26.1
By sex	women	27.1	25.6	4.0	13.9	29.3
	men	32.4	30.4	1.0	18.6	17.6
By financial situation	bad	18.8	25.0	6.3	37.5	12.5
	medium	28.8	28.8	4.0	10.4	28.0
	good	25.6	29.1	2.5	16.1	26.6
	very good	48.6	8.6	2.9	17.1	22.9

Source: Authors' elaboration

The survey results let us assume that negative migration tendencies, typical of Kielce, probably will not change in the near future. The reason why university students want to emigrate is their desire to improve their financial situation.

It can be assumed that the respondents' plans regarding the place of residence will not stop the process of city shrinking, quite on the contrary—they may intensify it. This will result in a further decrease in the number of residents, as well as the erosion of the social capital of the city. The outflow of young, well-educated residents will certainly have an impact on reducing entrepreneurship and the use of social and material infrastructure.

The institution of marriage is strongly rooted in Polish people's system of values. In Kielce and in the whole country, since the beginning of the 1990s, it has been possible to observe a continuous decrease in the number of new marriages. The mean age of getting married for the first time in Poland has been growing and, currently, it is nearly 30 years for men and 28 for women. The family plans of Kielce students were more optimistic in this respect. Over 94% of the respondents planned to formally enter into marriage. Over 50% planned to do that at the age of 24–26. This age range dominated in Poland in the 1990s. Nearly 1/4 of the respondents planned to get married at the age corresponding to the average for Poland, i.e. at 27–30 (Table 5.3). The age at which Kielce students planned to get married varied, depending on sex. Women were going to do that earlier than men (Table 5.4).

A significant factor that had an influence on the planned age at marriage was the respondents' evaluation of their financial status. The higher it was evaluated, the earlier the respondents planned to get married. Over 71% of the students who described their financial situation as very good planned to get married by the age of 26. Only 43% of those who described it as bad had similar plans (Table 5.4).

Table 5.4 The age at marriage planned by the respondents

Respondents		Planned age at marriage							
		Before 18	18–20	21–23	24–26	27–30	After 30	No such plans	I don't know
		%							
Total		0.3	1.3	13.1	50.4	24.3	4.5	5.3	0.8
By sex	women	0.4	1.8	15.0	53.8	20.1	2.2	5.5	1.1
	men	0.0	0.0	7.8	41.2	35.3	10.8	4.9	0.0
By financial situation	bad	0.0	0.0	18.8	25.0	25.0	6.3	18.8	6.3
	average	0.8	2.4	12.8	51.2	27.2	3.2	2.4	0.0
	good	0.0	1.0	13.1	50.3	24.1	5.0	6.0	0.5
	very good	0.0	0.0	11.4	60.0	14.3	5.7	5.7	2.9

Source: Authors' elaboration

The analysis of the students' age at which they were planning to get married shows that marriage and family were embedded more strongly in the minds of women than men. Men's plans more often concerned first of all their education, finding a job and only later marriage and family. It was an expression of their greater responsibility for the financial situation of the future family. A significant obstacle to getting married by young people studying in Kielce was their financial situation, as well as the growing desire to have a stable professional life and career. It can be assumed that delaying marriage will result in a decreasing birth rate in future years.

On average, Polish women have their first child at the age of 27. It is similar in Kielce, where students' plans in this respect do not diverge from national and regional trends. The age interval most often indicated by the respondents, as regards their plans to have the first child, was 27–30 (Table 5.5). Parental plans varied, depending on sex. Women planned to be parents earlier than men. Nearly half of female students and under one-third of male students planned to have the first child by the age of 26 (Table 5.5).

A significant factor was the students' evaluation of their own financial situation. The better it was, the earlier the parenthood plans were (Table 5.5). This probably resulted from the fact that the people assessing their financial situation the lowest were worried that they would be unable to raise their children in good conditions. Their own negative experience of financial shortages also had an impact on their intentions. Thus, it was not the future professional career or the consumptive model of life that stopped the students from planning the number of children, but the family's financial status, which was not the concern of the students who rated their financial situation higher (Table 5.5).

The age at which the respondents planned to have the first child was related to the age at which they planned to get married. They represented the phenomenon of "postponed parenthood", observed in Poland—young people first complete

Table 5.5 The age at having the first child, planned by the respondents

Respondents		Planned age at having the first child							
		Before 18	18–20	21–23	24–26	27–30	After 30	No such plans	I don't know
		%							
Total		0.0	0.3	4.8	36.0	42.4	8.0	8.0	0.5
By sex	women	0.0	0.4	5.5	39.6	41.4	4.8	7.7	0.7
	men	0.0	0.0	2.9	26.5	45.1	16.7	8.8	0.0
By financial situation	bad	0.0	0.0	0.0	25.0	31.3	12.5	25.0	6.3
	average	0.0	0.8	5.6	31.2	51.2	6.4	4.8	0.0
	good	0.0	0.0	4.5	38.7	41.2	8.0	7.5	0.0
	very good	0.0	0.0	5.7	42.9	22.9	11.4	14.3	2.9

Source: Authors' elaboration

Table 5.6 The number of children planned by the respondents

Respondents		Planned number of children				
		None	One	Two	Three	Four and more
		%				
Total		6.9	8.8	49.6	29.3	5.3
By sex	women	5.9	8.8	52.0	28.2	5.1
	men	9.8	8.8	43.1	32.4	5.9
By financial situation	bad	25.0	0.0	56.3	18.8	0.0
	average	4.8	5.6	47.2	32.0	10.4
	good	7.0	10.6	52.8	27.6	2.0
	very good	5.7	14.3	37.1	34.3	8.6

Source: Authors' elaboration

their education, next strive at economic stabilisation and—having achieved it, usually in their early thirties—they decide to have the first child (cf. Kotowska 2014). Such behaviour is typical of societies at the second demographic transition.

In 2019, the fertility rate in Poland stood at 1.41 child per woman and much lower in Kielce—1.21, which did not ensure generational replacement. Kielce students' plans were optimistic. Over 84% planned to have two and more children, about 9%—one child and about 7% did not intend to have children at all. Considering the respondents' sex, it was noticed that a larger percentage of men (nearly 10%) than women (nearly 6%) did not plan parenthood. On the other hand, there were more men (ca. 38%) than women (ca. 32%) willing to have three or more children (Table 5.6).

Similar to the age at which the students planned to get married and have children, a significant factor in planning the number of children was the evaluation of one's own financial situation. The students who rated it the lowest, usually (56%) planned to have two children. The situation in the remaining groups was similar, but the percentage of declarations in this respect was smaller (Table 5.6). The fear that it will be impossible to provide suitable living conditions for the children is visible when we analyse the respondents' declarations to have three, four and more children. About one-fifth of those who saw their financial situation as bad planned to have three children. Similar plans were reported by one-third of the respondents who considered it to be average, good or very good (Table 5.6). None of the students rating their financial situation as bad wanted to have four or more children (Table 5.6).

The figures regarding the planned number of children imply that Kielce students wanted and intended to have them. The number of children they planned (mean value 2.17) would ensure generational replacement. However, research shows (Kotowska and Giza-Poleszczuk 2010; Matysiak and Mynarska 2014) large discrepancies between young Poles' plans and their actual execution. Although the respondents usually wanted to have two children, the majority will probably have one, matching the family model characteristic of the second demographic transition.

Kielce students' attitude to family values

During the second demographic transition, the importance of marriage weakened for the benefit of cohabitation. The attitude to the latter changed from negative to positive. The number of relationships of this type in Poland is rising. The survey revealed that most respondents (ca. 61%) had a positive attitude to informal partnerships. A positive attitude to cohabitation was expressed more often by women than men (Table 5.7). In view of the students' financial situation, the largest percentage of those who expressed a positive or rather positive attitude to informal relationships was represented by the group of the lowest financial status. Among the respondents in a better financial situation, the percentage of students with a positive or rather positive attitude to cohabitation was smaller (Table 5.7). Negative and rather negative attitude to informal relationships was declared by nearly one-fifth of all respondents. It was more often expressed by men than women (Table 5.7). The smallest percentage of students expressing a negative and rather negative attitude to cohabitation was identified among the respondents who rated their financial situation the lowest. A larger percentage of respondents expressing negative attitudes to informal partnerships was identified in groups rating their financial status higher. Nearly every fifth respondent did not have an opinion regarding informal partnerships (Table 5.7).

The university students' attitudes to informal partnerships are liberal, similar to the attitudes of the Polish society as a whole (Janiszewska 2013). However, it does not indicate that traditional family norms and values are declining among students, particularly women and the students in the least advantageous financial situation, which is suggested in the second demographic transition theory. It is untrue because the percentage of respondents who are not planning marriage is

Table 5.7 Respondents' attitude to informal partnerships

Respondents		Attitude to informal partnerships				
		Negative	Rather negative	Rather positive	Positive	I don't have an opinion
				%		
Total		14.4	24.0	19.7	22.7	19.2
By sex	women	12.5	21.2	20.9	24.5	20.9
	men	19.6	31.4	16.7	17.6	14.7
By financial situation	bad	18.8	31.3	12.5	18.8	18.8
	average	16.8	22.4	20.0	16.0	24.8
	good	12.1	25.1	18.6	26.6	17.6
	very good	17.1	20.0	28.6	25.7	8.6

Source: Authors' elaboration

insignificant. It can be assumed that a positive attitude to cohabitation concerns only the pre-marital period: preparation to marriage, prolonging the engagement period. Probably, relationships which are alternatives to marriage (e.g. LAT, concubinage) are accepted less willingly.

One of the characteristic features of a society at the second demographic transition period is the increasing number of singles. Living the life of a single is advocated by about one-quarter of Poles (*Kontrowersje wobec…* 2008). The majority of respondents expressed a positive attitude to this style of living. It was much more popular among women than men. The largest percentage of students declaring a positive attitude to living the life of a single was identified among the respondents describing their financial status as low. The supporters of this style of living made up over 80% of all the students rating their financial situation as bad. About 16% of the respondents expressed a negative attitude to being a single. The percentages of female and male students with a negative attitude were similar. The percentages of those who criticised single life were not significantly diversified among students evaluating their own financial situation differently. About 15% of the respondents did not have an opinion about living single life (Table 5.8).

Kielce academic youth, particularly women and the respondents who rated their financial status the lowest, had considerably more liberal opinions about living single life than the remaining part of Polish society (*Kontrowersje wobec…* 2008). The positive attitude to functioning in the society as a single, without a family, is characteristic of the second demographic transition and strictly connected with the autonomisation of the status of an individual, which involves pursuit of life without commitments and striving for financial self-sufficiency.

As it was mentioned earlier, a distinctive procreative feature of the Polish society was a continuous increase in the number of extra-marital births (Szukalski 2004, Kotowska 2014). Most respondents had a positive attitude to having children without a formal or informal relationship. The percentage of women representing this opinion was larger than that of men (Table 5.9). The smallest percentage of respondents

Table 5.8 Respondents' attitude to single life

Respondents		Attitude to single life				
		Negative	Rather negative	Rather positive	Positive	I don't know
				%		
	Total	6.7	10.1	25.3	42.7	15.2
By sex	women	7.7	8.8	25.3	45.8	12.5
	men	3.9	13.7	25.5	34.3	22.5
By financial situation	bad	0.0	18.8	25.0	56.3	0.0
	average	7.2	9.6	28.0	31.2	24.0
	good	6.0	10.1	24.6	47.2	12.1
	very good	11.4	8.6	20.0	51.4	8.6

Source: Authors' elaboration

Table 5.9 Respondents' attitude to having children and not being in a relationship

Respondents		Attitude to having children and not being in a relationship				
		Negative	Rather negative	Rather positive	Positive	I don't know
				%		
	Total	10.1	21.1	25.6	28.5	14.7
By sex	women	8.4	20.5	26.4	30.0	14.7
	men	14.7	22.5	23.5	24.5	14.7
By financial situation	bad	6.3	18.8	6.3	31.3	37.5
	average	15.2	21.6	26.4	22.4	14.4
	good	7.5	19.1	29.1	29.6	14.6
	very good	8.6	31.4	11.4	42.9	5.7

Source: Authors' elaboration

expressing a positive attitude here was identified among the students who rated their own financial situation the lowest. In groups rating it higher, this percentage was larger (Table 5.9). A negative opinion about having children and not being in a relationship was expressed by nearly one-third of the respondents. Men expressed such an attitude more often than women. The respondents who rated their financial situation as very good and average usually did not approve of having children when not being in a relationship. In the remaining groups, this percentage was smaller. About 15% of the respondents did not have a clearly defined opinion (Table 5.9).

The academic youth taking part in the survey, especially women and people in the best financial situation, had a positive attitude to extra-marital births. It may be explained by social the changes occurring during the second demographic transition, such as the growing acceptance of women's freedom and autonomisation, e.g. as regards taking procreative decisions. It should be noted, however, that a sizeable group (over one-third) of the respondents, first of all men and people of a lower financial status, expressed more radical views and did not approve of having children and not being in a relationship.

A phenomenon observed in modern developed societies is not only having children and living single life, but also raising them. The number of single parents in Poland is growing. The social acceptance of this phenomenon is increasing. Back in the 1970s and 1980s, single parenthood was associated with social pathology (Siudem and Siudem 2008), while nowadays single parents raise considerably less controversy. However, the opinions about single parenthood varied among Kielce students. Nearly 42% declared a positive attitude to it, women more often than men (Table 5.10). Over half of the respondents approving of single parenthood evaluated their financial situation as very good. The lower it was rated, the smaller this percentage was (Table 5.10). Nearly 40% of Kielce students declared a negative attitude to single parenthood. Over half of the men and one-third of women represented this view. The largest percentage of respondents who expressed a negative opinion about single parenthood rated their financial status the lowest, while among those who evaluated it better—this percentage was lower. Every fifth respondent did not have an opinion regarding single parenthood (Table 5.10).

The respondents' attitude to single parenthood varied. A positive attitude was expressed first of all by women and people who rated their financial situation the highest. This group was characterised by greater economic independence and openness to new family patterns and the weakening of norms and values, which were important in a traditional family (Janiszewska 2013). A negative attitude to single parenthood was expressed primarily by men and people who rated their financial situation the lowest. It can be assumed that this group of respondents represented the traditional, Catholic part of the society, for whom only the full family environment is appropriate for raising a child.

Table 5.10 Respondents' attitude to single parenthood

Respondents		Attitude to single parenthood				
		Negative	*Rather negative*	*Rather positive*	*Positive*	*I don't know*
		%				
	Total	14.4	24.0	19.7	22.7	19.2
By sex	women	12.5	20.9	24.5	20.9	14.7
	men	19.6	16.7	17.6	14.7	14.7
By financial situation	bad	18.8	12.5	18.8	18.8	37.5
	average	16.8	20.0	16.0	24.8	14.4
	good	12.1	18.6	26.6	17.6	14.6
	very good	17.1	28.6	25.7	8.6	5.7

Source: Authors' elaboration

As it was mentioned earlier, the number of divorces in Poland and in Kielce increased as a result of re-evaluating social attitudes and the liberalisation of views on human freedom, individualism and self-fulfilment. However, most Kielce students expressed a negative attitude to divorces. Criticism towards divorces was more common among men than women. The respondents who described their financial status as bad, average and good, mostly represented negative attitudes. In the group whose members evaluated their financial situation as very good, the percentage of students with a negative attitude was smaller. Positive attitudes to divorce were expressed by ¼ of the respondents (Table 5.11). The group of respondents who rated their financial situation the highest included the largest percentage of students representing a positive attitude to divorces. In the remaining groups, this percentage was smaller. Every fifth respondent did not have an opinion regarding divorces (Table 5.11).

The university students participating in the survey, especially men and people who rated their financial situation the lowest, expressed a negative opinion regarding divorces. A positive attitude was expressed by only one-quarter of the respondents. They were usually women and people in the best financial situation. The results of the survey suggest that the predominant attitude to divorces among Kielce students was a negative one. This may result from the traditional perception of the family and a significant role of the Catholic religion, which assumes insolubility of marriage.

As shown by the results of earlier studies (*Kontrowersje wobec...* 2008, Janiszewska 2013), a great majority of young Poles (over 76%) have a positive attitude to premarital sex. A similar opinion was expressed by Kielce students. A significant majority declared a positive attitude to premarital sex. Differences of opinion depending on sex were insignificant. A positive attitude to premarital sex was expressed by 80% of both, women and men. The group of students who rated their financial status the highest included the largest percentage of respondents declaring a positive attitude to premarital sex. The lower the students rated their financial

Table 5.11 Respondents' attitude to divorces

Respondents		Attitude to divorces				
		Negative	Rather negative	Rather positive	Positive	I don't know
		%				
	Total	26.9	28.5	10.4	14.9	19.2
By sex	women	24.5	29.7	11.4	15.4	19.0
	men	33.3	25.5	7.8	13.7	19.6
By financial situation	bad	25.0	31.3	12.5	12.5	18.8
	average	35.2	22.4	8.8	11.2	22.4
	good	22.6	33.2	10.6	15.6	18.1
	very good	22.9	22.9	14.3	25.7	14.3

Source: Authors' elaboration

situation, the smaller the percentage of those with a positive view to premarital sex. A negative attitude to premarital sex was expressed by nearly 8% of the respondents, more often by women than men (Table 5.12). With regard to their financial status, the largest percentage of students disapproving of premarital sex was identified in the group rating their financial situation the lowest, and the smallest—in the group rating it the highest. Every tenth respondent did not have any opinion on this matter (Table 5.12).

A great majority of the respondents, regardless of their sex and financial situation, expressed a positive opinion about premarital sex. The survey results confirmed a significant change in the traditional perception of a relationship, i.e. the separation of sex life from procreation, as well as from marriage. This also shows that young people tend to adopt new cultural models, increasingly accepted by the Polish society (*Kontrowersje wobec...* 2008).

A considerable majority of young Poles have a positive opinion not only about premarital sex but also about contraception (*Kontrowersje wobec...* 2008, Janiszewska 2013). Kielce students who took part in the survey mostly described their attitude to contraception as positive (Table 5.13). The approval was expressed by slightly more women than men. The percentage of students approving of contraception depended on the evaluation of one's own financial situation. Negative and rather negative attitudes were expressed by less than 8%. The differences of opinions depending on the respondents' sex were insignificant. The students' evaluation of their own financial situation had a stronger impact than the difference in sex. Among those who rated it the highest, there was nobody who declared a negative attitude towards contraception. In the remaining groups, which rated their financial situation lower, the percentage of respondents with negative opinions on contraception ranged from ca. 6% to ca. 11% (Table 5.13). Every tenth respondent did not have an opinion about contraception (Table 5.13).

Table 5.12 Respondents' attitude to premarital sex

Respondents		Attitude to premarital sex				
		Negative	Rather negative	Rather positive	Positive	I don't know
		%				
	Total	4.0	3.7	20.0	59.7	12.5
By sex	women	3.7	4.8	20.9	58.2	12.5
	men	4.9	1.0	17.6	63.7	12.7
By financial situation	bad	12.5	6.3	12.5	50.0	18.8
	average	4.8	4.8	17.6	56.8	16.0
	good	3.5	3.0	23.1	59.3	11.1
	very good	0.0	2.9	14.3	77.1	5.7

Source: Authors' elaboration

Table 5.13 Respondents' attitude to contraception

Respondents		Attitude to contraception				
		Negative	Rather negative	Rather positive	Positive	I don't know
				%		
	Total	4.8	2.9	21.6	60.3	10.4
By sex	women	5.1	2.6	20.9	61.9	9.5
	men	3.9	3.9	23.5	55.9	12.7
By financial situation	bad	0.0	6.3	6.3	56.3	31.3
	average	6.4	4.8	24.8	50.4	13.6
	good	5.0	2.0	22.1	63.3	7.5
	very good	0.0	0.0	14.3	80.0	5.7

Source: Authors' elaboration

Members of the social group included in the study represented a liberal attitude to contraception. Regardless of their own sex and financial situation, a great majority of them expressed a positive opinion about it. Similar to the positive attitude to premarital sex, this is also a proof of the contemporary transformation of traditional sexual norms, earlier shaped mainly by the Catholic religion, where the aim of sex was procreation. Similar attitudes among young people have been indicated before by Janiszewska (2013) and Skibiński (2017).

An important part of the development of an adult person is marriage. Most young Poles express a positive attitude towards it (Bakiera 2008). Kielce students were even more enthusiastic. A great majority declared a positive attitude to formalising a relationship through the sacrament or a civil ceremony. The attitudes towards formalising a relationship depended on the sex. A positive attitude to marriage was expressed by a larger percentage of women than men (Table 5.14). A significant factor diversifying the attitudes was the assessment of one's own financial situation. Among the students who rated their financial situation the lowest, the percentage of those representing a positive attitude to formalising a relationship was the largest. In groups evaluating their financial situation higher, this percentage was smaller (Table 5.14). A negative attitude toward formalising a relationship was represented by nearly 9%; the percentage of men presenting this view was higher than the percentage of women. In the groups of respondents rating their financial situation as very good, good and average, the percentage of students objecting to formalising a relationship ranged from 6% to 12%. None of the students who evaluated their financial situation the lowest had a negative attitude. About 8% did not have an opinion on the matter (Table 5.14).

The respondents from Kielce, especially women and the persons who rated their financial situation the lowest, expressed a positive attitude to marriage. The way young people approached it depended on many factors. In Poland, it was mainly determined by tradition. Despite the fact that young people have a positive opinion about staying in informal relationships, marriage remains for them a common model of man and woman's coexistence (Bakiera 2008).

Table 5.14 Respondents' attitude to formalising a relationship through the sacrament or a civil ceremony

Respondents		Attitude to formalising a relationship				
		Negative	Rather negative	Rather positive	Positive	I don't know
				%		
Total		2.1	6.4	18.9	64.5	8.0
By sex	women	2.2	3.7	15.0	72.9	6.2
	men	2.0	13.7	29.4	42.2	12.7
By financial situation	bad	0.0	0.0	25.0	68.8	6.3
	average	3.2	8.8	17.6	59.2	11.2
	good	2.0	5.5	20.1	65.8	6.5
	very good	0.0	5.7	14.3	74.3	5.7

Source: Authors' elaboration

Table 5.15 Respondents' attitude to abortion

Respondents		Attitude to abortion				
		Negative	Rather negative	Rather positive	Positive	I don't know
				%		
Total		18.7	23.7	16.0	27.7	13.9
By sex	women	16.8	24.9	14.3	30.8	13.2
	men	23.5	20.6	20.6	19.6	15.7
By financial situation	bad	12.5	25.0	25.0	18.8	18.8
	average	16.0	24.0	14.4	31.2	14.4
	good	18.1	23.6	18.1	26.6	13.6
	very good	34.3	22.9	5.7	25.7	11.4

Source: Authors' elaboration

The issue to abortion raises many controversies. According to CBOS (Centre for Public Opinion Research) (*Opinie na temat dopuszczalności...* 2010), the admissibility of abortion strongly divided the Polish society. Every second adult Pole objected to the right to abortion and 45% claimed that it should be allowed. The opinions of the students participating in the survey also strongly varied. About 43% of respondents claimed that abortion is admissible. Some difference occurred depending on the respondents' sex. A slightly smaller percentage of women than men claimed that abortion is admissible (Table 5.15). The views differed more if we consider the evaluation of one's own financial situation. The largest percentage of students who accepted abortion were those who rated their financial situation the highest. In the groups rating it lower, this percentage was also lower. About 44% of the respondents claimed that abortion should be forbidden. In this group, there were more women than men. The smallest percentage of students regarding abortion as inadmissible were those who declared a very good financial situation. In the remaining groups, where the students rated their financial situation lower, this percentage was higher. Every seventh respondent did not have an opinion about the admissibility of abortion (Table 5.15).

The respondents' views concerning abortion agreed with the views of adult Poles. The most liberal were the students who rated their own financial situation the highest. In contrast, according to CBOS results (*Opinie na temat dopuszczalności...* 2010), the main factor determining the attitudes towards abortion were the worldview differences, especially the approach to religion and Church teachings. It can be assumed that they had an influence on the students' opinions as well.

Summary

Since the beginning of the transformation period, Kielce has been losing its population potential. In 1988–2019, the number of residents decreased by about 16,000. The average annual population decrease rate stood at 0.15%, which is typical of a shrinking city. Since the late 1980s, the rate of demographic processes has declined considerably, which is manifested by natural population loss and negative migration balance. Negative trends are also confirmed by GUS (the Central Statistical Office), forecasting further shrinking of the city. It is predicted that in 2050, Kielce will be inhabited by 138,000 people, which is about one-third less than today. The phenomena related to the second demographic transition, i.e. the decreasing fertility rate, falling number of new marriages and the increasing number of divorces may aggravate the population decline. The depopulation of the city will bring negative economic consequences, such as declining entrepreneurship and problems on the local labour market.

The results of the survey conducted among Kielce students lead to the conclusion that the shrinking of the city is unlikely to stop. Nearly half of the respondents declared their wish to leave the city in the near future. Implementing these plans would not only increase the negative migration balance, but also affect the population growth. It would also lead to decreasing the quality of the city's social capital.

A more optimistic picture of Kielce demography arises from the students' plans regarding the size and structure of their potential families. A great majority of young people intend to formalise their future relationship and have children. Most of them are planning to have at least two children, which would ensure a simple generational replacement. However, young people's plans very often diverge from reality. It is also confirmed by the students' plans as regards the age at having the first child. The research shows that postponing maternity results in low fertility and fertility rate among women.

Kielce students' attitudes to family values do not fully fit into the framework of the second demographic transition theory. They represent a duality: on the one hand, they are traditional, as regards marriage, single parenthood, divorces or abortion, and on the other—they are liberal with regard to premarital sex, cohabitation, staying single, or contraception. This dualism is particularly visible in approving of partnerships and strongly supporting the idea of traditional marriage at the same time. It may indicate that the students participating in the survey were characterised by moderate modernity and a large dose of traditionalism.

Many changes characteristic of the second demographic transition were evaluated by the students negatively. Nevertheless, they are becoming increasingly common among the Polish society (divorces, single parenthood). Transformation and modernisation processes have led to a considerable cultural transformation of metropolitan communities. The study described herein indicates that in medium-sized, shrinking cities, these processes are less intense.

Notes

1   According to the definition of a shrinking city, proposed by the International Research Network for Shrinking Cities, Kielce meets the criteria of such a city (Pallagst 2010). It is inhabited by more than 10,000 people and has been experiencing a loss of population at the rate of 0.15% annually. Negative migration and procreative tendencies have caused unfavourable changes in population sex and age structures, making the city and the whole Świętokrzyskie region a problem area, with regard to demography. Economic indicators (unemployment rate, professional activity, the number of firms per 1000 inhabitants) place Kielce at the last positions in the group of cities of the same size.
2   Unless another source is indicated, the statistical data used in this part of the article has been provided by the Central Statistical Office.

References

Abramowska-Kmon A., 2014. Przemiany struktur rodzin i gospodarstw domowych, [in:] I. E. Kotowska (ed.), *Niska dzietność w Polsce w kontekście percepcji Polaków*. Diagnoza społeczna 2013, Raport tematyczny, Ministerstwo Pracy i Polityki Społecznej, Centrum Rozwoju Zasobów Ludzkich, Warszawa, pp. 25–32.

Audirac I., Arroyo-Alejandre J., (eds.). 2010. Shrinking cities South/North, Florida State University, Universidad de Guadalajara, UCLA Program on Mexico, Profmex/World and Juan Pablos Editor.

Bakiera L., 2008. Postawy młodych dorosłych wobec małżeństwa, *Psychologia Rozwojowa*, 13 (3), pp. 67–68.

Beaujouan E., Brzozowska Z., Zeman K., 2016. The limited effect of increasing educational attainment on childlessness trends in twentieth-century Europe, women born 1916–65, *Population Studies*, 70 (3), pp. 1–17.

Coale A. J., 1973. The demographic transition reconsidered, Proceedings of the international population conference, International Union for the Scientific Study of Population, Liège, 1, pp. 53–72.

Cunningham-Sabot E., Audirac I., Fol S., Martinez-Fernandez C., 2013. Theoretical approaches of "shrinking cities, https://www.researchgate.net/publication/289122839 [accessed 27 April 2020].

Cunningham-Sabot E., Fol S., 2009. Shrinking cities in France and Great Britain: a silent process?, [in:] K. Pallagst et al., (eds), *The future of shrinking cities*, Berkeley, University of California: Center for Global Metropolitan Studies, Institute of Urban and Regional Development, and the Shrinking Cities International Research Network (SCiRN), IURD, pp. 17–27.

Downs A., 1997. The challenge of our declining big cities, *Housing Policy Debate*, 8 (2), pp. 359–408.

Duszczyk M., Fihel A., Kiełkowska M., Kordasiewicz A., Radziwinowiczówna A., 2014. Analiza kontekstualna i przyczynowa zmian rodziny i dzietności, Studia i Materiały, 2, UW, Warszawa.

Esteve A., Lopez-Ruiz L. A., Spijker J., 2013. Disentangling how educational expansion did not increase women's age at union formation in Latin America from 1970 to 2000, *Demographic Research*, 28, pp. 63–76.

Feng W., Quanhee Y., 1996. Age at marriage and the first birth interval – the emerging change in sexual behaviour among young couples in China, *Population and Development Review*, 22(2), pp. 299–320.

Frątczak E., Kozłowski W., 2005. Rodzinne tablice trwania życia, OW SGH, Warszawa.

Frątczak E., Pęczkowski M., 2002. Zmiany w postawach i zachowaniach reprodukcyjnych młodego i średniego pokolenia Polek i Polaków i ich wpływ na proces formowania związków, rodzin, gospodarstw domowych, SGH, Warszawa.

Frejka T., Jones G. W., Sardon J.-P., 2010. East Asian childbearing patterns and policy developments, *Population and Development Review*, 36 (3), pp. 579–606.

Herudzińska, M. H., 2012. Rodzina w świadomości społecznej: co (kto) tak naprawdę stanowi rodzinę? Społeczne (re)konstruowanie definicji rodziny, *Wychowanie w Rodzinie*, 6, pp. 15–41.

Janiszewska A., 2013. Małżeństwa vs związki nieformalne w opiniach młodych mieszkańców Łodzi, *Space – Society – Economy*, 12, pp. 185–211.

Jończy R., Rokita-Poskart D., Tanas M., 2013. Exodus absolwentów szkół średnich województwa opolskiego do dużych ośrodków regionalnych kraju oraz za granicę, Wojewódzki Urząd Pracy w Opolu, Uniwersytet Ekonomiczny we Wrocławiu, Opole.

Kocot-Górecka K., 2014. Zmiany demograficzne w Polsce po 1989 r. i ich wybrane konsekwencje, [in:] I. E. Kotowska (ed.), *Niska dzietność w Polsce w kontekście percepcji Polaków. Diagnoza społeczna 2013*, Raport tematyczny, Ministerstwo Pracy i Polityki Społecznej and Centrum Rozwoju Zasobów Ludzkich, Warszawa, pp. 9–19.

Kontrowersje wobec różnych zjawisk dotyczących życia małżeńskiego i rodzinnego, Komunikat z badań, 2008. Centrum Badania Opinii Społecznej, Warszawa.

Kotowska I. E., 1999. Drugie przejście demograficzne i jego uwarunkowania, [in:] Przemiany demograficzne w Polsce w latach 90. w świetle koncepcji drugiego przejścia demograficznego, Oficyna Wydawnicza Szkoły Głównej Handlowej, Warszawa.

Kotowska I. E., 2009. Zmiany modelu rodziny a zmiany aktywności zawodowej kobiet w Europie, [in:] I. E. Kotowska (ed.), *Strukturalne i kulturowe uwarunkowania aktywności zawodowej kobiet w Polsce*, Wydawnictwo Naukowe Scholar, Warszawa, pp. 15–56.

Kotowska I. E., (ed.), 2014. Niska dzietność w Polsce w kontekście percepcji Polaków. Diagnoza społeczna 2013, Raport tematyczny, Ministerstwo Pracy i Polityki Społecznej and Centrum Rozwoju Zasobów Ludzkich, Warszawa.

Kotowska I. E., Giza-Poleszczuk A., 2010. Zmiany demograficzno-społeczne i ich wpływ na rekonceptualizację polityki rodzinnej w kierunku równowagi w zakresie ochrony praw rodziny i poszczególnych jej członków, Polska na tle Europy, [in:] E. Leś, S. Bernini (eds.), *Przemiany rodziny w Polsce i we Włoszech i ich implikacje dla polityki rodzinnej*, Wydawnictwa Uniwersytetu Warszawskiego, Warszawa, pp. 31–68.

Kuong B., Perelli-Harris B., Padmadas S., 2016. The unexpected rise of cohabitation in the Philippines – Evidence for a negative educational gradient, pdfs.semanticscholar.org/c07 4/108249d601221e5d1f3326b712e9e41be58d.pdf, [accessed 20 July 2020].

Kurek S., Lange M., 2013. Zmiany zachowań prokreacyjnych w Polsce w ujęciu przestrzennym, Prace Monograficzne, 658, Uniwersytet Pedagogiczny, Kraków.

Kurkiewicz J., 1998. Modele przemian płodności w wybranych krajach europejskich w świetle drugiego przejścia demograficznego, Zeszyty Naukowe Akademii Ekonomicznej w Krakowie, Seria Specjalna: Monografie, 131.

Lesthaeghe R., 1983. A century of demographic and cultural change in Western Europe. An exploration of underlying dimensions, *Population and Development Review*, (9) 3, pp. 411–435.

Lesthaeghe R., 1991. The second demographic transition in western countries – an interpretation, Interuniversity Programme in Demography, Vrije Universiteit Brussels, IPD-Working Paper, 2, pp. 313–334.

Lesthaeghe R., 2010. The unfolding story of the second demographic transition, *Population and Development Review*, 36 (2), pp. 211–251.

Lesthaeghe R., 2020. The second demographic transition, 1986–2020: sub-replacement fertility and rising cohabitation – a global update, Genus, 76, https://doi.org/10.1186/s41118-020-00077-4, [accessed 20 July 2020].

Lesthaeghe R., Van de Kaa D. J., 1986. Twee demografische transities, [in:] R. Lesthaeghe, D. J. Van de Kaa (eds.), *Groei of Krimp? Van Loghum Slaterus*, Deventer: Mens en Maatschappij.

Lesthaeghe R., Lopez-Colas J., Neidert L., 2016. The social geography of unmarried cohabitation in the USA, 2007–2011, [in:] A. Esteve, R. Lesthaeghe (eds.), *Cohabitation and marriage in the Americas – Geo-historical legacies and new trends*, Springer Open, Switzerland, pp. 101–132.

Matysiak A., Mynarska M., 2014. Urodzenia w kohabitacji: wybór czy konieczność?, [in:] A. Matysiak (ed.), *Nowe wzorce formowania i rozwoju rodziny w Polsce. Przyczyny oraz wpływ na zadowolenie z życia*, Wydawnictwo Naukowe Scholar, Warszawa, pp. 104–129.

Mokomane Z., 2005. A demographic and socio-economic portrait of cohabitation in Botswana, *Societies in Transition*, 36 (1), pp. 57–73.

Musiał-Malago' M., 2018. Wybrane aspekty kurczenia się miast w Polsce, *Studia Miejskie*, 29, pp. 61–75.

Musiał-Malago' M., 2017. Przestrzenne zróżnicowanie procesu kurczenia się miast w Polsce, Prace Naukowe Uniwersytetu Ekonomicznego we Wrocławiu, 467, Wrocław.

Mykhnenko V., Turok I., 2008. East European cities – patterns of growth and decline, 1960–2005, *International Planning Studies*, 13 (4), pp. 311–342.

Mynarska M., 2014. Zmiany zachowań dotyczących rodziny w Polsce na tle innych krajów Europy, [in:] I. E. Kotowska (ed.), *Niska dzietność w Polsce w kontekście percepcji Polaków. Diagnoza społeczna 2013*, Raport tematyczny, Ministerstwo Pracy i Polityki Społecznej and Centrum Rozwoju Zasobów Ludzkich, Warszawa, pp. 19–25.

Nowak J., Wieteska M. (eds.), 2019. *Akademickość polskich miast*, Polski Instytut Ekonomiczny, Warszawa.

Opinie na temat dopuszczalności aborcji. Komunikat z badań, 2010. Centrum Badania Opinii Społecznej, Warszawa.

Pallagst K., 2010. Viewpoint. The planning research agenda: shrinking cities – a challenge for planning cultures, *Town Planning Review*, pp. 81–85.

Philipov D., Kohler H.-P., 2001. Tempo effects in the fertility decline in Eastern Europe: Evidence from Bulgaria, the Czech Republic, Hungary, Poland, and Russia, *European Journal of Population*, 17, pp. 37–60.

Podogrodzka M., 2013. Wzorzec płodności a wzorzec małżeńskości w Polsce po okresie transformacji, *Space – Society – Economy*, 12, pp. 167–183.

Pruszyński J., Putz J., 2016. Efekt drugiego przejścia demograficznego na strukturę społeczeństwa w Polsce i związane z tym wyzwania, *Gerontologia Polska*, 24, pp. 127–132.

Rokitowska J., 2017. Skłonność do migracji młodych ludzi na przykładzie gminy Oborniki Śląskie – potencjalne implikacje dla rozwoju lokalnego, *Ekonomia XXI wieku*, 1 (13), pp. 119–132.

Siudem A., Siudem I., 2008. *Profil psychologiczny osób samotnie wychowujących dzieci*, UMCS, Lublin.

Skibiński A., 2017. Uwarunkowania postaw prokreacyjnych i matrymonialnych młodzieży akademickiej województwa śląskiego, *Studia Ekonomiczne, Zeszyty Naukowe Uniwersytetu Ekonomicznego w Katowicach*, 309, pp. 17–27.

Slany K., 2002. *Alternatywne formy życia małżeńsko-rodzinnego w ponowoczesnym świecie*, NOMOS, Kraków.

Solga B., 2013. *Miejsce i znaczenie migracji zagranicznych w rozwoju regionalnym*, Instytut Śląski, Opole.

Stryjakiewicz T., Jaroszewska E., 2011. Kurczące się miasta (shrinking cieties) i strategie ich regenracji, https://www.researchgate.net/publication/308899213 [accessed 27 April 2020].

Stryjakiewicz T., Jaroszewska E., Marcińczak S., Ogrodowczyk A., Rumpel P., Siwek T., Slach O., 2014. Współczesny kontekst i podstawy teoretyczno-metodologiczne analizy procesu kurczenia się miast, [in:] Stryjakiewicz T. (ed.), *Kurczenie się miast w Europie Środkowo-Wschodniej*, Bogucki Wydawnictwo Naukowe, Poznań.

Szafraniec K., 2011. *Młodzi 2011*, Kancelaria Prezesa Rady Ministrów, Warszawa.

Szukalski P., 2004. Urodzenia pozamałżeńskie w Polsce, [in:] A. Warzywoda-Kruszyńska, P. Szukalski (eds.), *Rodzina w zmieniającym się społeczeństwie polskim*, Wydawnictwo Uniwersytetu Łódzkiego, Łódź, pp. 111–142.

Szukalski P., 2005. Wielkość i struktura rodziny a przejście demograficzne, Dylematy Współczesnych Rodzin, Roczniki Socjologii Rodziny XVI, UAM, Poznań, pp. 95–110.

Szukalski P., 2019. Wzrost dzietności w ostatnich latach: dlaczego najbardziej z niego korzystają duże miasta?, Demografia i Gerontologia Społeczna, Biuletyn Informacyjny, 5.

Van de Kaa D. J., 1978. Recent trends in fertility in Western Europe, NIDI-Working Paper, 11, Voorburg.

Van de Kaa D. J., 1987. Europe's second demographic transition, Population Reference Bureau, Washington, DC.

Van de Kaa D. J., 2003. 'Demographics in transition': an essay on continuity and discontinuity in valuechange, [in:] I. E. Kotowska, J. Jóźwiak (eds.), *Population of Central and Eastern Europe: challenges and opportunities*, Statistical Publishing Establishment, Warsaw, pp. 641–663.

Wiechman T., 2007. *Between spectacular projects and pragmatic deconstruction. The future of shrinking cities: problems, patterns & strategies of urban transformation in a global context*, Berkeley, CA.

Wu T., Martinez-Fernandez C., 2009. Shrinking cities: A global overview and concerns about Australian cases, [in:] K. Pallagst et al. (eds.), *The future of shrinking cities-problems, patterns and strategies of urban transformation in a global context*, IURD, Berkeley.

Zborowski A., Soja M., Łobodziński A., 2012. Population trends in Polish cities – stagnation, depopulation or shrinkage? *Prace Geograficzne*, 130, pp. 7–28.

6 Manufacturing in the post-industrial city

The role of a "Hidden Sector" in the development of Pécs, Hungary

Gábor Lux

Introduction

While the post-socialist transformation of regions and cities in Central Europe has brought about profound change in political life and socio-economic systems, recent studies have put renewed emphasis on the interplay between historically embedded legacies and new phenomena, emphasising the *longue durée* of systemic transformations (Lux and Faragó 2018; Gorzelak 2020; Gorzelak and Smętkowski 2020). Productive, institutional, cultural and even political legacies matter—sometimes more than we would like to admit. The majority of case studies concerned with urban transformation inadvertently deal with the eye-catching and the dramatic: spectacular successes and unqualified failures. The former often enter best-practice manuals without sufficient attention to special or even unique circumstances which had made them so (Lovering 1999); the latter, typically less studied, become cautionary tales in academia, and sometimes a testing ground for new regeneration schemes in policy.

Yet highlighting dramatic processes and outliers neglects the existence and problems of the several "in-betweens", the "mixed types": regions and cities which do not fit neatly into either category, exhibiting dubious development patterns or moderate success/failure. Cartographic visualisation—usually based on the horizontal division of space (Nemes Nagy 2009)—results in the secondary and tertiary aspects of spatial units "dropping off the map", while a stylised approach to the study of patterns and processes reduces space "into points devoid of any territorial dimension" (Capello 2016, 8). Furthermore, even cases which fit a specific analytical framework (e.g. metropolitan areas, Old Industrial Regions, shrinking cities, etc.) are rarely pure specimens, but complex entities with specific characteristics and contradictions. Even cities with strong specialisation in specific activities have secondary or tertiary branches which play a considerable role in employment and value creation, even if they are rarely mentioned.

This chapter uses the example of the manufacturing sector of Pécs, a Central European city better known for its culture to highlight the significance of what will be henceforth referred to as "hidden sectors"; economic activities which are obscured by other, locally dominant development patterns. Overshadowed or obscured by socio-economic phenomena commanding more attention, they exist in the background—either taken for granted, or, as in our case, neglected due to their poor fit with dominant development narratives. Hidden sectors exist

DOI: 10.4324/9781003039792-6

everywhere, even in developed core regions; however, the peripheries can be most sensitive to their problems. Hidden sectors can easily "fall through the gaps", either dwindling into irrelevance, or never "taking off".

Contemporary development in Central European manufacturing has mostly been studied on the examples of regions with a strong industrial specialisation, either deeply integrated into foreign direct investment (FDI) networks (e.g. Pavlínek, Domański and Guzik 2009; Wójtowicz and Rachwał 2014; Pavlínek 2015) or dealing with the consequences of industrial restructuring (Drobniak, Kolka and Skowroński 2012; Sucháček et al. 2012). Studies on peripheral branches such as textiles and light industries, or the industry of the peripheries have been a lot more scarce (Pickles et al. 2006; Molnár 2013), and are more often the subject of domestic publications. And yet cities with struggling industries are commonplace in Central Europe, and may be closer to the norm than the exception when it comes to smaller cities on the periphery. While re-industrialisation has been the prevailing trend of the decade after the 2008–2010 crisis, this, too, comes with uneven consequences.

Pécs, the city being investigated in this chapter, is a regional centre of 142,000 inhabitants in the region of Southern Transdanubia, Hungary. With a history stretching back to the Roman era, Pécs has long had to contend with the duality between its peripheral geographical position in Hungary's Budapest-centric space economy (Figure 6.1), and its higher aspirations in the field of culture and education. The present chapter will focus on the city's manufacturing industries, whose development has taken place in an adverse context. Pécs and the surrounding region lie outside Hungary's modern-day manufacturing core areas, which have

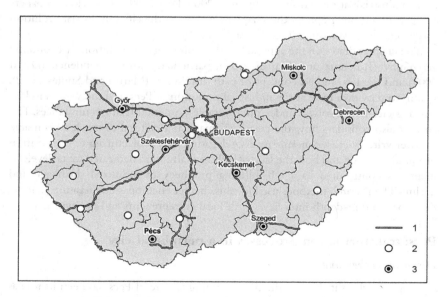

Figure 6.1 The location of Pécs within Hungary.

Legend: 1—highways, 2—county seats, 3—regional centres. Source: Author's elaboration.

been integrated into global production networks (GPNs) through FDI inflows. Likewise, recent re-industrialisation processes, gathering pace after the 2008–2010 crisis and particularly since 2012, have bypassed the city and the entire region, accompanied by a gradual decline in development rankings, and a decrease of both population (from 170,000 to 142,000 between 1990 and 2019) and employment numbers. Accordingly, Pécs can be fairly characterised as a cultural centre, an industrial periphery, and a shrinking city.

In academia, Pécs has been one of the more studied Hungarian urban centres. Particular attention has been dedicated to the city's transformation from mining under socialism into a city of culture, its bid for European Capital of Culture 2010, and the controversial aftermath of this programme. Papers have focused on the city's urban structure and functional transformation (Trócsányi 2011), processes of culture-based regeneration (Faragó 2012; Tubaldi 2014; Turşie 2015; Keresnyei and Egedy 2016), managing a cultural flagship event (Pálné Kovács 2013; Lähdesmäki, 2014; Pálné Kovács and Grünhut 2015; Németh 2016), the city's gateway role (Pap et al. 2013), and the tourism impacts of building a cultural economy (Aubert, Marton and Raffay 2015). Compared to this abundance, reflections on the city's development outside the cultural sphere have been scarce. The previous papers make occasional mention of the city's manufacturing industries, but mainly do so in passing, or as a problematised industrial heritage to be superseded by a more modern development model. The manufacturing industries of the city are "there" in the background, somewhere, but either taken for granted, or viewed with an implied scepticism regarding their growth prospects. Studies on the city's industry include a few broad overviews on economic and industrial transformation (Hrubi 2006a, 2006b; Rácz, Kovács and Horeczki 2020), a policy-oriented book on re-industrialisation (Hrubi 2009; Szerb 2009; Póla 2009), as well as two papers and part of a monograph by this chapter's author on the city's industrial evolution and economic governance (Lux 2010a, 2015, 2017).

The chapter draws on the approach and terminology of evolutionary economic geography (Boschma and Martin 2010), particularly path dependence (David 2007) and the less studied concept of path dissolution (Martin and Sunley 2012). It aims to provide a look at the industrial evolution of Pécs, and the way its industrial legacies have evolved under unfavourable post-industrial circumstances. The chapter also considers how the city's self-image as a cultural centre has been able to "overwrite" socio-economic reality, relegating its manufacturing to the position of a hidden sector, and how the lack of sufficient attention dedicated to its development has contributed to a self-fulfilling prophecy of industrial shrinkage and decline. The research methods used for this chapter encompassed literature analysis, fieldwork, and in-depth interviews with local entrepreneurs and policymakers.[1]

Pécs: transformation processes in a peripheral city

Historical background

Through much of its history, the distinguishing feature of Pécs has been its role as a cultural hub and episcopal seat, as well as a centre of high-skilled craftmanship. With a history stretching back to the Roman provincial seat of Sopianae, it became a bishopric in the mediaeval Kingdom of Hungary, serving as a south-western gateway towards Dalmatia, Italy and the Balkans. Pécs served as the seat of Hungary's first university (established 1367, destroyed during the Ottoman occupation, and re-established as an academy in 1833), emerging as one of the main hubs of renaissance culture and learning. Its role as a conduit for Italian influences was cut short by the Turkish conquest (1526/1543–1686), a period which saw it function as a military outpost and a hub of Balkans trade networks; and ended with the city's almost complete destruction by the liberation wars and the plague, which left the city in ruins, and mostly depopulated.

In the 18th and early 19th century, Pécs regained its position as a bishopric and clerical stronghold, purchasing its free royal city status in 1777. As the city was distant from the major river routes, it became specialised in complex, easily transported, high-value goods for luxury markets (Kaposi 2006). Light and food industries represented the first wave of industrialisation, developing into its iconic brands: watches (several smaller workshops forming what would now be called a cluster), cosmetic soaps, champagne (Littke—1859), glovemaking (the "Hamerli" manufacture—1861), church organs (Angster—1867), and decorative ceramics. The Zsolnay Porcelain Factory (1853) came to symbolise the city in Europe, with products ranging from household porcelain to lustrous roof tiles and distinctive art pieces, based on continuous technological innovation and contemporary design. This mixture of industry and culture, organised by prominent industrialists and supported by a flowering of local banks (Gál 2009), was joined by more typical industrial branches. The arrival of steamboat shipping (1833) and the railway network (1859) gave rise to coal mining and an ironworks, and a growing tradition of machine industry (Horváth 2006; Kaposi 2006). The city's "capitalist" industrial structure proved enduring, experiencing modest growth, but remaining stable until 1945. In this era, Pécs had a diversified economy, both in its mixture of industries, and its other city functions: public administration, higher education, trade, and multiple artistic schools drawn to the city's sub-Mediterranean climate and intellectual ambience (literature, fine and applied arts, and multiple prominent members of the Bauhaus movement).

Socialist industrialisation after the Second World War resulted in the city's radical and thorough socio-economic restructuring. Most of its traditional industries were deliberately shut down or downsized, and its cultural functions atrophied while the city would expand at a rapid pace due to the massive expansion of coal and uranium mining. The population of Pécs grew from 88,000 to 165,000 inhabitants between 1949 and 1970, and multiple new housing estates were constructed on its perimeter. The definitive economic geography handbook of the time (Markos 1962) describes Pécs as a settlement with developed mining, industry and culture, noting that 42% of its active workers were now employed in mining and industry. These figures masked a more sobering reality. While mining provided unusually high incomes for 18,000 coal and approximately 8000 uranium miners (precise figures conspicuously omitted from contemporary statistical yearbooks), the economy of the city took on a distorted, mono-structural character, and was effectively dominated by low-skilled work producing raw materials for distant steelworks, and the Soviet nuclear programme. Hidden unemployment among inhabitants who were not accounted for by official statistics but could not find a job in the mining-dominated economy (a problem mainly affecting married women), and shortages of housing, public services and household goods became overwhelming concerns.

The problems of industrial development, coupled with the rise of a reformist generation in the local party organs over the 1960s, resulted in an ambitious restructuring plan to modernise the local economy by halving mining employment by 1980, and re-specialising in new, more advanced "basic industries" (Lux 2010b). A rare example of bottom-up initiatives in a socialist state, the plans had no realistic chance of success even in a decade of careful economic reform, but nevertheless achieved some results. Traditional food and light industries were revived to create new jobs, and "branch-plant companies", the subsidiaries of Budapest-based firms were opened in the machinery, electronics and optical sector—the second pillar of the city's current manufacturing base.

Post-socialist development processes

The post-socialist transformation of Pécs resulted in the rapid unravelling of its mining complexes. Already massively unprofitable, both the coal and uranium mines were shut down over the 1990s, followed by comprehensive recultivation efforts, and the early retirement of the majority of workers. Structural change was almost as rapid and radical as the great expansion of mining over the 1950s and 1960s. "Branch-plant" manufacturing companies experienced severe difficulties in transition, as these units were not independent firms, but externally managed subsidiaries without independent management, R&D, or design functions. Unlike Hungary's north-western industrial cities, Pécs did not benefit much from the first wave of FDI inflows due to its geographic isolation, the lack of motorway access (only completed in 2010, and then towards Budapest instead of Hungary's western export markets—cf. Figure 6.1), and the proximity of the Yugoslavian conflict. Hence, three larger firms represented the majority of FDI: Hauni Hungaria (machinery), Hantarex (electronics, 1991–1994; later superseded by Nokia, 1994–2001; and by Elcoteq, 1998–2011), and British American Tobacco (tobacco). Successive investment waves also found little interest in the city's limited labour market. Through the steady decline of manufacturing, Pécs, previously in the middle of the national rankings, converged towards Hungary's de-industrialised southern periphery (Figure 6.2).

Partly through necessity, partly through a revival of earlier traditions, the cultural and political elite reimagined Pécs as a city of culture and higher education, contrasted in the public imagination with declining manufacturing, and a now-rejected, quickly forgotten mining past. The university emerged as the city's most important employer next to the local government, growing to 33,000 students and 6300 staff by the mid-2000s. The sizable public sector served as a buffer from the depression experienced by other industrial crisis regions, and offered a hopeful way out—in the medium term. References to complex cultural products and "cultural industry" started to appear in early revitalisation plans (Faragó, Horváth and Hrubi 1990; Pécs Városfejlesztési Koncepciója 1995), and eventually came to dominate

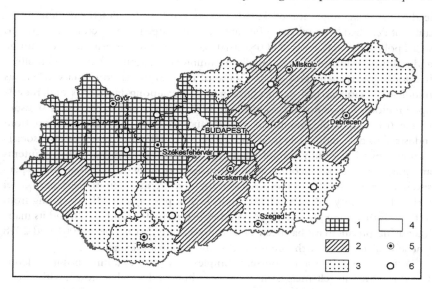

Figure 6.2 Pécs in the spatial structure of Hungarian manufacturing.
Legend: 1—integrated industrial counties, 2—re-industrialising counties, 3—industrial peripheries, 4—
service-based core area, 5—regional centres, 6—county seats. Source: Author's elaboration based on
regional typologies by Nemes Nagy—Lőcsei (2015), Lengyel et al. (2016), and Lux (2017).

development discourse. As the most important expression of these aims, a new,
two-pronged growth strategy was formulated (Borderless City 2005; The Pole of
Quality of Life 2005). Three industry/service clusters with good future potential
were identified in the health, environmental and cultural industries; furthermore,
the city submitted and won a bid for the 2010 "European Capital of Culture"
(ECoC) title (although shared with the vastly larger Essen and Istanbul). The
growth pole strategies are just as interesting for what they were missing as what
they actually contained: the city's actually functioning industries, or indeed references to manufacturing in general were conspicuously absent from the picture, and
from the programmes' future images.

The 2008 crisis had shaken the city economy particularly badly. Consumption-
oriented cultural functions had to contend with a severe drop of real demand,
damaging the already problem-riddled ECoC event. The university, one of the
cornerstones of the local economy, had undergone its own crisis, brought about by
the region's sharp demographic decline and the *de facto* introduction of tuition fees,
which had been particularly severe for the university's main focus, humanities and
social sciences. The number of students decreased from 33,000 to 20,000, a change
only partially compensated by its growing specialisation in international medical
students (growing from 1200 to 5000 between 2012 and 2019). The city's service-
based branches proved to have remarkably weak resilience; in years of severe auster-
ity, they proved dangerously dependent on transfers from the central government.

The 2011 shuttering of Elcoteq, the city's largest employer in manufacturing,
dealt a severe blow to electronic industry, and the inability to find a new investor

lead to the mass emigration of skilled workers. For two decades, the declining fortunes of Pécs had been relatively bearable due to a large public sector and generous social policy: the new crisis of 2008–2010 had exposed the severity of its economic and political problems. A now rapidly declining population, indebted and ineffectual local government, and general sense of depression placed the city where its middle-class citizens had never considered being: among Hungary's underdeveloped regions. This is not always borne out by the most outwardly visible statistics (Table 6.1)—unemployment had peaked and receded, and the transition from industrial to post-industrial city is considered a natural consequence of post-socialist modernisation—but it is reinforced by finer economic indicators. For all intents and purposes, the entire county is an industrial periphery (Lux 2010a), and a "persistent loser" of post-1970s industrial development trends (Nemes Nagy and Lőcsei 2015). The county (NUTS-3 level) ranks low in per capita GDP (declining from 11th to 16th among Hungary's 19 counties between 2000 and 2017), and its manufacturing productivity has been ranked persistently last between 2001 and 2018 (the annual 14.6% growth rate is close to the national average).

Post-transition Pécs is, of course, a complex entity with multiple possible identities: it is, fitting its self-image, still a city of culture and a college town, with a relatively high quality of life that is not always reflected in the "hard numbers". Considering the *genius loci* and its self-image, it is a post-industrial city with a vibrant cultural life. Between the 1970s and about 2004 (the year of EU accession), it would show many characteristics of industrial crisis regions, although this status has receded into the past after thorough de-industrialisation. Today, it is perhaps most accurately described as a shrinking city. The combination of economic decline and population loss puts it solidly within the group of cities found across the developed world which must contend with managing long-term decrease (Martinez-Fernandez et al. 2012; Richardson and Woon Nam 2014), a process now helped along by seemingly irreversible processes of cumulative causation (Hospers 2014).

Table 6.1 Main structural indicators of Pécs and Baranya county (1950–2019)

		1950	1960	1970	1980	1990	2000	2010	2019**
Population (1000)	Baranya	362.8	400	421.2	423.5	418.6	400.8	391.5	361.0
	Pécs	87.5	114.1	141.3	168.7	170.0	157.3	157.7	142.9
Employment (BA, %)	agric.	56.5	39.2	29.7	19.9	18.3	8.1	4.8	4.6
	industry	19.9	30.2	41.1	42.2	38.7	33.1	26.4	28.9
	services	23.6	30.6	29.2	37.9	43.0	58.8	68.8	66.5
Unemployment (BA, %)		-	-	-	-	4.2*	3.4	7.2	3.5
Investment (BA, %)	agric.	n/a	10.9	25.1	25.1	25.7	5.7	8.5	13.0
	industry	n/a	65.1	41.6	37.6	37.5	54.9	23.9	45.2
	services	n/a	24.0	33.3	37.3	36.8	39.4	67.6	41.8
Coal production (1000 t)		1,400	2,847	4,151	3,065	1,735	753	0	0

* Peaked at 13.5% in 1993.
** First two quarters. Source: Author's calculations and construction based on data from statistical yearbooks (Baranya Megye Statisztikai Évkönyve 1956, 1960, 1970, 1980, 1990, 2000, 2019).

Manufacturing in the post-industrial city

Despite its peripheral status and post-industrial character, even the industries of Pécs show a fairly diverse picture. The mining industries which had once made the city take on a mono-structural character are completely gone, leaving very little of their traditions except a housing estate named "Uranium City" (ironically, a much more pleasant place than it sounds). The remaining industries are small both individually and collectively. All of them are rooted to some extent in local productive traditions, and represent different forms of path development (Figure 6.3).

The first group of local industries concerns *FDI-based manufacturing activities*. This segment of the city's economy concentrates its largest employees outside the public sector and utilities. Even for Central Europe's geographic and economic peripheries, Pécs has a very low FDI presence. As Rácz, Kovács and Horeczki (2020) note, only 250 out of 9,000 registered companies in the city have foreign capital, and machine industrial firm Hauni Hungaria (1300 employees) is the only one with more than 1000 jobs—making Pécs unique in this respect among Hungary's eight large cities. Three other foreign-owned manufacturing companies employ over 250 staff (British American Tobacco—835, tobacco; Bader Gruppe—516, metal cabinets; Harman Professional Ltd.—432, electronics), and four more employ over 100.

With a few exceptions (most notably Hauni), these firms are specialised in low-to-medium-tech assembly functions, with low local added value, and strong external control over local management. They represent the lower range, or perhaps *peripheral form* of Central Europe's typical FDI-based manufacturing development path, and are the current manifestations of path-dependent evolution. The city's foreign-owned companies are built on previous knowledge sets, some in machinery, some in electronics, and some in food and light industries. Without good transport connections (Pécs only gained its highway connection in 2010), the investors of the 1990s were mostly attracted by available pools of skilled labour or privatisation opportunities. However, while the investing firms have brought modern equipment and efficient management methods to the city, their presence has not led to the sustainable modernisation, or even stabilisation of growth paths in these mature industries. No comprehensive efforts had been undertaken—by

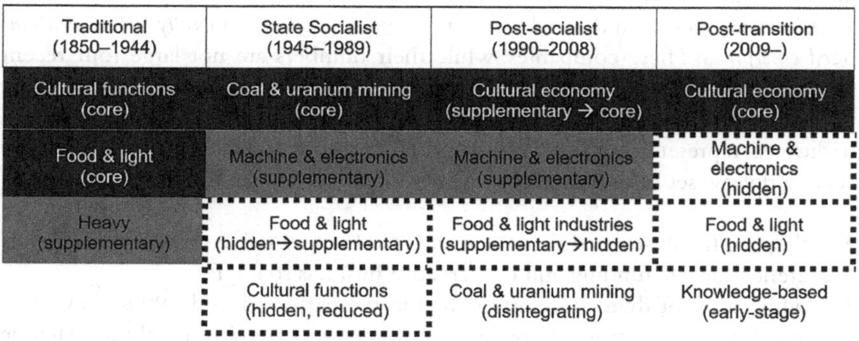

Traditional (1850–1944)	State Socialist (1945–1989)	Post-socialist (1990–2008)	Post-transition (2009–)
Cultural functions (core)	Coal & uranium mining (core)	Cultural economy (supplementary → core)	Cultural economy (core)
Food & light (core)	Machine & electronics (supplementary)	Machine & electronics (supplementary)	Machine & electronics (hidden)
Heavy (supplementary)	Food & light (hidden→supplementary)	Food & light industries (supplementary→hidden)	Food & light (hidden)
	Cultural functions (hidden, reduced)	Coal & uranium mining (disintegrating)	Knowledge-based (early-stage)

Figure 6.3 The industrial mix of Pécs in four periods.

Source: Author's elaboration.

either the public or the private sector—to update the knowledge sets of these branches, or even ensure their continued viability through strong vocational education. Manufacturing *in general* was considered a thing of the past, something not worth the effort and scarce resources to maintain; and development policy in the city has focused on allocating resources to cultural and education functions to the neglect of the productive sector. Consequently, the city's manufacturing base has been slowly shrinking, falling behind national trends in both employment and productivity.

Whereas the prevailing post-crisis trend in Central Europe's manufacturing sector has been local re-investment in existing sites and slowly increasing embeddedness (Pavlínek 2015; Rechnitzer 2016; Józsa 2016; Rechnitzer and Fekete 2019), the FDI-driven industry of Pécs has experienced an opposite process. Elcoteq, its dominant FDI plant became disembedded, and with its closure in 2011 due to external causes (the collapse of the parent firm), the formerly 5–8,000 jobs in electronics dwindled away as former employees moved to more dynamic regions in northwestern Hungary and abroad. Low-value added routine activities are easily uprooted: Elcoteq was eventually succeeded by smaller investment projects (most notably Harman), and a few medium-sized, locally owned companies, but these collectively represent much less employment and production potential than the original plant.

Without upgrading to higher value-added, more embedded functions, manufacturing branches relying on routine activities and cost advantages prove particularly footloose and "slippery". Moreover, even the scarce attractions which had once drawn these FDI projects are no longer present. With shrinking industries, the pool of skilled labour to draw on, and the knowledge sets which might be reused by new investors, are depleted due to retirement, out-migration, and insufficient replacement (due to a downsized vocational education system). While Pécs is at realistic risk of losing even some of its existing industries, it can offer precious little incentives to attract new ones. Neither the construction of a previously missing motorway (2010), nor the city's three well-equipped but under-used industrial estates have been able to change this situation. Indeed, all they show is that the 1990s approaches of investment promotion are no longer sufficient: cities must offer attractive industrial milieus, an active and skilled labour force, and investment-friendly local authorities to succeed.

A second group of industrial firms encompasses a group of *locally owned medium-sized enterprises*. These companies, while their numbers are not large (our recent study[2] identified 23 in Pécs and 16 more in nearby settlements), show a more promising picture than FDI plants. They are deeply rooted in the local industrial traditions, representing path-dependent endogenous development. While they cover a diverse set of industries, these firms show certain similarities as a group:

- they are family-owned firms characterised by slow, stable growth, with a preference for self-funding and conservative business strategies;
- while most of them were established in the early and mid-1990s, they reuse and develop traditional knowledge sets found in the city, mainly in machine industry, electronics, and light industrial activities;

- they are often specialised in niche markets, with relatively high exporting activity, reliance on skilled labour, and, in recent years, incremental innovation behaviour based on a mixture of continuous improvement and new product design. The group also includes Tier 2 automotive suppliers, although these companies try to hedge their bets by also developing own products which reduces their dependency on FDI-controlled value chains.

Our recent comparative study of domestic medium-sized manufacturing firms in Hungary (Kovács, Lux and Páger 2016) has concluded that Pécs had a higher than expected number of these companies, especially in comparison with more successful industrial cities. It seems safe to conclude that they have emerged in the absence of large-scale FDI, by reusing the legacy (manpower and knowledge sets) of defunct state socialist enterprises. The closest analogy to these companies is the German Mittelstand, a group of locally embedded family firms typically found in Germany's smaller cities and rural areas. The mid-sized firms of Pécs are, of course, separated from this much more robust and numerous group by multiple criteria, most prominently their multigenerational development (most are near their first generation change), but there are valid parallels with the generalised Mittelstand model (Lehrer and Schmid 2015; Welter et al. 2016; De Massis et al. 2017), most prominently the ability to mobilise endogenous capabilities to sidestep resource scarcity, high local embeddedness, and a reliance on local social capital. That is to say, medium-sized manufacturing companies in Pécs are not *yet* full Mittelstand-type firms, but many of them *may* develop in this direction, given sufficient time and opportunities.

It must be noted that, despite its moderate success and healthy development patterns, this group is also facing problems. The diversified industrial mix also implies a lack of effective specialisation outside machine industry (which has formed an effective and active regional cluster) and winemaking (whose production base lies outside the city in a rural wine region), leaving the local Small and Medium Enterprise (SME) sector fragmented, without sufficient critical mass. Other clusters, while they exist, are either very small (the remnants of glovemaking) or dominated by public actors which are not active in actual production functions (cultural cluster). Some industrial traditions, notably organ-building and ceramics, are rooted in one or two companies, and do not seem to generate new spinoffs or cluster-type networks. Some companies operate in branches exposed to intense cost-based competition from East Asia. As the experiences of shrinking light industries in Hungary show (Molnár 2013), holding together local production networks in the generational perspective is a formidable challenge. Access to skilled labour is a critical concern, and an area where interviewed firms and policymakers have been consistently ringing alarm bells. The resilience of the city's endogenous development path has been tested by the 2008 crisis, but it is yet to be tested by time. It is a question for the future whether it can withstand further deindustrialisation processes, or (paradoxically) stand its own if the city eventually succeeds at attracting larger FDI projects. Our research in FDI-dominated regions suggests that crowding-out and congestion effects are very much at play.

A third, prospective group of industrial firms encompasses *innovative, knowledge-based industries*. Innovative firms are the "Holy Grail" of regional development, and a commonly suggested panacea to the ills of underdevelopment and low competitiveness. But, as literature shows (Malory 1485), grails are elusive and hard to obtain. Knowledge-based industries require high capitalisation, and show immense agglomeration tendencies in a few global centres, with limited development potential on the periphery (Audretsch 1998; Kilar 2015; and in particular Kasabov 2011). In Pécs, two small groups of innovative firms are of particular note, both connected to the university. One is made up of minor IT start-ups and the subsidiaries of Budapest-based IT firms, and has very few connections to manufacturing. The second group represent university spin-offs active in the biomedical sector, founded by academic entrepreneurs. In-depth studies (Erdős and Varga 2012, 2013) have found that while these entities generate valuable knowledge, there are significant barriers to their commercialisation, particularly in a peripheral region distant from the major innovation hubs. Without a supportive business environment, adequate financing, and business know-how, the work of academic entrepreneurs is more likely to be purchased and put to use by firms in global core regions, promptly delocalised from their region of origin. Likewise, previous attempts to link up the university's medical knowledge base and local manufacturing companies (as in the 2005 growth pole programme) have met major hurdles. While local firms have sufficient capabilities to manufacture high-quality medical instruments, even relatively simple tools (e.g. surgical metal) must surpass significant barriers to entry posed by international quality standards—a practice benefitting large corporations. Nevertheless, biomed and broadly understood health industries may be one of the seeds for the city's future development, a rare area where it has comparative advantages—but the exploitation of this potential will require a concentrated, long-term development effort.

The institutional context of industrial change

During its transformation from industrial to post-industrial city, the institutional setting of Pécs had also changed profoundly. Despite significant personal continuity between the state socialist and post-socialist city elite, the leaders of the emerging new *urban regime* in Pécs accepted and eventually embraced the radical structural change of the local economy. The resulting regime, somewhat uncharacteristically among Central Europe's former industrial cities, does not have the traditional industrial and business affiliation of the typical "caretaker" and "developmental" regime types (Stone 1993; De Socio 2007). Rather, the city's political machine could be described as *an early-stage progressive regime*. As De Socio notes in his study of US urban regimes, progressive regimes prioritise civic functions, the service economy, multimedia services, high technology and, most notably for our case, culture-based development and the creative class.

> Progressiveness in this context more closely resembles the (...) middle-class arts-and-culture progressive regime type that focuses on historical preservation, quality of urban design, tourism, or environmental quality rather than a low-income, opportunity-expanding progressive regime prototype.
>
> (De Socio 2007, 358)

Concomitantly with the rise of the city's reorientation towards its cultural, consumption and higher education functions, interest groups representing manufacturing industries became marginalised in policymaking. Similar to De Socio's (2010) description of "sunset firms" in US Rustbelt cities, industry as a whole became "tainted" by the legacy of mining's failure: operating at a communicative disadvantage, they were persistently sidelined in the rush for priority-setting and resource mobilisation. For a long time, the local government had no formal investment/business promotion team at all, and only through the lobbying efforts of local entrepreneurs would this office be established—very late, well into the late 2010s. Likewise, Hungary's central government—regardless of its political affiliation—had either considered Pécs as a cultural city without need for manufacturing development (and, as under state socialism, not poor enough to be prioritised in development funding), or written it off as a hopeless case.

Absent a cohesive policy effort to organise the local economy, these responsibilities fell on the city's business community. The *Chamber of Commerce and Industry of Pécs and Baranya (PBKIK)* has emerged as the main lobbyist for local entrepreneurs, and is considered one of the more active actors in Hungary's network of economic chambers. Much like in the German model (Maenning and Ölschläger 2011), local chambers play a vital role in the reproduction of relational and social capital; in Hungary, they have been at the centre of bottom-up re-industrialisation initiatives. Póla (2020) lists seven areas where chambers should have a role in local manufacturing development:

1. improving the availability of skilled labour;
2. assisting generational change in enterprises;
3. interior market development;
4. fostering innovation and knowledge transfer;
5. increasing local added value;
6. fostering digitalisation and the spread of Industry 4.0 solutions;
7. developing reformed vocational education systems.

These aims, which form a comprehensive set of objectives to develop local and regional industrial milieus, have been in the focus of PBKIK, particularly in cluster-building, business matchmaking, and re-establishing an effective dual vocational education system after a long period of neglect. For decades, the Chamber's efforts had fallen on deaf ears; and it is only when the city's decline have become clearly visible that they have made some inroads.

The third actor in the city's institutional network is the *University of Pécs*. In spite of its history and prestige, the university has mostly been an absent player in Pécs's economic development. One reason for this is its *disciplinary structure*: like Hungary's other old provincial universities, it is specialised in medicine, the humanities and social sciences, lacking a strong technical faculty. The Faculty of Engineering and Information Technology (est. 1970) had only provided three-year education on the college level until 2004. It had first specialised in architecture and mechanical engineering, and has been very slow to develop beyond a small provincial college.

To this date, it is underdeveloped for the city's needs (IT as a universally applicable discipline; a sufficient number of mechanical engineering and electronics graduates). The second reason has to do with *institutional rigidity*. The university maintains limited relations with the local economy, and its attitudes have been described by interviewed economic stakeholders as "distant", "bureaucratic", "aristocratic", and "a useless humanities-based elite" (sic). There are strong organisational and cultural barriers that limit the university's knowledge transfer role—even beyond the aforementioned disciplinary mismatch. The University of Pécs has been far less successful in contributing to local and regional development than smaller, less prestigious but more entrepreneurial "mid-range" universities in Hungary and other Central European countries (Gál and Ptaček 2011, 2018; Rechnitzer 2016; Rechnitzer and Fekete 2019). Recognising this problem, the University's Szentágothai Research Centre (2012) has been established to provide a supportive environment for knowledge transfer and the commercialisation of innovation (particularly in biomed), but this institution is rather new, and its effects on regional development are yet to be properly surveyed.

Conclusions

While they may be out of the public eye, "hidden sectors" represent a substantial slice of post-transition regional structures in Central Europe. Beyond a non-negligible role in employment and value creation, they should be understood to contribute to the diversification and resilience of regions and cities. From the perspective of evolutionary economic geography, they contribute to the mixture of activities which characterise the modern city. Some of these sectors may contribute to new path creation (e.g. the seeds of medical industry in Pécs), some are relatively stable (e.g. the city's machine industrial cluster), while others are mature branches with limited growth potential, but still capable of renewal and a shift towards high value-added production (e.g. industries rooted in traditional craftsmanship). The neglect of specific activities can also lead to the decline, disintegration and eventual disappearance of productive traditions. Mining has entirely vanished from Pécs along with its specialised knowledge base, while the long-term future of the electronics industry is constrained by low upgrading performance. Neither traditions have strong successor industries.

Hidden sectors—typically mature branches taken for granted or experiencing decline—may move towards path dissolution and the loss of valuable development potential. A vicious circle emerges (Figure 6.4):

- Hidden sectors must often contend with a real or perceived declining position in local and regional economies due to external and internal factors. The consequences may involve slow shrinkage, or in more extreme cases dissolution (the falling apart of production systems needed to sustain competitive firms) and delocalisation (the flight of remaining companies to more advantageous investment locations).
- Hidden sectors typically experience policy neglect. They are at a narrative disadvantage in political agenda-setting due to their assumed failure or

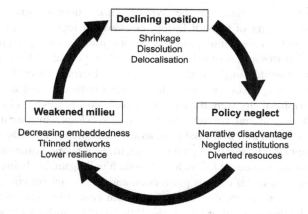

Figure 6.4 Hidden sectors and path dissolution.
Source: Author's elaboration.

diminished attractiveness, and may be written off as "out of fashion" or "unfix-able". Consequently, background institutions (education, training, business support, etc.) which are vital to the efficient operation of industrial milieus are often downsized or shuttered, introducing further difficulties for local firms. Public resources are spent on new, promising activities, and loans for SMEs may prove just as hard to secure.

- The neglect or marginalisation of hidden sectors usually results in a slow, often imperceptible erosion of the intangible factors which maintain a local indus-trial milieu. Yet one by one, these binding forces can become disconnected: firms become disembedded from their locality, business and knowledge net-works are thinned as actors and networking opportunities fade. In the end, the resulting milieus are weaker, less able to generate common value, and become more vulnerable to exogenous shocks—which produce new waves of decline.

Vicious circles, as it is well known, are hard to escape. Yet they do not represent the only possible future for hidden sectors.

In the example of Pécs, reducing the "idea" of a city into a pure concept (the former "mining city" vs. the new "city of culture"), and failing to account for sectors which did not fit this ideal picture has reduced the *diversity* of the local economy. The consequences can be felt in both concrete losses and (more importantly) opportunity costs, missed or underexploited potential. While external factors had played a role in the adverse de-industrialisation of the city and the surrounding city-region, many of the causes were endogenous, and can be traced back to poor policy decisions or public sector neglect. Whereas the abandonment of a previously diversified industrial mix under state socialism had created a mono-functional city, the abandonment of manufacturing and the milieus which would sustain it during post-socialism has contributed to a loss of economic potential and processes of urban shrinkage. Over the last decades, a wide gap has emerged between the city's post-industrial aspirations and the realities of its actual position in the post-industrial economy.

Yet there is no easy return to the "typical", more successful growth path followed by the majority of Central European industrial regions. In the peripheral context, FDI inflows are limited, and they mostly just reinforce peripherality. It is more likely that the way forward lies in the exploitation of endogenous growth potential found mainly in domestic companies, and strongly localised sources of development. Going beyond rhetoric, the city must rediscover that its former successes as a peripheral city were once found in high-quality products in niche markets; and that its cultural functions had once been closely intertwined with its industrial innovation and industrial production—not merely publicly funded cultural consumption. Therein lies a path forward, towards re-specialisation and future prosperity in a high value-added niche role which will, again, be different from the globally integrated spaces of mass production and mass employment. This is not a straight and wide road, nor does it come without costs. Pécs must accept that its destiny may be that of a minor city in the 100–110,000 inhabitant range, and that being a successful "city of culture" is untenable without also being a successful "city of industry": that is, its self-image and aspirations should be that of a diversified city, where culture and industry are not rival future images, but complementary parts of a unified vision.

Notes

1 Some of the empirical work for the chapter was undertaken in three previous research projects over the span of the last decade, with repeated inquiries providing an insight into ongoing evolutionary processes. These projects included OTKA grant #81789 ("Specific questions of institutionalising agglomerations within and beyond administrative structures"; 2010–2012); OTKA grant #K75906 ("The effects of industrial investments and transportation linkages on spatial structures"; 2010–2012); and NKFIH grant #K115577 ("The role of medium-sized enterprises in regional industrial competitiveness"; 2015–2019).

2 NKFIH grant #K115577 ("The role of medium-sized enterprises in regional industrial competitiveness"; 2015–2019).

References

Aubert, A., Marton, G. & Raffay, Z. (2015). Impacts of the European Capital of Culture of Pécs on the city's tourism. *Geographica Timisiensis*, 24 (1), 31–42.

Audretsch, D. B. (1998). Agglomeration and the location of innovative activity. *Oxford Review of Economic Policy*, 14 (2), 18–29. https://doi.org/10.1093/oxrep/14.2.18

Baranya Megye Statisztikai Évkönyve [Statistical Yearbook of Baranya County] (1956). Budapest: Központi Statisztikai Hivatal.

Baranya Megye Statisztikai Évkönyve [Statistical Yearbook of Baranya County] (1960). Budapest: Központi Statisztikai Hivatal.

Baranya Megye Statisztikai Évkönyve [Statistical Yearbook of Baranya County] (1970). Budapest: Központi Statisztikai Hivatal.

Baranya Megye Statisztikai Évkönyve [Statistical Yearbook of Baranya County] (1980). Budapest: Központi Statisztikai Hivatal.

Baranya Megye Statisztikai Évkönyve [Statistical Yearbook of Baranya County] (1990). Budapest: Központi Statisztikai Hivatal.

Baranya Megye Statisztikai Évkönyve [Statistical Yearbook of Baranya County] (2000). Budapest: Központi Statisztikai Hivatal.

Baranya Megye Statisztikai Évkönyve [Statistical Yearbook of Baranya County] (2019). Budapest: Központi Statisztikai Hivatal.

Borderless City: European Capital of Culture – Pécs, 2010. (2005). Pécs: Europe Centre Pbc.

Boschma, R. & Martin, R. (2010). The aims and scope of evolutionary economic geography. In Boschma, R. & Martin, R. (eds): *The Handbook of Evolutionary Economic Geography* (pp. 3–39). Cheltenham: Edward Elgar.

Capello, R. (2016). *Regional Economics* (2nd ed.). London and New York: Routledge.

David, P. A. (2007). Path dependence: A foundational concept of historical social science. *Cliometrica*, 1 (2), 91–114.

De Massis, A., Audretsch, D., Uhlaner, D., & Kammerlander, N. (2017). Innovation with limited resources: Management lessons from the German Mittelstand. *Journal of Production Innovation Management*, 35 (1), 125–146. https://doi.org/10.1111/jpim.12373

De Socio, M. (2007). Business community structures and urban regimes: A comparative analysis. *Journal of Urban Affairs*, 29 (4), 339–366. https://doi.org/10.1111/j.1467-9906.2007.00350.x

De Socio, M. (2010). Marginalization of sunset firms in regime coalitions: A social network analysis. *Regional Studies*, 44 (2), 167–182. https://doi.org/10.1080/00343400903095246

Drobniak, A., Kolka, M. & Skowroński, M. (2012). Transition and Urban Economic Resilience in Poland's Post-industrial Cities: The Case of Katowice. *Regions*, 286, 13–15. https://doi.org/10.1080/13673882.2012.10515116

Erdős, K. & Varga, A. (2012). The academic entrepreneur: Myth or reality for increased regional growth for Europe? In Geenhuizen, M. & Nijkamp, P. (eds.): *Creative Knowledge Cities: Myths, Visions and Realities* (pp. 157–181). Cheltenham: Edward Elgar Publishing.

Erdős, K. & Varga, A. (2013). The role of academic spin-off founders' motivation in the Hungarian biotechnology sector. In João, J.M., Ferreira, M.R., Rutten, R. & Varga, A. (eds.): *Cooperation, Clusters, and Knowledge Transfer. Universities and Firms Towards Regional Competitiveness* (pp. 207–224). Springer: New York. https://doi.org/10.1007/978-3-642-33194-7_11

Faragó, L., Horváth, G. & Hrubi, L. (1990). *Szerkezetátalakítás és Regionális Politika.* [Restructuring and Regional Policy] Budapest: Ts-2/2 Program Iroda.

Faragó, L. (2012). Urban regeneration in a 'city of culture': The case of Pécs, Hungary. *European Spatial Research and Policy*, 19 (2), 103–120. https://doi.org/10.2478/v10105-012-0017-4

Gál, Z. (2009). *The Golden Age of Local Banking: The Hungarian Banking Network in the Early 20th Century.* Budapest: Gondolat.

Gál, Z. & Ptaček, P. (2011). The role of mid-range universities in knowledge transfer in non-metropolitan regions in Central Eastern Europe. *European Planning Studies*, 19 (9), 1669–1690. https://doi.org/10.1080/09654313.2011.586186

Gál, Z. & Ptaček, P. (2018). The role of mid-range universities in knowledge transfer and regional development: The case of five Central European regions. In: Varga, A. & Erdős, K. (eds.): *Handbook of Universities and Regional Development* (pp. 279–300). Cheltenham: Edward Elgar. https://doi.org/10.4337/9781784715717.00023

Gorzelak, G. (ed.) (2020). *Social and Economic Development in Central and Eastern Europe: Stability and Change After 1990.* London and New York: Routledge.

Gorzelak, G. & Smętkowski, M. (2020): Regional dynamics and structural changes in Central and Eastern European countries. In Gorzelak, G. (ed.): *Social and Economic Development in Central and Eastern Europe: Stability and Change After 1990* (pp. 207–224). London and New York: Routledge.

Horváth, Gy. (2006). A lassú fejlődés történelmi folyamatai a Dél-Dunántúlon. [Historical processes of slow change in Southern Transdanubia]. In Hajdú, Z. (ed.): *Dél-Dunántúl. A Kárpát-medence Régiói 3* (pp. 27–62). Pécs–Budapest: Dialóg Campus.

Hospers, G.-J. (2014). Urban shrinkage in the EU. In Richardson, H.W. & Woon Nam, C. (eds): *Shrinking Cities: A Global Perspective* (pp. 47–58). London and New York: Routledge.

Hrubi, L. (2006a). A régió gazdaságának átalakulási sajátossága. [The characteristics of the region's economic transformation] In Hajdú, Z. (ed.): *Dél-Dunántúl. A Kárpát-medence Régiói 3* (pp. 192–237). Pécs–Budapest: Dialóg Campus.

Hrubi, L. (2006b). A régió ipara. [The region's industry] In Hajdú, Z. (ed.): *Dél-Dunántúl. A Kárpát-medence Régiói 3* (pp. 238–255). Pécs–Budapest: Dialóg Campus.

Hrubi, L. (2009). Tudás és újraiparosítás a dél-dunántúli régióban. [Knowledge and reindustrialisation in the Southern Transdanubian Region] In Fodor, I. (ed.): *A Régiók Újraiparosítása: A Dél-Dunántúl Esélyei.* (pp. 61–68). Pécs: MTA Regionális Kutatások Központja.

Józsa, V. (2016). *A Vállalati Beágyazódás Útjai Magyarországon.* [Pathways Towards Corporate Embeddedness in Hungary] Budapest: Dialóg Campus.

Kaposi, Z. (2006). *Pécs Gazdasági Fejlődése 1867–2000.* [The Economic Development of Pécs, 1867–2000] Pécs: Pécs–Baranyai Kereskedelmi és Iparkamara, Pécs.

Kasabov, E. (2011). Towards a theory of peripheral, early-stage clusters. *Regional Studies, 45* (6), 827–842. https://doi.org/10.1080/00343401003724651

Keresnyei, K. & Egedy, T. (2016). A pécsi kreatív osztály helyzetének értékelése statisztikai és empirikus kutatások alapján. [Evaluating the position of the creative class in Pécs on the basis of statistical and empirical studies]. *Tér és Társadalom, 30* (1), 57–78. https://doi.org/10.17649/TET.30.1.2730

Kilar, W. (2015). Settlement concentration of economic potential represented by IT corporations. *Geographia Polonica, 88* (1), 123–141. https://doi.org/10.7163/GPol.0009

Kovács, Sz., Lux, G. & Páger, B. (2016). Medium-sized manufacturing enterprises in Hungary: A statistical survey. *Studia Miejskie, 24,* 59–71.

Lähdesmäki, T. (2014). Discourses of Europeanness in the reception of the European Capital of Culture events: The case of Pécs 2010. *European Urban and Regional Studies, 21* (2), 191–205. https://doi.org/10.1177/0969776412448092

Lehrer, M. & Schmid, S. (2015). Germany's industrial family firms: Prospering islands of social capital in a financialized world? *Competition & Change,* 19 (4), 301–316. https://doi.org/10.1177/1024529415581970

Lengyel, I., Szakálné Kanó, I., Vas, Z. & Lengyel, B. (2016). Az újraiparosodás térbeli kérdőjelei Magyarországon. [The spatial question marks of reindustrialisation in Hungary] *Közgazdasági Szemle,* 53 (6), 615–646.

Lovering, J. (1999). Theory led by policy: The inadequacies of the 'new regionalism' (Illustrated from the case of Wales). *International Journal of Urban and Regional Research,* 23 (2), 379–395.

Lux, G. (2010a). From industrial periphery to cultural capital? Restructuring and institution-building in Pécs. In Sucháček, J. & Petersen, J.J. (eds.): *Development in Minor Cities: Institutions Matter* (pp. 103–126). Ostrava: VŠB – Technical University of Ostrava.

Lux, G. (2010b). Periférikus fejlődés, szerkezetátalakítási törekvések: Baranya megye és az államszocialista iparpolitika [Peripheral development and restructuring initiatives: Baranya county and state socialist industrial policy]. *Közép-Európai Közlemények,* 3 (3), 161–169.

Lux, G. (2015). Minor cities in a metropolitan world: Challenges for development and governance in three Hungarian urban agglomerations. *International Planning Studies,* 20 (1–2), 21–38. https://doi.org/10.1080/13563475.2014.942491

Lux, G. (2017). *Újraiparosodás Közép-Európában* [Reindustrialisation in Central Europe]. Budapest–Pécs: Dialóg Campus.

Lux, G. & Faragó, L. (2018). Conclusion: An evolutionary look at new development paths. In Lux, G. & Horváth, Gy. (eds.): *The Routledge Handbook to Regional Development in Central and Eastern Europe* (pp. 309–319). London and New York: Routledge.

Maenning, W. & Ölschläger, M. (2011). Innovative milieux and regional competitiveness: The role of associations and chambers of commerce and industry in Germany. *Regional Studies*, 45 (4), 441–452. https://doi.org/10.1080/00343401003601917

Malory, T. (1485). *Le Morte d'Arthur.* Westminster: William Caxton.

Markos, Gy. (1962). *Magyarország gazdasági földrajza.* [The Economic Geography of Hungary] Budapest: Közgazdasági és Jogi Könyvkiadó.

Martin, R. & Sunley, P. (2012). The place of path dependence in an evolutionary perspective on the economic landscape. In Boschma, R. & Martin, R. (eds.): *The Handbook of Evolutionary Economic Geography* (pp. 62–92). Cheltenham: Edward Elgar.

Martinez-Fernandez, C., Audirac, I., Fol, S. & Cunningham-Sabot, E. (2012). Shrinking cities: Urban challenges of globalization. *International Journal of Urban and Regional Research*, 36 (2), 213–225.

Molnár, E. (2013). Egy zsugorodó iparág újrapozicionálásának kérdőjelei: Magyarország cipőgyártása a rendszerváltás után. [Questions of repositioning a shrinking industry: Hungarian footwear industry after transformation.] *Tér és Társadalom*, 27 (4), 95–114. https://doi.org/10.17649/TET.27.4.2577

Nemes Nagy, J. (2009). *Terek, Helyek, Régiók: A Regionális Tudomány Alapjai.* [Spaces, Places, and Regions: The Foundations of Regional Science]. Budapest: Akadémiai Kiadó.

Nemes Nagy, J. & Lőcsei, H. (2015). Hosszú távú megyei ipari növekedési pályák (1964– 2013). [Long-term industrial development paths in counties (1964–2013)] *Területi Statisztika*, 55 (2), 100–121.

Németh, Á. (2016). European Capitals of Culture – Digging deeper into the governance of the mega-event. *Territory, Politics, Governance*, 4 (1), 52–74.

Pálné Kovács, I. (2013). Pécs, as the victim of multi-level governance: The case of the project 'European Capital of Culture' in 2010. *Urban Research & Practice*, 6 (3), 365–375. https://doi.org/10.1080/17535069.2013.827907

Pálné Kovács, I. & Grünhut, Z. (2015). The "European Capital of Culture - Pécs": Territorial governance challenges within a centralised context. In Schmitt, P. & Well, L.V. (eds.): *Territorial Governance across Europe: Pathways, Practices and Prospects* (pp. 81–94). New York and London: Routledge.

Pap, N., Gonda, T. & Raffay, Z. (2013). Pécs, a possible gateway city. *Forum Geografic*, 12 (2), 178–186. https://doi.org/10.5775/fg.2067-4635.2013.208.d

Pavlínek, P. (2015). Foreign Direct Investment and the development of the automotive industry in Central and Eastern Europe. In Galgóczi, B., Drahokoupil, J., Bernaciak, M. & Pavlínek, P. (eds.): *Foreign Investment in Eastern and Southern Europe After 2008. Still a Lever of Growth?* (pp. 209–255). Omaha: University of Nebraska.

Pavlínek, P., Domański, B. & Guzik, R. (2009). Industrial upgrading through Foreign Direct Investment in Central European automotive manufacturing. *European Urban and Regional Studies*, 16 (1), 43–63. https://doi.org/10.1177/0969776408098932

Pickles, J., Smith, A., Buček, M., Roukova, P., & Begg, R. (2006). Upgrading, changing competitive pressures, and diverse practices in the East and Central European apparel industry. *Environment and Planning A: Economy and Space*, 38 (12), 2305–2324. https://doi.org/10.1068/a38259

Pécs, the Pole of Quality of Life. (2005) Pécs: Local Government.

112   *Gábor Lux*

Pécs Városfejlesztési Koncepciója [Pécs City Development Concept] (1995). Pécs: Magyar Tudományos Akadémia Regionális Kutatások Központja Dunántúli Tudományos Intézete.

Póla, P. (2009). Az újraiparosítás intézményi feltételei. [The institutional conditions of reindustrialisation] In Fodor, I. (ed.): *A Régiók Újraiparosítása: A Dél-Dunántúl Esélyei* (pp. 69–78). Pécs: MTA Regionális Kutatások Központja.

Póla, P. (2020). Az iparvállalatokat támogató helyi intézményrendszer. [Local institutions in support of industrial firms.] In Lux, G. (ed.): *Hazai középvállalatok és regionális fejlődés* (pp. 145–161). Budapest: Ludovika.

Rácz, S., Kovács, S. Z. & Horeczki, R. (2020). Pécs fejlődési pályája. [The development path of Pécs.] In Rechnitzer, J. (ed.): *Nagyvárosok Magyarországon* (pp. 248–260). Budapest: Dialóg–Campus.

Rechnitzer, J. (2016). *A Területi Tőke a Városfejlődésben – A Győr-kód* [Territorial Capital in Urban Development: The Győr Code] Budapest–Pécs: Dialóg Campus.

Rechnitzer, J. & Fekete, D. (2019). *Együtt Nagyok: Város és Vállalat 25 Éve* [Great Together: 25 Years of the City and the Company] Budapest: Dialóg-Campus.

Richardson, H.W. & Woon Nam, C. (2014). Shrinking cities. In Richardson, H.W. & Woon Nam, C. (eds): *Shrinking Cities: A Global Perspective* (pp. 1–7). London and New York: Routledge.

Stone, C.M. (1993). Urban regimes and the capacity to govern: A political economy approach. *Journal of Urban Affairs*, 15 (1), 1–28. https://doi.org/10.1111/j.1467-9906.1993.tb00300.x

Sucháček, J., Krpcová, M., Stachoňová, M., Holešinská, L. & Adamovský, J. (2012). Transition and resilience in Czech post-industrial towns: The case of Ostrava and Karviná. *Regions*, 286, 17–19. https://doi.org/10.1080/13673882.2012.10515118

Szerb, L. (2009). Tradicionális és új iparágak fejlesztési lehetőségei Magyarországon és a Dél-Dunántúlon. [The development possibilities of traditional and new industries in Hungary and Southern Transdanubia.] In Fodor, I. (ed.): *A Régiók Újraiparosítása: A Dél-Dunántúl Esélyei* (pp. 45–60). Pécs: MTA Regionális Kutatások Központja.

Trócsányi, A. (2011). The spatial implications of urban renewal carried out by the ECC programs in Pécs. *Hungarian Geographical Bulletin*, 60 (3), 261–284.

Tubaldi, M. (2014). *An attempt at socio-economic regeneration through culture in a Central European city: The case of Pécs. Discussion Papers 94*. Pécs: Institute for Regional Studies, Centre for Economic and Regional Studies, Hungarian Academy of Sciences.

Turşie, C. (2015). Re-Inventing the centre-periphery relation by the European Capitals of Culture. Case-studies: Marseille-Provence 2013 and Pecs 2010. *Eurolimes*, 19, 71–84.

Welter, F., Baker, T., Audretsch, D. B. & Gartner, W. B. (2016). Everyday entrepreneurship: A call for entrepreneurship research to embrace entrepreneurial diversity. *Entrepreneurship Theory and Practice*, 41 (3), 311–321. https://doi.org/10.1111/etap.12258

Wójtowicz, M. & Rachwał, T. (2014). Globalization and new centers of automotive manufacturing: The case of Brazil, Mexico, and Central Europe. *Prace Komisji Geografii Przemysłu*, 25 (1), 81–107.

7   Socio-economic development in Bratislava during post-socialism

Pavol Korec and Slavomír Ondoš

Introduction

The socio-economic transition of post-socialist cities is a gradual complex process of changes affecting all domains of the society. The crucial elements were effective immediately at the end of the 1980s with the sudden and unexpected removal of the former socialist regime. In political domain, based on personal freedom and responsibility, a corridor was open towards democracy. Competition between varying interests and positions of stakeholders in the city replaced earlier hierarchical government without pluralism. In the economic domain, the free market was non-existent and economic institutions were mimicking many of the effects organically emerging from the competition (Sýkora 1993a; Musil and Illner 1994; Smith et al. 2008).

Socialism brought consequences in economic bases, social structures and inner spatial structures of the cities. Overdevelopment of industry, absence of private entrepreneurship, state monopoly in the bank sector, the industrial construction of housing complexes, and functional and physical underdevelopment in central areas became typical features of the cities in Central and Eastern Europe before 1989 (Węcławowicz 1992; Smith 1996; Korec 2003). Economically, socially and structurally neglected urban environments offered a good background for changes arriving during the 1980s. Collapse of socialism at the end of the 1980s and the appearance of a new political and socio-economic regime finally started an extensive urban transition.

Several conditions influenced the individual course and intensity of transition after 1989 in Bratislava, compared to other post-socialist metropolises. First, necessary to mention is a strategic setting in Central Europe and a unique micro-location regarding the environmental conditions, with a neglected potential of Danube for urban development. Second, on 1 January 1993 Bratislava became the capital of a new independent state for the first time in its history (if we omit the period between 1939 and 1945, during the Second World War, which was not a standard regime). Finally, globalisation and its consequences have brought drastic changes in economy of the city and caused changes in its social structure.

DOI: 10.4324/9781003039792-7

The changes were extensive and dynamic immediately after 1989. When Professor Walter Thomi, head of the Department of Geography at the University of Halle-Wittenberg, visited the city in 1997, he described it as "a unique laboratory of urban geography".

The aims of this chapter are: to review the basic features of the geographical setting, to draw attention to complex historical development of the city, to analyze changes in demographic conditions after 1989, to analyze a new economic base and change in use of city's territory in chronological order, and to point out the peculiarities of the process of suburbanisation, which changes the urban structure.

Main dimensions of socio-economic change in post-socialism

Many post-socialist countries achieved basic reforms of the political system over the first months after the collapse of socialism. The key aspect for understanding post-socialist urban change is the distinction among: a brief period changing basic principles of political and economic organisation; a medium-term period of population's behaviour adaptation and adoption of habits and cultural norms in new environment; a long-term period of transition of multiple spheres affecting broader societal changes with more stable patterns of urban morphology, land use and residential segregation. Understanding and interpreting post-socialist urban restructuring reflects the interactions among the following three aspects of post-socialist transition: the institutional change that created a general societal framework for transition; changes of the social, economic, cultural and political practices exhibited in the usual life of people, firms and institutions and resulting in social restructuring; and, the dynamics of urban change (Sýkora and Bouzarovski 2012).

With respect to urban change, a main part of the political change was the return to self-governance and the later shift of control from central (state) to local (community) level, which generally strengthened the role of the local community and eroded the power of the central state (Kovács 2000). The end of centrally state-planned economy has created processes and spatially well documented effects in immediate and deep change in structure of economy and areas of cities (Mantey and Kępkowicz 2018). Cities were under the sudden pressure of post-socialist change in shortly after 1989 (Smith et al. 2008; Stenning and Hörschelmann 2008; Sýkora and Bouzarovski 2012; Golubchikov et al. 2014; Jayne and Ferenčuhová 2015).

In opinions of several urban geographers in Central Europe (Sýkora 1993a; Węcławowicz 1997; Matlovič 1998) there were the following six trends in developing post-socialist cities which started immediately after 1989:

- Return of the self-government, the shift of an absolute control over the space from central to local institutions;
- Return of land rent and increasing number of actors competing for space;
- Transition in the economic base of the city;
- Increase of social and spatial differentiation and the changing rules of population's spatial distribution from political to economic criteria;
- Change of urban landscape;
- Architecture.

Cities concentrate a large part of the population and economic activity within an area, as a result of dynamic interactions between the economic activity and demographic growth. Differences between various urban systems in history and in different parts of the world knowledge play a key role in transition of economic activity. In economic theory knowledge is represented by human capital. For developed cities, sectors based on a knowledge-intensive work determine degree of their ability to compete with other cities. The growing role of cities in the knowledge economy model relates with the advantages of spatial proximity creating extra benefits of an effective collaboration between stakeholders (Van Winden et al. 2007).

City centres, and especially their historical cores, have moreover become a major topic for geographers (Musil and Illner 1994; Wolaniuk 1997; Sýkora 1993b; Matlovič 1995; Kovács and Dövényi 1998; Ira 2003; Ondoš and Korec 2004). According to Musil and Illner (1994), the centre of Prague in the late 1980s was in a critical situation, caused by unsuitable equipment, several buildings being demolished and insufficient maintenance. Wolaniuk (1997) mapped the institutions in the centre of the Polish city of Łódź and development in six timeframes between 1891 and 1993. According to her, the centre of Łódź does not meet the functional requirements and is predominantly occupied by retail and services. According to Kovács and Dövényi (1998), in post-war Hungary, public expenditure on urban development has been directed to housing estates on the periphery, which has caused the centers to decline. After the political change influenced by the market economy and the returning land rent's value, the city centres underwent an extensive functional modification and construction. Ira (2003) points to revitalisation in the morphological structure, commercialisation, fragmentation and expansion of sacral elements in the functional structure in the historical core. Ondoš and Korec (2004) describe how many (443) buildings were fundamentally reconstructed in the centre of Bratislava on an area of approximately six square kilometers between 1990 and 2001. A further (69) new buildings appeared during the time. On the example of Prague, Sýkora (1993b) explains how gentrification "changes the face of post-socialist city center". Later, especially with the new millennium, suburbanisation became one of the key themes of urban geography in post-socialist Europe (Ouředníček 2006, 2007; Sýkora and Ouředníček 2007; Novák and Sýkora 2007).

The growing role of suburbanisation in metropolitan development is not unique to the capital cities as other major cities in post-socialist countries follow a similar path. Suburbanisation should thus be considered as one of the crucial topics in the study of urban change in post-socialist cities. During the decade from 1997 to 2006, post-socialist Prague experienced profound changes of spatial organisation, with suburbanisation bringing radical reorganisation of metropolitan space (Novák and Sýkora 2007). But an interesting and attention catching process of change was furthermore happening in neighbourhoods to which some inhabitants escaped from the large housing estates. The escape out of modernist housing estates was not as massive as contemporary sociology predicted, although no failure of these neighbourhoods has taken place towards ghetto-like structural elements known from the western metropolises.

Since any larger construction did not appear for many years after the institutional gap at the end of the 1980s, the main arena of development has moved further towards urban periphery in the form of suburbanisation, and closer to the city centres as gentrification of inner-city blocks (Haase et al. 2012; Špačková and Ouředníček 2012). Both have been described as the processes changing both materiality and society of these places, inevitably bringing conflict and class replacement as the newly arriving inhabitants practically destroyed what had been in place and old communities had no effective means to oppose what is happening with their environment and local regime. Still, these processes were temporary and from the perspective of a city as a whole they brought new quality connected with the larger, more intensive and high-density place.

In the early 1990s several authors drew attention to cities of Central and Eastern post-socialist Europe, which would no longer be protected from competition. With the urban development, the key national centres were facing a crucial problem of their future role and prosperity in general within Europe. Major political, economic and cultural centres are no more isolated by economy and political borders competing with other European cities. Post-socialist cities must cope with a new reality and search for their own trajectory not only in national but at least in continental settlement network (Hall 1993; Grimm 1994; Vaishar 2000).

Concerning the state territory, the capital city of Slovakia is eccentrically located as shown in Figure 7.1. Geographically at the edge, despite its dominance in political, economic, cultural and social spheres, its role as the national capital was not questioned. Practically all positive factors influencing regional economic development clustered advantageously in the Bratislava metropolitan region after 1989. Shortly after the fall of socialism the city wins extraordinarily dominant position within the regional structure of Slovakia which is seen negatively in other regions. Proximity to the Western Europe economically and politically amplifies these processes (Korec 2003).

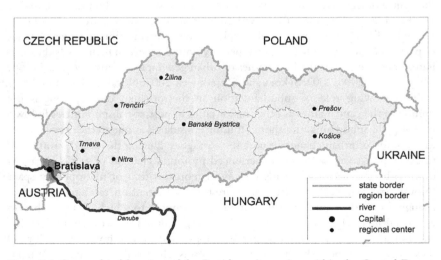

Figure 7.1 Geographical location of the Bratislava city region within the Central Europe and national territory.

Source: Authors' elaboration.

Geography, history and local governance design

Located in the centre of Europe, close to the contact of four major landscape elements played a major role at the beginning of city's history and it plays a crucial role in its development even today. The Alps and Carpathians, with lowlands of Vienna and Danube basins meet each other in the city's vicinity. The role of the contact space of both mountain ranges was strengthened by the Danube river, which penetrates through lower parts of both mountain ranges right in this area. Main transport links stretching from surrounding lowlands in northwest and southeast meet in the point where the Danube intersects the mountain ranges. As a consequence, crossing points of European routes located nearby since early historic times have frequently had military, strategic, economic and commercial functions (Korec and Galasová 1994).

In the immediate area, three large ethnic groups, German, Slavic and Hungarian, met in the historical evolution. Historical development, relationships ranging from collaboration to confrontation often alternated between the existing state departments in the area of the city. For the first time, locality gained significance at the beginning of the tenth century, when it became one of the three centres of the Great Moravia, the first joint state unit of old Czechs and old Slovaks (Štefanovičová 2011). The economic base of the city has begun to take form in the medieval era, when it developed mainly in three economic sectors—trade, crafts and agriculture. New development incentives occurred in the second half of the sixteenth century when due to the Turkish takeover of Budín, Pest and Székesfehérvár, the city became a temporary capital and coronation city of the Hungarian Kingdom in 1536. In 1563, the coronation of Maximilian II (a member Austrian House of Habsburg) took place here. Until 1790, further ten Hungarian kings, one queen and seven royal wives had this privilege in Bratislava. On June 25, 1741, Maria Theresa was crowned the Queen of Hungary in Bratislava (Szabó 2020).

The Hungarian administration, religious offices and judiciary moved in during the 16th and 17th centuries. Moreover, there were others like education, culture, religious activities that witnessed a growth in the city. In that period, the city started to form itself into a multi-functional regional center with predominant non-production activities, a crossroads of ethnicities and cultures.

The period between the 16th and 18th centuries was a "golden age" of Bratislava, as noted by historians, planners and geographers. During this period, the city's area grew, as numerous aristocratic palaces and church buildings were built (Kunec 2019; Szabó 2020). The establishment of Czechoslovakia in 1918 is a major event for its further development. Based on the 1910 population census, the population of Bratislava was predominantly of German (41.9 % of total) and Hungarian ethnicities (40.5 % of total) in the early 20th century. The third ethnic group was represented by Slovaks with only 14.9 % of the total population. The political forces at the foundation of Czechoslovakia were well aware of the strategic and political significance of Bratislava and therefore insisted from the beginning of discussions on determining the border between Czechoslovakia, Hungary and Austria on its unconditional connection (Holec 2020; Kováč 2020).

At the contact point with Hungary and Austria, the cadastral border of Bratislava is identical with the state border. The capital city is only 65 kilometres from Vienna, the capital of Austria, and 193 kilometres from Budapest, the capital city of Hungary. Political composition of Europe after the Second World War had resulted in two politically, economically and socially different regions just close to Bratislava. The border between the Eastern communist bloc and Western capitalist Europe was an effective barrier, called "iron curtain". Contacts between these regions ceased and remained sparse. Despite being in the geographical centre of Europe, political and economic positions of Bratislava turned to peripheral. Marginality and an immediate presence of borders with the "political enemies" influenced the economy and inner spatial structure of the city (Korec 2003).

Political and economic situation rapidly changed after the fall of communism in 1989. An optimistic prognosis appeared shortly in the context of integrating Europe and Bratislava was ranked within the top lucrative places where investors should start their business. An evaluation of 473 European regions, carried out by research and consulting institute Empirica AG in Bonn, Germany found the region of Vienna, Bratislava and Györ to be the most promising economic area in Europe from the point of view of capital returns in 1993 (Finka and Žigray 2005). A new continent's "golden triangle" of productivity was formed in the territory encompassing three nations (Korec 2003).

As a result of democratic processes, former Czecho-Slovakia was divided into newly formed independent states, Slovakia and Czech Republic as of 1 January 1993. Today both countries, together with Poland and Hungary, form the core of Central European region and closely collaborate within the "Visegrad Group". Bratislava became the capital city on January 1, 1993. It has around 430,000 inhabitants and has two levels of self-government. The first one represents the level of the city headed by the lord-mayor (in Slovak called "primátor"). Moreover, there is a city council. The lord mayor is elected every four years directly by citizens. The second level of self-government, the local level, is represented by 17 city boroughs or municipalities, headed by mayors (in Slovak called "starosta"). At the level of major urban agglomerations, however, the power decentralisation meant often the weakening of city governments and increasing power and competences of the districts and communities. This fragmentation of self-government of large cities resulted in difficulties as far as the elaboration and implementation of comprehensive urban development programs is concerned and raised the sustainability question in governance (Bennett 1998).

Processes of population development

Within twenty years of the fall of socialism, major changes appeared in population trends, too. After 1989, the city's size and its demographic structures started to change soon. Between 1991 and 2001 the city declined in population for the first time between two official censuses after the first historical census in 1869, with the population growth index reaching to 0.97 (Table 7.1). The population decrease continued in the following decade, with the index even lower than before. The most dynamic population growth in history of Bratislava started with the beginning of the 20th century.

Table 7.1 Development of Bratislava's population after 1869

Year	Thousand	% of SR	Index	Year	Thousand	% of SR	Index
1869	46.5	1.9	–	**1950**	192.9	5.6	1.39
1880	48.0	1.9	1.03	**1961**	241.8	5.8	1.25
1890	52.4	2.0	1.09	**1970**	285.4	6.3	1.18
1900	61.5	2.2	1.17	**1980**	380.3	7.6	1.33
1910	73.5	2.5	1.19	**1991**	442.2	8.4	1.16
1921	93.2	3.1	1.27	**2001**	428.7	8.0	0.97
1930	123.8	3.7	1.33	**2011**	411.8	7.6	0.96
1940	139.0	3.9	1.12	**2019***	435.3	8.0	1.06

Sources: Retrospective lexicon of CSSR municipalities 1850–1970, FSÚ Praha, 1978; Statistical office of the Slovak Republic (1992–2020).
* End of year balance instead of census.

The yearly population increase declined quickly from 8000 inhabitants in the 1970s and early 1980s to 5000 in 1987 and 2000 in 1991. Negative population development continued after 1991. Due to the stable value of mortality (about 9 deaths per 1,000 inhabitants), the natural population decrease occurred in 1995 for the first time in the city's modern history. Migration of the population followed the trends of natural population increase after 1989. The number of emigrants was stable from 1975 to 2001 reaching about 3000 to 4000 persons a year. However, the number of immigrants declined quite intensively after 1987 and even more until 1997, from approximately 7000 in the late 1980s to 3000 in 1997. Migration decline and decrease of population occurs in 1997 for the first time in modern history after the Second World War. In 1997, the number of migrants declined by 393 and the natural decrease was a little higher than migration decrease, reaching 500. Population declined by 893 in 1997. With 1997, the city's population size started decreasing (Korec 2003). The reasons for this development were mainly the following: (1) cessation of construction of large residential complexes and a transition to a standard housing market; (2) new demographic behaviour of the population; and (3) suburbanisation.

Following trends in population's natural and migration change, the age structure of the population has changed, too. During the 1991–2001 decade the number of children (up to 14 years) decreased from 103,000 to 60,000 and their share within the city's population decreased from 23.2% to 14.0 % (Table 7.2). The size of the population in productive age (aged between 15 and 59 years) increased in the same period from 261,000 to 286,000 (the portion increased from 59.1% to 66.7%). The number of inhabitants in higher age categories (60 years and more) increased from 78,000 to 83,000 and with their share on population increasing from 17.7% to 19.4%. The population development was alarming above all in a sharp decline of the young age categories. Over the next decade, the decline in the child component slowed markedly. The share of the productive age population fell slightly, and the growth in the number of older age categories remained high. Suburbanisation

Table 7.2 Development of age structure of Bratislava's population after 1989

Age	1991		2001		2011		2019*	
Group	Thousand	%	Thousand	%	Thousand	%	Thousand	%
0-14	102.8	23.2	59.9	14.0	52.5	12.7	70.9	16.3
15-59	261.3	59.1	285.8	66.7	270.6	65.7	255.0	58.6
60+	78.1	17.7	83.0	19.4	88.8	21.6	109.4	25.1
Total	442.2	100.0	428.7	100.0	411.8	100.0	435.3	100.0

Source: Statistical office of the Slovak Republic (1992–2020).
* End of year balance instead of census.

Table 7.3 Development of ethnic structure of Bratislava's population after 1851

	Share of total (%)				Population
Year	Slovak	German	Hungary	Others	Thousand
1851	17.9	74.6	5.9	1.6	42.2
1890	16.6	59.9	19.9	3.6	52.4
1910	14.9	41.9	40.5	2.7	78.2
1921	32.9	36.3	29.1	1.7	93.2
1930	48.5	26.5	15.3	9.7	123.8
1950	90.2	0.2	3.5	6.1	192.9
1991	90.9	0.3	4.6	4.2	442.2
2011	90.9	0.2	3.4	5.5	411.8
2019*	89.7	0.4	3.4	6.4	435.3

Sources: Retrospektívny lexikón obcí ČSSR 1850–1970, FSÚ Praha, 1978; Statistical office of the Slovak Republic (1992–2020).
* End of year balance instead of census.

processes have intensively impacted on the age structure development, too. Parents aged between 30 and 45 years with children have been moving out for rural villages in the vicinity.

Bratislava was once a multi-ethnic city. Until Czechoslovakia's establishment in 1918, the portion of Slovaks was stable at below 20.0%. Back in 1910, only 14.9% of the population declared their Slovak ethnicity (see Table 7.3), while the portion of Germans and Hungarians was roughly the same, hovering at around 40.0%. The situation in the ethnic structure of population changed sharply after 1918 and especially after the end of the Second World War. The share of German, Hungarian and "Czechoslovak" ethnic groups was similar in 1921, but Germans" and Hungarians' expulsion after the war brought a dominance of "Czechoslovak" ethnicity in the 1950 census (Table 7.3). In the 1960 census a Slovak ethnicity was introduced. Since the war the share of Slovak ethnicity remains stable at above 90.0%, the share of Germans has dropped to almost zero and the share of Hungarians is stable at around 3.5%. A tragic feature of the ethnic structure development is that before the Second World War as many as 12.0% of Jews lived in the city. As a result of the policy of "fascist Slovak state" at the time of the Second World War, only 0.4% of Jews remained after the end of the war (Hofreichter and Janoviček 2018).

Processes of economic transition

Development of economic structure of Bratislava since 1918 was complicated. Above all, the disruption of relations between Vienna and Budapest meant a loss for the city. Before 1918, the regional economy was closely linked with the larger domestic market of the Austrian-Hungarian state. Despite not having sovereignty, after 1918 the country was one of the two basic parts of the sovereign Czechoslovakia. After that, the city started to develop as an economical and cultural, and later even a political-administrative centre of the Slovak nation. In 1919, the government founded the first nation's university, "Comenius University in Bratislava", and in 1920, "Slovak National Theatre" was established here as the first professional theatre. In the same year, the "Slovak Bank", existing as early as in 1879 moved to Bratislava from Ružomberok. Also, some new industrial plants and service activities started to operate here and Bratislava launched a development trajectory towards a multi-functional city.

The key formation of the city's economic base took place in the 1948–1989 period. Under the socialist regime, its developing economic base absorbed few more impulses. The principal ones were the following: (1) the 1948–1980 industrialisation period which contributed 31 new factories and the reconstruction of existing old industrial plants with production expansion; (2) the strengthening of political, administrative, economic and cultural roles; and (3) the establishment of a common economic market of the socialist states in Europe. Some unsatisfactory features in the extensive industrial development were present even before 1989. The three frequent problems were: (1) industry had an inappropriate sector structure, generated workplaces with lower qualification requirements; (2) extensive land near the city centre was occupied by industry with uneconomically low intensity of use; (3) industrial production had a negative influence on the air quality (Korec 1997).

As Dostál and Hampl (1994) emphasised, considerable numbers of work opportunities in the cities of socialist countries and in their economies were partly determined by the growing bureaucracy in political and economic organisations of the state. Bratislava as a city has been developing for more than seven hundred years. However, the forty years of socialist development of the city were decisive for forming both its economic base and its inner spatial structure.

The sectoral structure of the city's economy has gained a new form soon after 1989. The international element in its economy strengthened particularly in the non-productive sectors of the economy, especially the financial sector, trade and entrepreneurship, real estate business, information technologies and others. The share of the workforce employed in industry and construction sectors started to fall in the city's economy structure. However, we should mention that two Bratislava's industrial plants are still the leaders of the national economy. The *Volkswagen* automotive company is top of the chart of Slovakia's non-financial companies according to annual sales and the petrochemical company *Slovnaft* reaches the fifth position in the same chart.

Table 7.4 illustrates a sharp decline in the share of the secondary sector of the economy after 1989 and an increase in the third sector's portion. In the first period of economic transition a decrease in employment appeared in the industry sector. In 1985, there were 69,000 jobs in the manufacturing sector (26.5% of the total number of employees) and only 49,000 in 1998 (15.9% of total). Jobs in the construction sector decreased from 47,000 (16.7%) to 25,000 (8.0%). The financial sector increased the number of its employees from less than 2000 (0.7%) in 1985 to 14,000 (4.5%) in 1998 and later to 20,000 (5.9%) in 2018. In thirty years, the city became a new financial centre in central Europe. After 1989, other non-productive sectors of the economy (wholesale and retail sector, real estate and trade services and public administration) recorded a remarkable increase in jobs as illustrated in Table 7.4.

Table 7.5 documents the development of selected indicators of Bratislava's economic weight in Slovakia after 1989. Despite the sharp employment decline in Bratislava's manufacturing sector, the city maintained its dominance over national economy (9.0% of the total national industry's employment). This is mainly the result of workforce with both of the above-mentioned plants (*Volkswagen* with 15,000 in 2018 and *Slovnaft* with 2500 employees). The financial sector has been one of the key sectors generating urban changes within the city. Bratislava as a leading financial centre of the country concentrates 70.8% of the total employees in

Table 7.4 Employment structure in Bratislava by the main economic sectors

	1985	1998	2018
Agriculture	2.0	0.6	0.1
Primary sector	2.0	0.6	0.1
Industry	26.5	15.9	13.1
Construction	16.7	8.0	3.4
Secondary sector	43.1	23.9	16.5
Wholesale and retail services	11.1	18.3	22.2
Transport and logistics	8.9	10.1	7.5
Accommodation and gastronomy	1.5	2.1	2.3
Real estate, trade services	9.5	15.5	29.1
Finance and insurance	0.7	4.5	5.9
Public administration	3.5	5.9	10.3
Health and social services	5.4	5.7	6.1
Tertiary sector	54.9	75.6	83.4

Sources: Municipal statistical authority Bratislava (1986, 1999); Statistical office of the Slovak Republic (2020).

Table 7.5 Selected indicators of Bratislava's weight in Slovakia (in %)

	1985	1998	2018
Population	8.1	8.3	7.9
Total employment	11.6	14.0	20.9
Employment by sector			
Industry	8.6	7.4	9.3
Construction	15.8	12.0	17.9
Transport and ITC	27.0	18.3	31.9
Trade	10.4	21.4	26.3
Finance and insurance	21.6	37.4	70.8
Education	15.3	14.9	15.0
R&D	27.6	42.6	50.5

Sources: Municipal statistical authority Bratislava (1986); Statistical office of the Slovak Republic (1999, 2020).

the Slovak financial sector in 2018. Table 7.5 illustrates an increase in the research and development sectors, too. In 2018, more than one half of employees in research were concentrated in Bratislava.

The privatisation of state-owned economic institutions was among the earliest tasks after the collapse of socialism and market mechanisms were established. Along with new businesses, many surviving companies resurrected from former national enterprises or cooperatives. Small businesses were born quickly and relatively easily across the country. Many former state-owned companies were sold to the first generation of small entrepreneurs, often former employees or managers having the know-how and necessary starting private capital and personal connections for continuation of their activities and survival depending on competitiveness.

Explanation of factors triggering transition into the current stage of development correlates with the influence of the European Union. Opening of economy towards the Western partners and the standardisation of the institutional context enabled full connection with global value chains and synchronisation of local development with the global business. It is a historical coincidence that transition from totalitarian communism towards liberal capitalism happened as the West capitalist model of production entered an accelerated stage of transition into dematerialisation and digitisation, as one experiences it today. These processes were first shadowed by the post-communist transition, as observed from the local temporary perspective. Real transition was about to bring much more fundamental change to how people compete and collaborate.

Economic restructuring in Bratislava became heavily dependent on the international capital, allocating its production capacities with the primary motive of labour costs effectiveness. Slovakia has been among the new member states with the lowest wage levels within the common market, offering advantages for Western and later global actors active in Europe. Manufacturing is still the major sector generating employment opportunities in non-metropolitan regions. Labour costs in Slovakia before joining the European Union in 2003 were 3.1 euros per hour. In Hungary it was 3.8, in the Czech Republic 3.9 and in Poland 4.5 euros per hour. Over the same period, in Germany labour costs were 26.3 euros, 24.4 euros in France and 23.9 euros in the United Kingdom (Korec and Popjaková 2019). Slovakia soon experienced growth in a cluster of automotive and supporting industries, producing automobiles and consumer electronics at the local level of intensity never experienced before.

Changing inner spatial structure

The inner spatial structure was developed with an irregular pattern before 1989. Urban territory grew towards northeast and east directions from the city centre before 1970. There have been three major obstacles, two in the physical landscape—the mountain range of the Small Carpathians and the Danube river—and a political one represented by the state border with Austria. With respect to the future development of the city, one needs to emphasise the following features of its inner spatial structure in 1989:

- The central city was functionally and physically underdeveloped. In reality, it was a cluster of devastated buildings, without any reconstruction and suitable exploitation efforts;
- Large industrial zones with low land use intensity close to the city centre;
- Extensive mono-functional housing complexes at the periphery, which offered only limited opportunities and standards of services to their inhabitants;
- Unfinished main road network, which would support local exchange within the city, regional and transit transport;
- Artificial barrier to the city development at the east side of the city, stretching south-north and represented by petrochemical plant "Slovnaft", the airport, freight railway station in Bratislava-Vajnory and a natural protected wetland of Jurský Šúr;
- Large-scale integration of surrounding villages: 13 villages joined city in 1946 and 1971, cadastral territory increased from 58.5 km² in 1945 to 367.8 km² in 1989. The city's territory consists of intensively used agricultural land (21.0% of city's area) and forests (40%). There is large share of classical rural type of neighbourhoods within the city.

The inner spatial structure has changed rapidly since 1989. The following six clearly visible processes of urban change dominated here up to 2004 (the integration into the European Union):

- Commercialisation and revitalising of the city centre;
- Industrial decline and revitalisation of production areas located within urbanised areas of the city;

- Commercialisation by implantation of various business, trade and services activities into originally mono-functional housing estates;
- Intensification of the territory, construction of office buildings in central parts and construction of residences with high standard of living in green areas;
- Construction of hyper-markets and large shopping and services centres, along main roads and in large residential zones;
- Intensification of land located along highways crossing the city and other six main radial roads.

All these processes have been associated with radical change of land use patterns in terms of replacing existing activities with new and economically more efficient ones and, simultaneously, with physical upgrading.

There is a connection between inner-city revitalisation and the growing integration of these places to the world economy. The physical and social upgrading of these neighbourhoods is progressing by the corporate and commercial expansion of the global market. New corporate headquarters business and commercial centres, hotels and tourist facilities have flooded the city centres all around the metropolises of East Europe (Kovács 2000). Revitalisation and economic and social upgrading of areas near the city centre had special prerequisites.

The Danube river with its wide banks created by the old town and expanding the city centre has become, thanks to a centrally open space, the main target of capital investment. There were three qualitatively different areas in close to the city centre, which were suitable (predestined) for revitalisation. Attractive but unused areas were located along the left bank of Danube. Large business centres with a share of residential function (*River Park*, *Zuckermandel*, *Eurovea I* and *II*) grew quickly after accession of the Slovak Republic into the European Union in 2004 on the Danube bank. Another complex (*Vydrica*), located close to the Old town of the city, is under construction (documented in Figure 7.2).

The old industrial plants from the early twentieth century represent another attractive area close to the city centre. In the former large manufacturing area of Mlynské Nivy, new business projects such as *Twin City A*, *B* and *C*, *Nivy Tower I* and *II*, *Ister Tower*, *Panorama Tower I* and *II* and *Panorama Business* and large residential complex *Sky Park* are expanding and will soon change the skyline of the city. Construction development in this area is a typical example of the "manhattanisation" process in large cities of countries in Central and Eastern post-communist Europe (Figure 7.2). The use of intelligent technologies in the above-mentioned construction projects saves operating costs and environment. Offices in these facilities are fully occupied by powerful global companies. The third attractive area near the city centre is the area along the motorway passing along the right bank of the Danube river in the city neighbourhood of Bratislava-Petržalka. Here, we find new large business centers of *Aupark Tower*, *Digital Park I* and *II*, *Einsteinova Business Center*, *Tatrabanka Tower* and *Einpark* (in Figure 7.2).

Typical practically mono-functional residential areas appeared in various dimensions around close by former rural settlements during the period of socialism. Their population ranges from 7000 in Bratislava-Lamač or 12,000 in

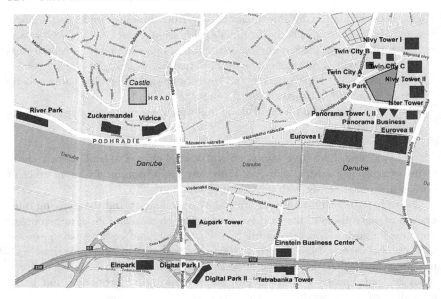

Figure 7.2 Central city with important commercial development projects of latest years along Danube and the main roads similar to those in Bratislava-Petržalka.

Source: Authors' elaboration.

Bratislava-Devínska Nová Ves to larger 40,000 residents in Bratislava-Ružinov and Bratislava-Dúbravka, up to enormous 125,000 inhabitants in the most internationally well-known Bratislava-Petržalka. The two key processes, commercialisation and intensification started operating in these residential areas immediately after 1993. In several of these socialist residential neighbourhoods, a decline in population occurred after 1993. These large residential complexes have become attractive for foreign retail chains. From the beginning of the millennium, hypermarkets and supermarkets *Tesco, Kaufland, Billa, Lidl* brands arrived here. Originally monofunctional residential complexes turned into multi-functional ones with representation of trade, services and finance and became more attractive for living.

Suburbanisation

Suburbanisation is a major development process of cities and regions in the post-socialist development period, changing social and physical environment from a rural type directly to a suburban type. Often it is associated with migration of city residents to the hinterland in search for more space and higher quality of housing. Loss of agricultural land and traffic jam on roads are main symptoms of suburbanisation. In the suburban zone, loss of agricultural land is not yet considered a problem. Villages have large quantities of unused arable land, vineyards and orchards available in result of earlier abandonment.

Four quantitatively and qualitatively distinct phases of suburbanisation could be identified in Bratislava region's recent evolution. The first phase (until 1996) only

Table 7.6 Population structure by ethnicity (2001) and share of the core's population in suburban districts

District	Ethnicity (%)			Urban (%)	
	SK	HU	Other	2001	2011
Malacky	97.0	0.3	2.8	40.7	39.0
Pezinok	97.3	0.4	2.2	63.2	60.9
Senec	76.9	20.2	2.9	28.3	25.6
Dunajská Streda	14.1	83.3	2.7	39.8	37.8

Source: Statistical office of the Slovak Republic (2002, 2012).

showed the first signs of forthcoming suburbanisation. Between 1996 and 2001, a gradual expansion of suburbanisation caused the migration of wealthier residents to newly built family houses on vacant plots on the outskirts of rural municipalities. Between 2002 and 2009, a construction hugely accelerated involving commercial investments and sprawling residential areas often completely segregated from the original local settlement. After 2009, the suburbanisation growth saw a stagnation due to economic crisis (Šveda 2016).

Distinctly specific socio-cultural zones around Bratislava exist among suburban belt segments. Living in each of these parts has its specifics as Table 7.6 and Figure 7.3 illustrate, regardless of transport accessibility from the city.

Initiation of a specific cross-border suburbanisation (stretching to Austria and Hungary) followed the accession Slovakia to the European Union and so-called Schengen area. These steps ensured free mobility of European citizens across national borders, which added large open areas for development of a functional city in Austria and Hungary for the residential market. The adjacent border regions of Austria and Hungary have had a comfortable transport accessibility from/to the city until today, with lower road traffic, and even lower prices of real estates. Residential suburbanisation in the municipalities beyond national borders has been in progress. Number of Slovak citizens living in Austrian and Hungarian municipalities is increasing and more distant villages are affected, too.

About ten years ago local studies found that suburbanisation associated with only three Austrian municipalities (Kitsee, Berg and Wolfsthal) and one Hungarian municipality (Rajka). Eight years later, there were fourteen municipalities in Austria, where the share of Slovaks was higher than 5.0%, and in Hungary there were seven such municipalities. Table 7.7 lists ten border municipalities with the highest proportion of Slovak residents. These municipalities are outlined in Figure 7.3. It is realistic to expect that suburbanisation in the Austrian and Hungarian territories will continue in the forthcoming years despite the fact that land and property prices in both countries have been rising over the past years, according to the market disproportion created. The Covid-19 crisis of 2020 brought new relations in suburbanisation in the Austrian and Hungarian borders. During the

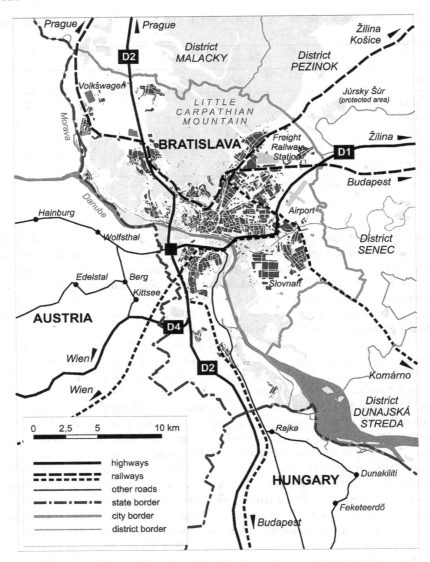

Figure 7.3 Internal spatial structure and main suburban development directions with expanding communities.

Source: Authors' elaboration.

lockdown, Slovaks living in Austria and Hungary were forced to declare their permanent residencies in order to even cross the state border.

In reality, suburbanisation did not exist during socialism. The largest proportion of urban residents lived in modernist public housing estates. The proportion of those living in large mono-functional housing estates reached 82.0% in Bratislava in 1989. The underlying social political values of the socialist regime did not favour

Table 7.7 Communities in the Austrian and Hungarian border area with
the largest share of Slovak citizens (as of 2018)

	Population	Slovaks	%
Kitsee (AT)	3,162	1,328	42.1
Feketeerdő (HU)	586	213	36.4
Rajka (HU)	3,006	846	28.1
Edelstal (AT)	747	188	25.2
Wolfsthal (AT)	1,415	336	23.7
Berg (AT)	1,152	231	19.9
Máriakálnok (HU)	1,723	316	18.3
Hainburg a.d. Donau (AT)	8,755	1,453	16.6
Dunakility (HU)	1,894	261	13.8
Levél (HU)	1,908	232	12.2

Sources: Statistical Database of Statistics Austria (2019); Hungarian Central
Statistical Office (2019); calculation by Korec (2020).
AT—Austria, HU—Hungary.

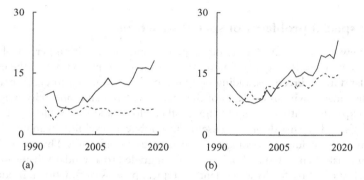

Figure 7.4 Population's migration into (full line) and out of (dashed line) Bratislava area (in
‰)—the metropolitan region (a), and separately the urban core of Bratislava (b).
Source: Statistical Office of the Slovak Republic (2020). Elaborated by authors.

status hierarchy and prestige. The egalitarian situation created suitable conditions
for suburbanisation process. Many residents in the city relatively quickly gathered
their economic resources using new market opportunities and living in housing
estates was no more comfortable for them.

Development of suburbanisation immediately after the fall of socialism was tur-
bulent in Slovakia, similarly to other Central European countries. Among others,
local attributes of suburbanisation in Bratislava include the following: rural communities gradually turned into suburban ones, actions of local governments ranging
between potential profits, risks and personal profits, activities of investors, development of housing prices and their differences between the city and suburbia, and
finally planning regulation seeking to balance preserving landscape values and economic pressure of capital. Figure 7.4 illustrates the development of housing construction in the metropolitan region and especially in the suburban belt over the
past twenty years affected by the impacts of the 2009 global economic downturn.

Many families contributing to suburban development have already returned to the city, mainly because of the unacceptable time-loss and stress caused by commuting. Early interests in private ownership of housing in a quiet rural environment slowly make room for more rational consideration of costs, disadvantages taken in their usual life, both in terms of financial costs and time-loss. The suburban belt is relatively poorly equipped by the public transport infrastructure, making the city more attractive again as the development projects across the inner city and outer city quarters still reachable by public transportation lines made living in the city comparatively more comfortable again. A considerable burden for the future of sustainable development in suburbs is symptomatic single function residential growth which misses modern amenities, practical accessibility of services and even a good-quality environment. large-scale suburban complexes of the north-east Bratislava region in Pezinok and Senec districts (Ivanka pri Dunaji, Chorvátsky Grob, Slovenský Grob, and many others today extending as far as Trnava region) concentrated thousands of inhabitants around old villages, along radial roads between them and Bratislava during previous decades.

Some special problems of spatial structure

Many new functional areas were developed during the transition period all across the city's territory. Residential functions intensified and spread towards the edge of urban land in all directions, adding networks of family housing to peripheral parts and later small towns and villages of the region. Central and inner-city neighborhoods expanded with activities bringing office blocks and towers along arterial or ring roads, and crossroads offering useful automobile accessibility in an increasingly problematic city due to morning and late afternoon traffic jams. The inner city has been gradually renovated and public spaces upgraded to a standard shape, improving quality of living for local residents. A typical process of functional mixing has redeveloped many formerly residential blocks into multi-functional zones, returning the city back to its more traditional shape and quality. More density and intensity of use has been well adapted and served by original infrastructure, especially a well-developed dense public transport system.

Large vacant territories are spread in the east suburbs and accessible locations, where the vast area of the former chemical industry operated for decades until the early 1990s. Degraded post-industrial areas are an economic and environmental issue, because valuable locations avoid development due to the toxic pollution infiltrated in soil and underground water. Few separate locations have been identified with soil toxicity, historical or recent. One of the exclusive locations at the Mlynské Nivy, which is under current development into a new high-rise center, has an unsolved problem underground. Pollution of Danube sediments remains here since the Second World War, when military destroyed the former Apollo oil refinery, and only partial solutions cleaned the sites of new commercial complexes, including *Eurovea* mall and residences.

Many more territorially defined problems wait in large premises of the former chemical plant *Istrochem*, in the east of the city. To the present day, the area is empty, slowly turning into green but toxic wilderness behind concrete protection walls.

Preliminary development visions for these locations were soon abandoned even in conditions of profits potentially generated from commercial and residential streets, which would otherwise have enough space for tens of thousands of inhabitants in the area, readily accessible by trams from both north and south, which is no longer the case anywhere else in the city. Instead, developers focus on suburbs without any infrastructure, with the exception of inadequate peripheral roads in the south and west of the city.

Gentrification used to have a limited area and source of demand in the same time compared with the bigger neighboring cities. Upgrading process with negative effects on local communities was present in Bratislava as early as the middle of the 1990s. City has only a few locations suitable for similar discovery of dense inner city, with traditional urban qualities of the late 19th or the early 20th centuries. Located near the historical central city, these soon became economically inaccessible due to competition with retail and consumer services, bars and coffee houses, negatively perceived due to increase in non-stop tourism. Residential function has been nearly completely pushed out of the Old Town. However, the neighbouring ring of inner-city neighborhoods became an attractive alternative. Owners of the properties have seen the value of their former rented apartments multiplying in comparison to the price they paid for owning their apartments.

The residential real-estate market has soon restored a steep price gradient, making the apartments in relatively small historical houses attractive as an investment opportunity and subject to high demand because of the immediate supply of all central city amenities. Young families of professionals participating in the new and profitable service economy have been increasingly preferring to live in inner-city blocks, meaning that there is no need for them to commute. Central neighbourhoods have saturated the existing supply. The owners overbuilt their residential buildings with vertical extensions in large numbers, and territorially extended towards the east, where older housing estates from the 1950s and the 1960s offer similar qualities of residential environment. The gentrification process is ongoing. New attractive services arrive serving a demand not generated by their earlier inhabitants. Housing has been predominantly owner-occupied since the early 1990s; therefore, no similar spreading problem to what was happening with inaccessibility of the households in renting sector happens here.

The development of housing construction in Bratislava region over the past two decades is shown in Figure 7.5. The global downturn after 2009 has clearly afflicted housing construction in the Bratislava region. Investment property inflow signaling potential price speculation is present only in the urban core of the region.

Conclusion

The socio-economic transition between a socialist and a capitalist city in principle took place according to the standard process in Bratislava after 1989. Recent capital city status and extraordinary development momentum merge into a unique trajectory of the "Bratislava type". The city and its metropolitan area have started to dominate in the national regional structure. Unique heritage of earlier decades slowly changed under the influence of developments in the local and regional

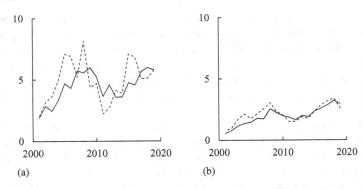

Figure 7.5 Housing construction in thousands of finished units (full line) and started units (dashed line) in Bratislava area—the metropolitan region (a), and separately the suburban belt of districts Malacky, Pezinok and Senec (b).

Source: Authors' elaboration on the basis of the materials from Statistical Office of the Slovak Republic (2020).

economy. Growth of a truly modern city has started and continues. Today, the city has a mixture of functions generated mainly by business service economy represented by the global corporations.

Underdeveloped in respect to the potential already recognised by professionals it fully accelerated its growth in the transition period. It had a vulnerable geographical location directly at the outer border of Europe's east bloc during the four decades between 1948 and 1989. This peculiarity blocked any economically sustainable development and while geopolitical interferences remained effective. Therefore, since 1989 the city is free for development dynamics first time in modern history regardles of how it unfolds next. There is a connection between revitalising inner-city neighborhoods and the growing integration of these places to the global economy. Corporate and commercial expansion in the competitive market drives the physical and social upgrading of these neighborhoods.

Corporate headquarters, business and commercial centres, hotels and tourist facilities have flooded the central city of Bratislava in a manner similar to that observed in metropolises in Central Europe. Revitalisation, economic and social upgrading of areas near the city centre had special prerequisites in Bratislava. There were three qualitatively different areas close to the centre, which were seemingly suitable for revitalisation. Large business centres with residential functions quickly mushroomed overlooking all of these attractive property locations.

Attractiveness for development of economic activities has increased during the social and economic transition. Metropolitan region of a country should be seen in terms of innovation space with possibility of participation for other regions of the country. Capital cities played a dominating role in earlier stages of economic development. But capital cities have acquired a new quality in the post-industrial stage, and in conditions of dominating global market.

The end of socialism stopped isolating former socialist countries and accelerated differentiation in their regional systems. Cities and their regions have, in general, joined two groups, the winners and the losers of transition within the emerging market. Differences in regional systems of the post-socialist countries have grown considerably. Regions in the west clearly joined the club of winners. Today, Bratislava region represents a growth pole even at the scale of the European Union's east part. Statistically, Bratislava region is considered the eighth richest within the Single Market by Eurostat. In addition, the city has been competing economically with the international network of cities of Central Europe, and so far competing successfully even of its story continues.

Acknowledgement

The authors prepared contents for this chapter with support of the project APVV 15–0184 "Intergenerational social networks in an aging city, continuity and innovation 2016–2020".

References

Bennett, R. J. (1998). Local government in post-socialist cities. In: Enyedi, G. (Ed.) *Social change and urban restructuring in central Europe*. Budapest: Akadémiai Kiadó, 35–54.

Dostál, P., Hampl, M. (1994). Changing economic base of Prague: towards new organizational dominance. In: Barlow, M., Dostál, P., and Hampl, M. (Eds.) *Development and administration of Prague*. Amsterdam: University of Amsterdam, 29–46.

Finka, M., Žigray, F. (2005). *Relationship among ideal European region and real European regions on the example of border region Bratislava – Vienna. Selected meta-scientific and applied aspects.* Bratislava: Faculty of Architecture, Slovak University of Technology in Bratislava.

Golubchikov, O., Badyina, A., Makhrova, A. (2014). The hybrid spatialities of transition: Capitalism, legacy and uneven urban economic restructuring. *Urban Studies*, 51(4), 617–633.

Grimm, F. (1994). *Zentrensysteme als Träger der Raumentwicklung in Mittel- und Osteuropa. Beiträge zur regionalen Geographie, 37.* Leipzig: Institut für Länderkunde, 128 p.

Haase, A., Grossmann, K., Steinführer, A. (2012). Transitory urbanites: New actors of residential change in Polish and Czech inner cities. *Cities*, 29(5), 318–326.

Hall, P. (1993). Forces shaping urban Europe. *Urban Studies*, 30(6), 883–898.

Hofreichter, R., Janoviček, P. (2018). *Z Prešporku do Sôlnohradu, Strednou Európou proti prúdu času.* Bratislava: Slovart, 256 p.

Holec, R. (2020). *Trianon, triumf a katastrofa.* Bratislava: Marenčin PT, 367 p.

Ira, V. (2003). The changing intra-urban structure of the Bratislava city and its perception. *Geografický časopis*, 55(2), 91–107.

Jayne, M., Ferenčuhová, S. (2015). Comfort, identity and fashion in the post-socialist city: Materialities, assemblages and context. *Journal of Consumer Culture*, 15(3), 329–350.

Korec, P. (1997). New tendencies of manufacturing development in Bratislava. In: Kovács, Z., and Wiessner, R. (Eds.) *Prozesse und Perspektiven der Stadtentwicklung in Ostmitteleuropa.* Passau: L.I.S.Verlag, 131–140.

Korec, P. (2003). Bratislava's perspective in Europe. Hallesches Jahrbuch für Geowissenschaften. *Reihe A: Geographie und Geoökologie*, 25, 7–17.

Korec, P., Galasová, S. (1994). Geografická poloha Bratislavy v nových hospodársko-politických podmienkach. *Geografický časopis*, 46(2), 75–86.

Korec, P., Popjaková, D. (2019). *Priemysel v Nitre: globálny, národný a regionálny context.* Bratislava: Univerzita Komenského, 210 p.

Kovács, Z. (Ed.). (2000). *Special issue on post-socialist urban transition in East and Central Europe.* Kluwer Academic Publishers.

Kovács, Z., Dövényi, Z. (1998). Geographical features of urban transition in Hungary. *Geographica Pannonica*, 2(1), 41–46.

Kováč, D. (2020). Hranice Slovenska v česko-slovenskom zahraničnom odboji na mierovej konferencii v Paríži. *Historická revue*, 31(6), 12–17.

Kunec, P. (2019). Vo víre vojen a medzinárodnej politiky. Vláda cisára Leopolda I. v rokoch 1657–1687. *Geografická revue*, 30(2), 11–17.

Mantey, D., Kępkowicz, A. (2018). Types of public spaces: The polish contribution to the discussion of suburban public space. *The Professional Geographer*, 70(4), 633–654.

Matlovič, R. (1995). Morfologické procesy v historickom jadre Prešova v svetle Conzenovej urbánno-morfologickej koncepcie. *Geographia Slovaca*, 10, 141–149.

Matlovič, R. (1998). Geografia priestorovej štruktúry mesta Prešov. *Geografické práce*, 8(1), Prešov: Prešovská univerzita, 260 p.

Musil, J., Illner, M. (1994). Development of Prague: Long-term trends and scenarios. In: Barlow, M., Dostál, P., Hampl, M. (Eds.) *Development and Administration of Prague*. Amsterdam: Instituut voor Sociale Geografie, 149–171.

Novák, J., Sýkora, L. (2007). A city in motion: Time-space activity and mobility patterns of suburban inhabitants and their structuration of the spatial organization of the Prague metropolitan area. *Geografiska Annaler*, 89(2), 147–167.

Ondoš, S., Korec, P. (2004). Prehľad a interpretácia stavebného vývoja centra Bratislavy v post-socialistickom období. *Geografický časopis*, 56(2), 153–168.

Ouředníček, M. (Ed.) (2006). *Sociální geografie Pražského městského regionu*. Praha: Univerzita Karlova v Praze, 156 p.

Ouředníček, M. (2007). Differential suburban development in the Prague urban region. *Geografiska Annaler: Human Geography*, 89(2), 111–125.

Smith, D. M. (1996). The socialist city. In: Andrusz, G., Harloe, M. and Szelenyi, I. (Eds.) *Cities after socialism: Urban and regional change and conflict in post-socialist societies*. Oxford: Blackwell, 70–99.

Smith, A., Stenning, A., Rochovská, A., Świątek, D. (2008). The emergence of a working poor: Labour markets, neoliberalisation and diverse economies in post-socialist cities. *Antipode*, 40(2), 283–311.

Stenning, A., Hörschelmann, K. (2008). History, geography and difference in the post-socialist world: Or, do we still need post-socialism? *Antipode*, 40(2), 312–335.

Sýkora, L. (1993a). City in transition: The role of rent gaps in Prague's revitalization. *Tijdschrift voor Economische en Sociale Geografie*, 84(4), 281–293.

Sýkora, L. (1993b). Gentrifikace: měníci se tvář vnitřních měst. In: Sýkora, L. (Ed.) *Teoretické přístupy a vybrané problémy v současni geografii*. Praha: Univerzita Karlova v Praze, 100–119.

Sýkora, L., Ouředníček, M. (2007). Sprawling post-communist metropolis: Commercial and residential suburbanisation in Prague and Brno, Czech Republic. In: Razin, E., Dijst, M., Vásquez, C. (Eds.) *Employment deconcentration in European metropolitan areas: Market forces versus planning regulations*. Dordrecht: Springer, 209–233.

Sýkora, L., Bouzarovski, S. (2012). Multiple transformations: Conceptualising the post-communist urban transition. *Urban Studies*, 49(1), 43–60.

Szabó, I. (2020). *Príbehy Prešporka v osemnástom storočí*. Bratislava: AlleGro Plus, 124 p.

Špačková, P., Ouředníček, M. (2012). Spinning the web: New social contacts of Prague's suburbanites. *Cities*, 29(5), 341–349.

Štefanovičová, T. (2011). Veľká Morava, vznik a vývoj. *Historická revue*, 22(8), 34–43.

Šveda, M. (2016). Život v Bratislavskom suburbia: prípadová štúdia mesta Stupava. *Sociológia*, 48(2), 139–171.

Vaishar, A. (2000). Globalizace a její vliv na rozvoj Brna. *Geografický časopis*, 52(1), 77–87.

Van Winden, W., Van den Berg, L., Pol, P. (2007). European cities in the knowledge economy: towards a typology. *Urban Studies*, 44(3), 525–549.

Węcławowicz, G. (1992). The socio-spatial structure of the socialist cities in East-Central Europe: the case of Poland, Czechoslovakia and Hungary. In: Lando, F. (Ed.) *Urban and rural geography*. Venice: Cafoscarina, 129–140.

Węcławowicz, G. (1997). The changing socio-spatial patterns in Polish cities. In: Kovács, Z., and Wiessner, R. (Eds.) *Münchener Geographische Hefte, 76, Prozesse und Perspektiven der Stadtentwicklung in Ostmitteleuropa*. Passau: L.I.S. Verlag, 75–82.

Wolaniuk, A. (1997). Spatial and functional changes in the city centre of Łódź. In: Liszewski, S., and Young, C. (Eds.) *A comparative study of Łódź and Manchester: Geographies of European cities in transition*. Łódź: University of Łódź, 137–158.

8 Brno in transition

From industrial legacy towards modern urban environment

Josef Kunc and Petr Tonev

Introduction

At the beginning of the 1990s the countries of Central Europe set off on their journey to political, social, and economic transformation. Concerning their economies, the goal was the same in all the countries: To change their central planning into market economy (Butschek, 1992; Holman, 2000; Nagy, 2001; Spěváček, 2002). The setting of the democratic conditions was difficult and it is still disputable whether some of the post-soviet authoritarian regimes could be called democratic (Žídek, 2017). Economic recession connected with the decline of industrial productivity in tens of percents was at the beginning of 1990s the inevitable consequence of the desire for rapid economic stabilisation and transformation (Enyedi, 1996; Hamilton et al., 2005; Garcia-Ayllon, 2018). Concerning the Czech Republic, or the earlier Czechoslovakia (until 1993) respectively, the decline of economy was due to unavoidable structural adaptations, e.g. the cancelling and reforming of unproductive agricultural and industrial units as well as excessive arms industries (Žídek, 2019). The collapse of the Council of Mutual Economic Assistance (CMEA), which had managed foreign trade, especially of exports to the Soviet Union and the German Democratic Republic, meant a great burden during the first years of transformation (Fischer and Gelb, 1991; Schmidt et al., 2015).

The Czech Republic, as well as nearby Slovakia, Hungary, and Poland, recovered quite quickly from the shock transformation therapy and reached very good results in both external and internal economic relations (levelling output balance, reorienteering the foreign trade to the 'West', the considerable reduction in unemployment, regulated inflation, and others). Basic structural changes in Central European economies saw the removal of a substantive part of manufacturing activities and increased employment in the service sector and its duration took less than one decade. The situation became more or less stable by the end of the 20th century (Holman, 2000, Steinführer, 2003; Hirt, 2013; Žídek, 2017). The dynamics of the changes was empowered by a very negative change of the reproduction behaviour, which was connected with the consequences of transformation as well as the inevitable start of the second demographic transition (Steinführer, Haase, 2007; Steinführer et al., 2010; Kabisch, Haase, 2011). These changes came to their climax at the end of 1990s as well.

Brno is the second-largest city in the Czech Republic, with almost 400,000 inhabitants. As such it ranks—and not only within the Central European

DOI: 10.4324/9781003039792-8

area—among European medium-sized cities, i.e. those up to 500,000 inhabitants (OECD, 2013). It ranks among the 100 biggest cities of the European Union. The Brno metropolitan area is the third-largest in the country (after Prague and Ostrava metropolitan areas). In its region Brno is really dominant and monocentric due to its size, the second-largest settlement not having even 15,000 inhabitants. It is possible to consider the other towns and villages Brno satellites.

The historical, economic, transport and administrative ties to Vienna were broken by the emergence of Czechoslovakia in 1918. It became one of the successor states of the Austro-Hungarian monarchy. In the interwar period, Czechoslovakia, or rather the Czech part, was considered a very developed industrial country, sometimes ranked among the 10 most industrialised countries in the world (Žídek, 2017). This included the territory of Bohemia, Moravia, Silesia (or at least the southeastern part of Silesia), Slovakia, and, until 1939, also Subcarpathian Russia. From 1948, Czechoslovakia became a state with a socialist establishment for 40 years and, together with other countries, also part of the sphere of influence of the Soviet Union (the so-called Eastern Bloc). Czechoslovakia ceased to exist on 31 December 1992, when it was divided into the Czech Republic and the Slovak Republic on 1 January 1993.

For more than a hundred years Brno's main connection has already been to Prague and, since the separation, in a slightly limited way, to the capital of Slovakia, Bratislava. Former significant North–South orientation (Vienna—Brno—Ostrava/Wroclaw) has been weakened in favour of the West–East orientation (Berlin—Prague—Brno—Bratislava—Budapest/Vienna), which is strengthened by important national and international transport corridors (Figure 8.1).

The aim of the chapter is to present the transformation of the economic base of the second-largest city in the Czech Republic, Brno, which has taken place in the past 30 years. Emphasis is placed on explaining the origin and development of industrial traditions and history, which are indelibly inscribed in the functional and spatial structure of the city and continue to affect it. Nevertheless, it is possible to point out, especially in the last two decades, the economic, political and spatial changes and the emerging polarity between the dynamic sectors of services, high-tech industry and innovation, and the declining traditional industries and the gradual transformation of Brno into a modern urban environment.

Methodically, the text is based on professional domestic and foreign literature reflecting, in particular, the economic transformation and changes in the industrial orientation of post-socialist cities. Emphasis is also placed on local studies focused on the industrial history and presence of Brno. The basic methodological approaches are analysis, comparison, abstraction, deduction, interpretation and synthesis, which are supported by descriptive statistics, illustrative maps created in ArcGIS and own author's photo documentation.

Industrial tradition in Brno and the first years of the transition period

Industrial production in the Czech Republic has a long and diverse history. Iron was produced in furnaces from the beginning of the 17th century, but the real development of manufacture production dates back to the turn of the 18th and

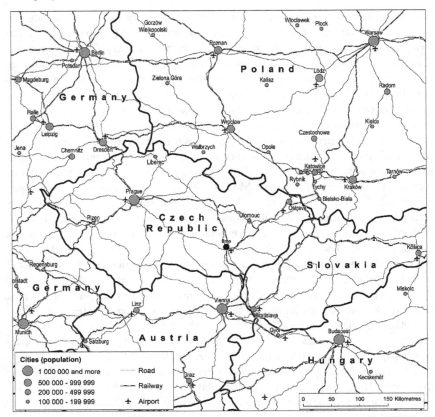

Figure 8.1 Map of Brno in the settlement and transport system of Central Europe.
Source: Authors' elaboration.

19th centuries. The textile industry was the one that developed most rapidly. Simultaneously, first sugar beet refineries emerged and the production of porcelain was started. Industrial and technical revolution made its way forward hand in hand with the invention of the steam engine which became the basis for the most important sector of the Czech industry—engineering. Concerning industry, the Czech Lands were the most developed part of the Habsburg monarchy. Different types of production naturally concentrated in underhill regions with sufficient supplies of fuel and water as well as in towns where the possibility of making use of the newly forming working-class workforce was rather high (Mareš, 1983; Toušek, Mulíček, 2003; Kunc, 2006). One of the most important industrial centres of the whole monarchy was Brno.

The history of industry in Brno goes back to the middle of the 18th century and is connected with the first textile manufactories. Towards the end of the 18th century, Brno was regarded the biggest textile centre in Central Europe (it was given the title of the "Austrian (Moravian) Manchester"; just like Łódź was called the "Polish Manchester") and focused mostly on wool production. In 1814, one of the first engineering factories in the Czech Lands was built in Šlapanice u Brna. Some 25 years later, the factory moved to Brno and this event started Brno engineering. The position of industry in the town is documented by the survey of 1869 (first census in Austria-Hungary) when almost one-third of its 75,000 inhabitants were employed in industry. At the turn of the 19th and 20th centuries, 38,000 people worked in Brno industrial factories, one-third of which were textile manufacture workers, whereas engineering employed only 6,600 people. The number of employees in engineering only reached the same level of the number of workers employed in the textile industry in 1918 (Mareš, 1983; Kunc, 1999; Kuča, 2000).

Engineering production became the dominant for of industrial production in Brno during the 20th century. In 1930, 22,000 people worked in this branch (15,500 in the textile industry) and the number of engineering workers also grew significantly after 1945. After 1948, Brno industrial factories became a part of the centrally controlled economy system of socialist Czechoslovakia. In those times the percentage of industrial workers compared to the overall population of Brno was well above the national average (Mareš, 1983). The aim of socialist economists and planners was to concentrate and rationalise the investments (Enyedi, 1996). For these reasons the industrial production in Brno was concentrated into big manufacturing premises with four biggest engineering companies employing two-thirds of the workers in the mid-1960s, and by the end of the 1980s still more than one-third of all the employees in the city (Kunc, 1999). By the end of the 1980s, almost one-quarter of the 100,000 industrial workers in Brno commuted to work from the towns and villages of the wider functional region of Brno (Mulíček et al., 2016). After Prague and Ostrava, Brno was the third-largest Czechoslovakian industrial centre.

According to Mareš (1983), Brno's share of the industrial employment in the Czech Socialist Republic of that time was 4.5%, and its manufacturing share was 3.5%. Concerning the industry branches, engineering and metallurgic industries were dominant (74% workers, 65% production) with textile industry still having its significance. The value of annual industrial production exceeded 17 billion Czech crowns. More than one-quarter of industrial production was exported mostly to the Soviet Union and other states of the CMEA, although many high-quality industrial products found its way to both developing and developed countries in all the continents of the world. Table 8.1 shows the number of industrial employees in Brno as of 31 December 1989. It is obvious that the position of engineering was still outstanding. The decline of the number of textile companies' employees had been a long process; on the other hand, electrical engineering was gaining significance in Brno after 1945 (especially high-voltage current electronics). The number of employees in Brno industries at the end of 1989 was floating just below the level of 100,000 personnel (Federal Statistical Office, 1990; Marek et al., 1997). About one-third of all the industrial workforce in Brno were employed by four engineering companies: Zetor, Zbrojovka, Královopolská strojírna, and První brněnská strojírna.

During the 1990–1997 period there came substantial changes in the industry branch structure employment in Brno. The number of employees in the insignificant primary sector (agriculture and forestry) decreased approximately by one-third (from 7,000 to 4,000), in the secondary sector (industry and construction) it

Table 8.1 Employees in industry branches of national economy as of 1989 in Brno and the Czech Republic

Industry branch	Number of employees			Index of specialisation*
	Brno		Czech Republic	
	Total	%	%	
Fuel	941	0.4	3.9	0.10
Energetics	3,335	1.3	1.1	1.22
Metallurgic	158	0.1	2.8	0.02
Chemical and rubber-making	2,911	1.2	2.3	0.51
Engineering	53,464	21.4	11.5	1.86
Electrical	9,928	4.0	2.7	1.48
Glass and construction materials	1,607	0.6	2.4	0.26
Wood and furniture	1,483	0.6	1.5	0.41
Metalworking	2,199	0.9	2.2	0.41
Paper-making	2,430	1.0	0.6	1.77
Textile, clothing, leather	9,360	3.8	5.4	0.70
Printing	1,180	0.5	0.3	1.45
Food-processing	4,967	2.0	1.4	1.40
Other industrial production	3,457	1.4	1.1	1.24
Industry total	**97,420**	**39.1**	**40.4**	**0.97**
Economy total	249,453	100.0	100.0	–

* Ratio of Brno % and Czech Republic %.
Source: Federal Statistical Office (1990). Labour force balance in CSFR (31.12.1989); Our own calculation.

decreased by approximately 30% (from 123,000 to 86,000) and vice versa the number of employees in the tertiary sector (retail and services) increased during this period by nearly 30% (from 120,000 to 170,000). The decrease of employment in the secondary sector was exclusively caused by its industrial part (civil engineering kept its share) where, from the original 97,000 workplaces in 1989, only 60,000 workplaces remained in 1997 (62% of the original number) (Kunc, 1999; Toušek, 2003).

The streamlining of industrial companies took place in two rounds: first part of redundant workers moved to non-manufacturing sectors, and the second part of them entirely departed from the labour market (the emergence of unemployment; retirement). By comparing the share of workers in Brno and in the Czech Republic we can find out that up to 1997 the share of the industry in the Czech Republic decreased by 8.5 percentage points (from 40.4 to 31.9%) whilst in Brno it declined by nearly 16 percentage points (from 39.1 to 23.2%). Only five biggest industrial companies employed more than a thousand employees by the end of 1997 and the total number of their personnel was only 10,500. By the end of 2000, the non-manufacturing sector employed already more than 70% of the Brno workforce. Fundamental structural changes and branch shifts had been finished (see Table 8.2) (Kunc, 1999; Toušek, 2003; Toušek and Mulíček, 2003).

Post-socialist de-industrialisation was fully on its way in the 1990s, which is clear from Table 8.3. Whilst industrial companies clearly dominated in 1989, the ranking of the biggest employers turned in favour of the tertiary sector over the course of just one decade. From available data we know that in 1989 there were 23 industrial companies in Brno which employed more than 1,000 employees while in 2000 only five companies had more than 1,000 employees (Toušek and Vašková, 2003). The same changes took place in the types of industrial branches which ranked the most important companies.

During the first transformation years Brno industry got into the scope of foreign capital. The biggest foreign company which entered Brno industry was Swiss—Swedish group Asea Brown Boveri (ABB) operating mainly in robotics,

Table 8.2 Development of employment in Brno according to sectors 1965–2000

National economy sector	Share of the total employment (%)			
	1965	1989	1996	2000
Primary	0.9	2.1	1.6	0.6
Secondary	60.6	49.8	33.2	28.9
Tertiary	38.5	48.1	65.2	70.5

Source: Hájek (1973); Mulíček (2010); Our own processing.

Table 8.3 Most important employers in Brno in 1989 and 2000

1989			2000		
Name	Number of employees	Branch	Name	Number of employees	Branch
Agrozet Zetor	10,261	engineering	The University Hospital Brno	4,524	healthcare
Zbrojovka	9,415	engineering	Czech Railways	3,329	transportation
Královopolská strojírna	6,557	engineering	Masaryk University	3,074	education
První brněnská strojírna	6,355	engineering	Brno Public Transport Authority	3,045	transportation
Chemont	2,608	engineering and chemical	Brno City Municipality	2,824	municipal service

Source: Toušek and Vašková (2003).

power, heavy electrical equipment, and automation technology areas. By the end of 1992 this company signed a joint venture contract with the engineering company První brněnská strojírna, a. s. and gradually gained a decisive share in it. Concerning the bigger companies, we can mention the Nová Mosilana, a.s. company, an important producer of 100% woollen textiles, who in 1994 made a joint venture company with the biggest textile group in the branch, the Italian Gaetano Marrzollo e Figli owning 90% of the shares. The last big company with a decisive share of foreign capital was the producer of healthcare equipment BMT Brněnská medicínská technika, a.s., that emerged through the privatisation of the Chirana Brno state company and the purchase of 70% of its shares by the German MMM Mnichov firm (Kunc, 1999; Toušek, 2003).

Changes in the city economy and policy

According to many authors, "industrialism" is a way in which industrial towns function and are organised. This way of functioning is deep-rooted and has different levels of intensity; it often prevails in the post-industrial phase of their development. The industrial production in the cities not only played the major role in the economic production system, but also served as the system for social reproduction and establishing the cultural forms. The way in which an industrial city was socially organised was significantly connected with the distribution of the shared sources, often in close connection with the industrial production (Byrne, 2002; Hamilton et al., 2005; Kunc et al., 2014b). Industrial society, with the working culture at its centre, clashed during the transformation period with the post-industrial city, which put on the stage new privileged groups of citizens and new preferred topics, as well as new cultural forms (Cudny, 2014).

If we take into consideration the city as a whole, with its distinctive and traditionally formed history and collectively formed knowledge, we can mark the city identity crisis as one of the outcomes of the changes in Brno's industrial structure. This identity crisis is closely connected with the "death" of the term "collectivist city" (Short, 2000). A collectivist city is characterised by a clearly defined role in a centrally planned national economy, with only few tools for its autonomous development, with a great deal of public services and investments, a high level of public estates sharing, and collective consumption of the space. Sýkora (2001) adds to the above-mentioned, that the concept of industrial production as the engine for the city economy seemed as the thesis with strong inertia even during the transformation period and it had a deep influence on public views and the activities of the city political representatives.

Toušek and Mulíček (2003) declare that there exist several important facts needed for the correct understanding of the city development during the 1990s transformation period, which followed the socialist period. In a similar way, Sýkora (1999) points out that the post-socialist cities of the 1990s were not quickly and fully transformed into capitalist cities. Their development after the fall of communism showed a number of specific features which in those days could not be generalised into the model of a transitional city.

The city of Brno never belonged into the category of single resource industrial cities—its industrial base has always been rather diverse, although the four biggest industrial companies (see above) employed roughly one-third of the industrial workforce. That is the reason why the development trajectory of the city is rather different from the trajectories of those cities with only one dominant sector (e.g. Ostrava and coal mining). The city development policy was in a directive manner and on the long-term basis influenced by the needs of the industry. According to Illner (1992), the influence of "industrial paternalism", the connection between the decision-making and manufacturing spheres (Pavlínek and Smith, 1998), was not as significant in Brno as it was for other Czech big industrial cities (e.g. Ostrava, Plzeň, Liberec). The industry of the socialist era was also supported by a relatively high-quality research and development base whose scale declined dramatically in the first years of the transformation (Steinführer, 2003; Mulíček and Toušek, 2004).

As early as 1991 the tertiary sector became dominant in Brno with more than a 50% share of the overall employment. The tertiary sector, including the dramatically developing retail trade and gradually formed new scientific-research and innovation bases linked to the high-tech production in the modern industrial and science-technology parks and zones, became the basis of the post-industrial transformation. Changes in the labour market logically became the most visible feature of the post-socialist restructuring of the city economy in the 1990s. (Dawson and Burns, 1998; Nagy, 2001; Szczyrba, 2005; Kunc et al., 2013; Križan et al., 2019).

The transition from the industrial to the post-industrial phase, especially in the case of traditional industrial cities, is connected not only with the shifts in the relative importance of individual sectors of the local economy, but also mainly with the deep institutional and cultural effects of this process. Byrne (2002) states that in the context of an industrial city it is not possible to limit the use of the term "industrial" solely to the sphere of manufacturing. In addition to this fact we must take into account other processes caused by industrial changes. As Musil (1993) points out, post-socialist societies underwent a dual transformation, both political and technological. Industrial restructuring is, and not only in the case of Brno, a good example of the dual character of the social transformation. The decline of the importance of industry for city life, the process which took several decades in the cities of Western Europe since the 1960s (Pacione, 2001; Le Gales, 2002), appeared later (since the early 1990s) and has been undergone much faster (about 20 years) in post-socialist cities.

Hand in hand with the decline of the traditional industrial branches, with the collapse of the centrally managed industry, and with the elimination of the biggest industrial companies, Brno city started to fulfil their new role of a manager, who is fully responsible for the economic development (Mulíček and Olšová, 2002). Moreover, since 1990s both the economically developed and the post-socialist city were viewed as an entrepreneurial concept which can be successful on the condition when comparative advantages and inner reserves are mobilised. The entrepreneurial concept of a city highly influenced the attempts to define the new role of a city in the changing economic environment and emphasise the priorities of economic policies (Amin, 2000; Sýkora and Bouzarovski, 2012).

Transitional development in the 1990s also brought important changes in the forms of city corporatism. The status of traditional industrial branches representatives, operating in groups of lobbyists influencing decision-making processes in cities, gradually weakened in favour of new players. The partners of the local municipal council, such as developers or groups of lobbyists, became more visible players of the economic transformation within the public and private sectors partnership processes (Bertaud, 2006; Brade et al., 2009). Nevertheless, the influence of the structures connected with the industry was still strong during the transformation period, or more precisely, it was stronger than it should have been compared to the declining importance of the traditional industrial branches. The historically viewed concept of the industrial production as the driving engine of the city economy during the transition period was still regarded as the element with high inertia and an ongoing deep influence on public attitude and the activities of the city of Brno political representatives. (Mulíček and Toušek, 2004).

The restructuring of the city economy during the transition period was most visible due to labour market changes (Kunc, 1999; Mickiewicz, 2005). The growing (long-term) unemployment strongly influenced the local authorities' attitude towards the issues of Brno industry transformation. The city economic strategy concerning the short-term view was focused on solving the workforce market issues, whereas the technological and structural change was the aim within the long-term horizon. In reality, the city functioned in a way where politically motivated measures which smoothened the effects of economic transformation often prevailed, and still prevail, over a strategic modernisation of the local economy (Toušek and Mulíček, 2003; Hirt et al., 2016).

The entrepreneurial city concept strongly influenced the attempts to define the city's new role in the changing economic environment as well as the attempts to emphasise the priorities of the economic policy. The limited city authority within the field of fiscal policy limits the city economic strategy to activities related to the construction and marketing of the newly emerging infrastructure. On the other hand, we have to realise that the concept of territorial tender and competition is rather controversial. According to many authors (Budd, 1998; Begg, 1999; Mickiewicz, 2005; Wiest, 2012; Garcia-Ayllon, 2018) there is no competition among the cities but among the individual companies located within the cities. The task of the city is thus to secure convenient conditions that would ensure a higher level of company competitiveness.

Spatial and social aspects of de-industrialisation

Industrial production was in the case of Brno, as well as in the case of other cities in Central Europe, the major decisive factor concerning the spatial organisation and the functional-spatial layout of the city. Despite the considerable manifestation of post-socialist de-industrialisation which culminated with its loss of economic dominance, the industrial heritage in Brno still leaves a highly significant trace which shapes the basic form of the city. The present spatial structure still resembles the urban patterns dating back to the times of the early and little coordinated phase of the industrialisation in the second half of the 19th century

(Toušek and Mulíček, 2003; Kunc et al., 2018). During the first phases of the industrialisation, the spatial needs of the industry were determined by the textile and engineering industries dependence on water, which was essential for the operation of the steam engines. That is the reason why the oldest industrial compounds are concentrated linearly in the structure of the present-day city, despite their out-of-control and additive development in the old days (Kunc, 2006). This structure had not significantly changed in the days of socialist industrialisation—the post-war industrial development continued mainly via further concentration of the factory buildings in their original locations (Kuča, 2000).

Changes in the spatial structure of industrial manufacturing are the result of many interactive processes. According to Pacione (2001), the post-industrial city development is generally characterised by a fragmentation of the traditional city form. The decentralisation of industry, the process taking place in Western Europe since the 1970s started in the Czech cities during the transition period of the 1990s (Turok and Mykhnenko, 2007; Kunc et al., 2014a). Despite the significant inertia of the industrial heritage, as mentioned above, the economic changes of the last two decades and their consequences will have great impact on the functional city structure and will influence its morphogenetic picture in the long-term horizon.

The decrease and rationalisation of manufacturing in industrial factories, the elimination of overemployment or ineffective use of outsourced services (catering, cleaning, security etc.) and the overall modernisation of the manufacturing process led to less intensive use of many traditional industrial operations. Structural changes connected with the flow of the workforce into the tertiary sector caused unprecedent dynamics in the service and business functions of the transitioning cities (Bertaud, 2006; Mulíček et al., 2016). An important positive effect on the city economy was also brought about by the liberalisation of the real estate market. The socialist real estate market was not functional and deeply deformed the use of different areas. Quite early, at the end of the 1990s, the real estate market liberalisation caused significant differentiation in the value of specific landplots with regard to their location within the city, their accessibility, and the image of the city borough (Sýkora, 2001; Garcia-Ayllon, 2018).

It can be stated, with a certain amount of simplification, that the polarity between the dynamic secondary sectors and the declining traditional branches made its mark onto the spatial structure of the city. Centrally located industrial compounds and premises fell behind while the developing suburban zones attracted new investments (Mulíček and Olšová, 2002). Today the location of new (often foreign) investments is mostly influenced by above-standard transport accessibility, quality of the workplace environment, structure of inhabitants with high levels of education, as well as the legal status of the particular landplot (form and kind of ownership, conditions of the rent, etc.). Gradually, the higher added value of new industrial companies and subjects operating services for companies as well as higher services, became positively evident in the post-socialist industrial environment; nevertheless, a large proportion of assembly factories remained in business (Birch et al., 2010; Kunc et al., 2018).

The above-mentioned facts resulted in the weakening of the industrial, and gradually the business, service, and institutional functions of the inner city in favour of the new economic structures localised on the outskirts of the city or within the city suburbs (Hutton, 2010; Taubenböck et al., 2013; Kunc et al., 2014b). The relocation of manufacturing activities to newly established industrial zones and technology parks which were connected to administrative and logistic centres, and the construction of vast retail premises (supermarkets, hypermarkets, DIY stores, and shopping centres) became both economic and spatial city-forming phenomena. With respect to the number of inhabitants, size of the area, and Brno's own morphostructure, we cannot speak about a typical change of a monocentric structure into a polycentric structure, as happened in the case of many post-socialist cities (Brade et al., 2009; Malý, 2016; Mulíček and Malý, 2018).

On the other hand, Brno avoided a very negative sociodemographic feature, depopulation and shrinking, which was so typical for the post-socialist development of many cities in the Central and Eastern Europe (Steinführer, 2003; Maas, 2009). A number of cities lost its exceptional economic and social status of a prosperous city located in the mining area, with dominant mining, and the iron and steel industry (e.g. Ostrava, Karviná, Katowice, Sosnowiec, Bytom) (Klusáček, 2005; Krzysztofik et al., 2012, 2013; Rumpel and Slach, 2014; Nekolová et al., 2016; Wolff et al., 2018), many other cities struggled with the growing trends of suburbanisation and the outflow of inhabitants beyond their administrative borders (Ott, 2001; Ouředníček, 2003; Schmidt et al., 2015).

The cities of the former German Democratic Republic (Leipzig, Dresden, Chemnitz, Magdeburg, Rostock and others) went through a specific development. Their inhabitants viewed the western part of the newly united Germany as a tempting immigration area (Kabisch, 2004; Wirth and Lintz, 2006; Steinführer and Haase, 2007). Turok and Mykhnenko (2007) proved in their study that whilst 78% of the western cities grew, 82% of the cities in Eastern Europe had been struggling with depopulation until 2005. In this respect, shrinking became an important feature of the transition in Central European countries.

The impacts of transition changes parallelly accompanied by significant demographic changes in the context of the second demographic transition did not leave Brno aside. The highest number of inhabitants from the beginnings of the 1990s has not been reached. Nevertheless, the relative population stability without extreme divergences provided the stimulating feature for Brno in the transition period. The development of the number of inhabitants is shown in Table 8.4.

Besides the modern industrial zones, logistic-administrative grounds and premises, and shopping campuses, there emerged new substantial spatial elements in the

Table 8.4 Development of Brno population during 1900–2021

Number of inhabitants (in thousands) in years											
1900	*1910*	*1921*	*1930*	*1950*	*1961*	*1970*	*1980*	*1991*	*2001*	*2011*	*2021*
177	217	238	284	299	324	344	371	388	376	386	382

Source: ČSÚ (2016). Historical lexicon of the Czech Republic municipalities 1869–2011; ČSÚ (2021) Number of inhabitants in the Czech Republic municipalities as of 1.1. 2021.

city, the brownfields (Kunc et al., 2014a; Nekolová et al., 2016). Due to all kinds of transition changes and a rather critical view of the municipal authority and the inhabitants, concerning the quality of life in the city area, brownfields became an ambivalent element of the development. On the one hand, they represent the potential and the possibility to reshape the abandoned areas, often located in the vicinity of the city centre, into a lively city organism; on the other hand, they represent the economic, social, environmental and visual stain on the face of the city, which is due to be dealt with by the public sector, ideally in cooperation with the private sector (Litt et al., 2002; De Sousa, 2006; Kunc et al., 2014b). Brownfields developer projects are thus a rather visible activity and in some cases the city became its own developer.

In Brno the historical development of industry and other manufacturing activities connected with the construction of the railway from the South-East and Vienna and its continuation via the city centre up North along the river Svitava created three important compounds of brownfield sites (see Figure 8.2). The first of them can be found in the South and South-East area, which borders the

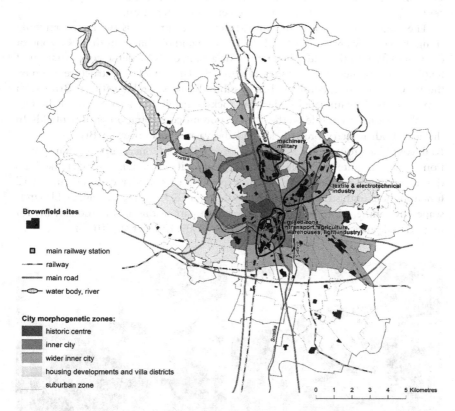

Figure 8.2 Spatial location of brownfield sites in Brno and the morphogenetic zones of the city.

Source: Mulíček (2010), authors' elaboration.

historical core. These are the compounds and premises of former textile factories and other manufacturing mills whose location in the vicinity of the city centre excludes the continuation of sophisticated manufacturing activities in the future. The usability of these locations is, on the one hand, limited to non-manufacturing activities, on the other hand this location is and will be highly prominent in the context of potential company headquarters, as well as office and administrative premises (Kunc and Tonev, 2008, 2009; Kunc et al., 2014b). An example of the revitalisation is an ongoing construction of a modern administrative complex, and in the future also a mixed housing and shopping complex called Nová Vlněna (New Vlněna, see also Figure 8.5) by a Dutch developer company CTP Invest in the place of the former textile company Vlněna. The plan is to create 85,000 square metres of office and business area where well-equipped and furnished classical offices as well as shared co-working premises will find their place. By 2030, the investment will rise to 160 million euros and up to 10,000 people should work here. Today, mostly IT firms and multinational companies such as Moravia IT, Avast, or Oracle have their residences within the complex which is still under construction. The basic benefit of this area is its location in the vicinity of the city centre, the main railway station, and the central coach station.

The second large area of brownfield sites concentration is the zone stretching along the river Svitava in the NE–SW direction and reaching out from the zone of the inner city into the wider inner city. The area is, due to a dense concentration of textile mills, engineering works, and chemistry plants in a linear structure given by the flow of the river, considered as possibly the most problematic in Brno with respect to the contaminated underlying rocks and soil. The area North of the historical core can be considered the third important concentration of brownfields. In this area both military brownfields and the vastest area covering Brno factory—Královopolská strojírna (37.5 hectares)—are located. Concerning the transportation, this area is quite accessible due to the greater city circle, it has sufficient facilities due to technical infrastructure and it is possible to consider extensive industrial functions here (industrial or logistics parks) (Figure 8.3). Taking the favourable landscape environment into account, the northern part of this area is also suitable for recreational purposes (Kunc and Tonev, 2008, 2009; Kunc et al., 2014b).

Figure 8.3 Urban renewal and revitalisation of the former engineering giant Zbrojovka Brno into a new city district.

Source: Josef Kunc archive.

In Figure 8.4 you can see the spatial distribution of other important types of developing zones—industrial zones, science-technology parks, business incubators, and shopping centres. Although the industrial zones Černovická terasa (Cernovice Terrace) and Areál Slatina (Slatina Site) are located within the zone of the wider inner city (zone 3), which stretches from the city centre down to the South-East, where both these zones lie, they are considerably distant from the historical core. By contrast, the Český technologický park (Czech Technology Park) is closer to Brno centre, but it is located on the border of zones 4 and 5. Two more zones are located behind the administrative border of Brno city (within the cadastral areas of Modřice and Šlapanice). Business incubators in Brno are concentrated (with just a few exceptions) within two locations—(i) the vicinity of Czech Technology Park (in the direct neighbourhood of Brno University of Technology); and (ii) the newly constructed university campus of Masaryk University. The relation of these "business subjects" to the academic sphere is more than obvious.

As mentioned above, and as Figure 8.5 proves, administrative centres naturally accumulated within the inner city zone, often in the immediate background of the

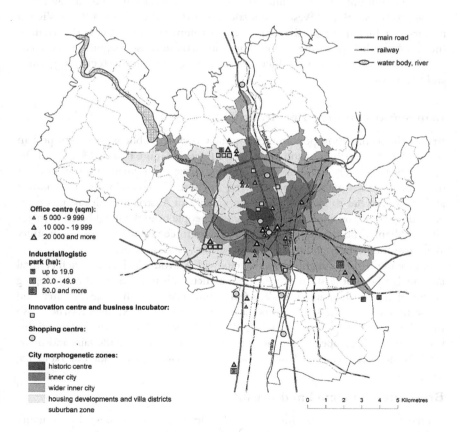

Figure 8.4 Spatial stratification of developing zones in Brno and the morphogenetic zones of the city.

Source: Mulíček (2010), authors' elaboration.

Figure 8.5 Shopping centre BrE Vaňkovka Gallery and a modern office complex Nová
    Vlněna in the immediate vicinity of the historic core.

Source: Josef Kunc archive.

historical core. Concerning the developing zones, the important location line features are commonly transport lines, in case of Brno specifically the railway tracks in the North–South and Western directions and the still unfinished Brno–Vienna motorway, which intersects the Prague–Brno–Olomouc motorway in the south of the city. As in the case of the business incubators, there is also a significant connection between certain types of office centres and existing science-technology parks and industrial zones.

Brno—modern urban environment

Brno is the second most populous city in the Czech Republic with a metropolitan area of 700,000 inhabitants and in the wider region of more than 1,000,000 inhabitants. Over the years, Brno has developed into one of the main innovation cities in Europe and has proven to be attractive for both Czech and foreign workforce and investors. Due to its location in the vicinity of Austria and Slovakia, Brno is particularly well positioned to attract talented people from the neighbouring major cities of Vienna and Bratislava. Brno and its workforce provide a great opportunity as an investor-friendly and multicultural business environment.

Currently, Brno is a highly attractive city with a favourable location, rich cultural life and, last but not least, an excellent infrastructure. It is a dynamically developing centre of high-tech industry, trade, science, information technology, research and innovation, with business incubators and scientific centres of excellence. Brno is also the exhibition centre of the Czech Republic. As the seat of one of the largest exhibition grounds in Europe, it hosts both local and international trade fairs and many other events every year. The basic characteristics of Brno are provided by Table 8.5.

Brno interconnected and liveable

Concerning its internal transportation, the city of Brno operates a high-quality and functional public transport system, which includes trams, trolleybuses, buses, and trains. This system enables people to get to work in an efficient and

Table 8.5 Brno—basic characteristics

380,000—the number of inhabitants

700,000—the number of inhabitants in Brno Metropolitan Area

150,000—people commuting to the city on a regular basis

250,000—total workforce

65,000—university students

140%—GDP per capita—EU 28 = 100%

1,320 €—the nominal monthly salary

Source: Brno City Profile (2019).

Table 8.6 Land transport linked to the surrounding cities

City	Distance (km)	Car (h)	Train (h)
Vienna	120	1:30	1:30
Bratislava	130	1:20	1:30
Prague	200	2:00	2:30
Katowice	260	2:30	4:15
Budapest	270	3:00	4:00
Munich	400	5:00	7:00
Leipzig	460	4:30	6:00
Berlin	550	6:00	7:30

Source: Brno City Profile (2019).

economical manner. Brno is well interconnected externally by main arterial motorways and railways that lead towards Prague and Germany in one direction and to Slovakia, Hungary and Austria in another direction (see Table 8.6). Brno is also connected internationally via its own Brno-Tuřany airport with a growing number of destinations and direct flights to London (2:10 h), Milano (1:25 h) and Berlin (1:10 h).

According to the newest Quality of Life Index—2020 Brno was ranked 86 worldwide for the quality of life and surpasses the capital Prague, which came in at 93. However, Brno came in 28th place in Europe and 4th in Eastern Europe, beaten only by Tallin, Ljubljana and Vilnius. The parameters of the survey include purchasing power, safety, health care, consumer prices, property-to-income ratio, traffic commute time, pollution and climate. Brno scored highly for safety, health care and climate. It was moderate for purchasing power, property-to-income ratio and pollution. It scored low in traffic commute times and very low in consumer prices. The quality of life in Brno is supported by the great range of cultural and leisure opportunities in and around the city (Numbeo, 2020).

Brno cultural, sporty and entertaining

Brno offers a wide range of sports, cultural and social activities and has numerous attractions and places of interest, including its stunning castle of Špilberk and many other historical monuments. One of the most famous functionalist residential villas in the world is located in Brno: Villa Tugendhat (inscribed on the list of UNESCO heritage sites since 2001). Another six UNESCO sites are situated within 100 km of Brno. Brno has also been known as a city with a long tradition of diverse cafés which have been an essential part of social and cultural life in Brno since the 18th century and in recent years also as a city of dozens of modern restaurants and trendy bars.

Concerning sports events, Brno has a long tradition of hosting motorbike and other motorsports racing events on the Masaryk circuit (established in 1930) including the famous MotoGP, which is one of the most prestigious and challenging events of the racing season.

Brno educated and entrepreneurial

Nowadays, Brno has about 65,000 students enrolled in 12 universities and 34 faculties (8,000 students are foreigners), many of which have long histories and are renowned for the high quality of their education. There are more than 16,000 new graduates each year, most of whom speak at least one European language. This figure denotes the highest density of graduates per capita in the Czech Republic.

The labour market in Brno is stable and currently offers around 15,000 job seekers. The total workforce numbers about 250,000 people. The availability of IT, economics and business students is one of the main factors attracting investors to the city. Business services are booming in Brno, with over 30 major international centres delivering a complex range of high-end IT, finance and business support functions for the largest global players. The sector employs more than 22,000 people in the city and represents one of the largest employers in the region, offering top-level careers for young professionals in the city.

The science and research sector in Brno is very important, with around 20,000 employees and more than 12,000 researchers. There are five centres of excellence in Brno, the most famous of which are: CEITEC (the Central European Institute of Technology), CzechGlobe (the Global Change Research Institute of the Czech Academy of Sciences), and FNUSA-ICRC (the International Clinical Research Centre).

625,000 m² is the current total modern office area in Brno. Approximately 80% of the area meets the requirements for A-class office space and more than 70% of the space is occupied by the companies from the IT sector. In terms of office rental, Brno is currently the most expensive among regional cities with more than 400,000 inhabitants in the Central European region, surpassing almost twice as large markets as Krakow and Wroclaw.

With almost 320,000 m² of retail space in the shopping centres, Brno has one of the highest shopping centre densities in the Czech Republic. National and international retail chains with a number of world-famous fashion brands are represented (Brno City Profile, 2019).

Conclusion

The history of Brno industry goes back to the middle of the 18th century and is connected with the first textile manufactories and later factories, which, at the beginning of the 20th century, were significantly surpassed by the engineering industry. For more than 200 years, the industry in Brno not only played a major role in the economic production system, but also served as the system for social reproduction and establishing the cultural forms. Industrial production was, in the case of Brno, the major decisive factor concerning the spatial organisation and the functional-spatial layout of the city. It is clear that thanks to the industrial heritage and the persistent spirit of industrial traditions, the city owes its prosperity and dynamic development in the 19th and 20th centuries.

It was not until the transitional developments in the 1990s that significant changes were brought about, and the position of representatives of traditional industries in interest groups influencing decision-making processes in cities gradually weakened in favour of new actors. Nevertheless, in the transition period, the influence of industry-related structures was still strong, respectively stronger than it should be in comparison with the declining importance of traditional industries.

In the last 20 years, the industrial, but also the commercial, service and institutional functions of the inner city have been gradually weakening at the expense of new economic structures located on the outskirts of the city. The transfer of production activities to newly formed industrial zones, technological and innovation parks connected to administrative and logistics centres, and the construction of large-scale retail entities have become an economic and spatial urban phenomenon.

Despite the considerable manifestation of post-socialist de-industrialisation which culminated with the loss of economic dominance, the industrial heritage in Brno still leaves a highly significant trace that shapes the basic form of the city. Despite this, Brno has managed to break free from the more or less rigid industrial legacy in recent decades and has become a city with a modern urban environment.

Acknowledgement

This chapter has been elaborated in the scope of the project MUNI/A/1248/2019 funded by the Masaryk University Brno.

References

Amin, A. (2000). The Economic Base of Contemporary Cities. In Bridge, G., Watson, S. (Eds.). *A Companion to the City*. Oxford: Blackwell, 115–129.

Begg, I. (1999). Cities and Competitiveness. *Urban Studies*, 3 (5/6), 795–809.

Bertaud, A. (2006). The Spatial Structures of Central and Eastern European cities. In Tsenkova, S., Nedovic-Budic, Z. (Eds.). *The Urban Mosaic of Post-Socialist Europe Space, Institutions and Policy*. New York: Verlag-Springer, 91–110.

Birch, K., MacKinnon, D., Cumbers, A. (2010). Old Industrial Regions in Europe: A Comparative Assessment of Economic Performance. *Regional Studies*, 44 (1), 35–53.

Brade, I., Gunter, H., Karin Wiest, K. (2009). Recent trends and future prospects of socio-spatial differentiation in urban regions of Central and Eastern Europe: A lull before the storm? *Cities*, 26, 233–244.

Brno City Profile (2019). *Brno City Profile. A Guide to the Business Services Sector.* [on-line]. Available at: https://brno2050.cz/wp-content/uploads/2019/11/Profil-m%C4%9Bsta-Brna.pdf

Budd, L. (1998). Territorial Competition and Globalisation: Scylla and Charybdis of European Cities. *Urban Studies*, 35 (4), 663–685.

Butschek, F. (1992). *External Shocks and Long-Term Patterns of Economic Growth in East-Central Europe.* Vienna: Mimeo.

Byrne, D. (2002). Industrial culture in a post-industrial world: The case of the North East of England. *City*, 6 (3), 279–289.

ČSÚ (2016). *Historický lexikon obcí České republiky 1869–2011.* [on-line]. Available at: https://www.czso.cz/documents/10180/20538302/13n106cd1.pdf/cf538eaa-7f70-49f6-8e76-dc88932650ef?version=1.0

ČSÚ (2021). *Počet obyvatel v obcích České republiky k 1. 1. 2021.* [on-line]. Available at: https://www.czso.cz/csu/czso/pocet-obyvatel-v-obcich-k-112021

Cudny, W. (2014). The influence of the "Komisarz Alex" TV series on the development of Łódź (Poland) in the eyes of city inhabitants. *Moravian Geographical Reports*, 22 (1), 33–43.

Hájek, Z. (1973). *Demografie Brna.* Praha: Academia.

Dawson, J., Burns, S. (1998). European Retailing: Dynamics, Restructuring and Development Issues. In Pinder, D. (Ed.) *The New Europe – Economy, Society and Environment.* Chichester: Wiley, 157–176.

De Sousa, C.A. (2006). Urban Brownfields Redevelopment in Canada: The Role of Local Government. *The Canadian Geographer*, 50 (3), 392–407.

Enyedi, G. (1996). Urbanization under socialism. In Andrusz, G., Harloe, M., Szelenyi, I. (Eds.). *Cities After Socialism: Urban and Regional Change and Conflict in Post-Socialist Societies.* Oxford: Blackwell, 100–118.

Federal Statistical Office (1990). *Labour force balance in CSFR* (31. 12. 1989). Praha: FSO.

Fischer, S., Gelb, A. (1991). The Process of Socialist Economic Transformation. *Journal of Economic Perspectives*, 5 (4), 91–105.

Garcia-Ayllon, S. (2018). Urban transformations as indicators of economic change in post-communist Eastern Europe: Territorial diagnosis through five case studies. *Habitat International*, 71, 29–37.

Hamilton, I.F.E., Dimitrovska Andrews, K., Pichler-Milanović, N. (2005). *Transformation of cities in central and eastern Europe. Towards globalization.* Tokyo-New York-Paris: United Nations University Press.

Hirt, S. (2013). Whatever Happened to the (Post)Socialist City? *Cities*, 32, 29–38.

Hirt, S., Ferenčuhová, S., Tuvikene, T. (2016). Conceptual forum: the "post-socialist" city. *Eurasian Geography and Economics*, 57 (4–5), 497–520.

Holman, R. (2000). *Transformace české ekonomiky: v komparaci s dalšími zeměmi střední Evropy.* Praha: CEP-Centrum pro ekonomiku a politiku.

Hutton, T.A. (2010). *The New Economy of the Inner City. Restructuring, regeneration and dislocation in the twenty-first-century metropolis.* London and New York: Routledge.

Illner, M. (1992). Municipalities and industrial paternalism in a "real socialist" society. In Dostal, P. et al. (Eds.). *Changing Territorial Administration in Czechoslovakia: international viewpoints.* Amsterdam: Instituut voor Sociale Geografie, 39–47.

Kabisch, N., Haase, D. (2011). Diversifying European Agglomerations: Evidence of Urban Population Trends for the 21st Century. *Population, Space and Place*, 17, 236–253.

Kabisch, S. (2004). Revitalisation Chances for Communities in Post-Mining Landscapes. *Peckiana*, 3, 87–99.

Klusáček, P. (2005). Downsizing of Bituminous Coal Mining and the Restructuring of Steel Works and Heavy Machine Engineering in the Ostrava Region. *Moravian Geographical Reports*, 13 (2), 3–12.

Križan, F., Bilková, K., Barlík, P., Kita, P., Šveda, M. (2019). Old and New Retail Environment in a Post-Communist City: Case Study from the Old Town in Bratislava, Slovakia. *Ekonomický časopis/Journal of Economics*, 67 (8), 879–898.

Krzysztofik, R., Kantor-Pietraga, I., Spórna, T.A. (2013). A Dynamic Approach to the Typology of Functional Derelict Areas (Sosnowiec, Poland). *Moravian Geographical Reports*, 20 (2), 20–35.

Krzysztofik, R., Runge, J., Kantor-Pietraga, I. (2012). Governance of Urban Shrinkage: A Tale of Two Polish Cities, Bytom and Sosnowiec. In Churski, P. (Ed.). *Contemporary Issues in Polish Geography*. Poznań: Bogucki Wydawnictwo Naukowe, 201–224.

Kuča, K. (2000). *Brno - vývoj města, předměstí a připojených vesnic*. Brno: Baset.

Kunc, J. (1999): Změny v průmyslu města Brna a jejich vliv na situaci na trhu práce. *Acta Facultatis Studiorum Humanitatis et Naturae Universitatis Prešoviensis. Prírodné vedy. Folia Geographica 3*, XXXII, 175–184.

Kunc, J. (2006). Historie a současnost průmyslové výroby na Moravě - regionální aspekt ekologického ohrožení krajiny. *Národohospodářský obzor*, 6 (3), 42–49.

Kunc, J. et al. (2013). *Časoprostorové modely nákupního chování české populace*. Brno: Masarykova univerzita.

Kunc, J., Martinát, S., Tonev, P., Frantál, B. (2014a). Destiny of urban brownfields: spatial patterns and perceived consequences of post-socialistic deindustrialization. *Transylvanian Review of Administrative Sciences*, 10 (41E), 109–128.

Kunc, J., Navrátil, J., Tonev, P., Frantál, B., Klusáček, P., Martinát, S., Havlíček, M., Černík, J. (2014b). Perception of urban renewal: reflexions and coherences of socio-spatial patterns (Brno, Czech Republic). *Geographia Technica*, 9 (1), 66–77.

Kunc, J., Tonev, P. (2008) Funkční a prostorová diferenciace brownfields – příklad města Brna. *Regionální studia*, 2 (1), 30–37.

Kunc, J., Tonev, P. (2009). Rozvojový potenciál města Brna (rozvojové zóny a inovační podnikání). In *Geodny Liberec 2008, výroční mezinárodní konference ČGS*. Liberec: Technická univerzita v Liberci, 78–86.

Kunc, J., Tonev, P., Martinát, S., Frantál, B., Klusáček, P., Dvořák, Z., Chaloupková, M., Jaňurová, M., Krajíčková, A., Šilhan, Z. (2018). Industrial legacy towards brownfields: historical and current specifics, territorial differences (Czech Republic). *Geographia Cassoviensis*, XII, (1), 76–91.

Le Gales, P. (2002). *European Cities - Social Conflicts and Governance*. Oxford: Oxford University Press.

Litt, J.S., Tran, N.L. Burke, T.A. (2002). Examining Urban Brownfields through the Public Health "Macroscope". *Environmental Health Perspectives*, 110 (2), 183–193.

Maas, A. (2009). Shrinking Cities in the Czech Republic? The Case Study of Brno. *Moravian Geographical Reports*, 17 (3), 27–40.

Malý, J. (2016). Impact of polycentric urban systems on intra-regional disparities: A micro-regional approach. *European Planning Studies*, 24, 116–138.

Marek, D., Toušek, V., Vančura, M. (1997). Transfonnace průmyslové výroby. In Kolejka, J. et al. (Ed.). *Životní prostředí - Brno 1996*. Brno: Magistrát města Brna, 31–35.

Mareš, J. (1983). Charakteristika průmyslu v území aglomerace. In Bína, J., Folk, Č. et al. (Eds.). *Geoekologie brněnské aglomerace*. Studia Geographica 83, 199–204.

Mickiewicz, T. (2005). *Economic Transition in Central Europe and the Commonwealth of Independent States*. New York: Palgrave Macmillan.

Mulíček, O. (2010). Prostorově diferencovaná dynamika maloobchodu v kontextu transformace města. In *Geografie pro život ve 21. století: Sborník příspěvků z XXII. sjezdu České geografické společnosti*. Ostrava: Ostravská univerzita v Ostravě, 513–516.

Mulíček, O., Malý, J. (2018). Moving towards more cohesive and polycentric spatial patterns? Evidence from the Czech Republic. *Papers in Regional Science*, 98 (2), 1177–1194.

Mulíček, O., Olšová, I. (2002). Město Brno a důsledky různých forem urbanizace. *Urbanismus a územní rozvoj*, 5 (6), 17–21.

Mulíček, O., Osman, R., Seidenglanz, D. (2016). Time-space Rhythms of the City – The Industrial and Post-industrial Brno. *Environment and Planning A*, 48 (1), 115–131.

Mulíček, O., Toušek, V. (2004). Changes of Brno Industry and their Urban Consequences. *Bulletin of Geography (socio-economic series)*, 3, 61–70.

Musil, J. (1993) Changing urban systems in post-Communist societies in central Europe: Analysis and prediction. *Urban Studies*, 30 (6), 899–905.

Nagy, E. (2001). Winners and Losers in the Transformation of City Centre Retailing in East Central Europe. *European Urban and Regional Studies*, 8 (4), 340–348.

Nekolová, J., Hájek, O., Novosák, J. (2016). The Location of Economic Activities in A Post-Communist City: Twenty Years of Land Use Change in the Ostrava Metropolitan Area (Czech Republic). *Transformations in Business & Economics*, 15 (3C), 335–351.

Numbeo (2020). Quality of Life Index by City 2020. [on-line]. Available at: https://www.numbeo.com/quality-of-life/rankings.jsp

OECD (2013). *Regional Statistics and Indicators: Definition of Functional Urban Areas for the OECD Metropolitan Database* [Online]. Available at: http://www.oecd.org/gov/regional-policy/Definition-of-Functional-Urban-Areas-for-theOECD-metropolitan-database.pdf.

Ott, T. (2001). From Concentration to De-concentration - Migration Patterns in the Post-socialist City. *Cities*, 18 (6), 403–412.

Ouředníček, M. (2003). Suburbanizace Prahy (The suburbanisation of Prague). *Sociologický časopis/Czech Sociological Review*, 39 (2), 235–253.

Pacione, M. (2001). *Urban Geography: A Global Perspective*. London: Routledge.

Pavlínek, P., Smith, A. (1998). Internationalization and Embeddedness in East-Central European Transition: The Contrasting Geographies of Inward Investment in the Czech and Slovak Republics. *Regional Studies*, 32 (7), 619–638.

Rumpel, P., Slach, O. (2014). Shrinking cities in central Europe. In Koutský, J., Raška, P., Dostál, P., Herrschel, T. (Eds.). *Transitions in Regional Science – Regions in Transition: Regional research in Central Europe*. Praha: Wolters Kluwer.

Schmidt, S., Fina, S., Siedentop, S. (2015). Post-socialist Sprawl: A Cross-Country Comparison. *European Planning Studies*, 23 (7), 1357–1380.

Short, J.R. (2000). Three Urban Discourses. In Bridge, G., Watson, S. (eds.) *A Companion to the City*. Oxford: Blackwell, 18–25.

Spěváček, V. (2002). *Transformace české ekonomiky*. Praha: Linde.

Steinführer, A. (2003). Socio-spatial Structures Between Persistency and Change. Historical and Contemporary Perspectives on Brno. *Sociologický časopis/Czech Sociological Review*, 39 (2), 169–192.

Steinführer, A., Bierzyński, A., Großmann, K., Haase, A., Kabisch, S., Klusáček, P. (2010). Population Decline in Polish and Czech Cities during Post-socialism? Looking Behind the Official Statistics. *Urban Studies*, 47 (11), 2325–2346.

Steinführer, A., Haase, A. (2007). Demographic change as a future challenge for cities in east central Europe. *Geografiska Annaler: Series B, Human Geography*, 89B (2), 183–195.

Sýkora, L. (1999) Changes in the Internal Spatial Structure of Post-communist Prague. *GeoJournal*, 49 (1), 79–89.

Sýkora, L. (2001): Proměny prostorové struktury Prahy v kontextu postkomunistické transformace. In Hampl, M. et al. (Eds.). *Regionální vývoj: specifika české transformace, evropská integrace a obecná teorie*. Praha: Univerzita Karlova v Praze, 127–166.

Sýkora, L. and Bouzarovski, S. (2012). Multiple Transformations: Conceptualising the Post-communist Urban Transition. *Urban Studies*, 49 (1), 43–60.

Szczyrba, Z. (2005). *Maloobchod v ČR po roce 1989: vývoj a trendy se zaměřením na geografickou organizaci*. Olomouc: Univerzita Palackého v Olomouci.

Taubenböck, H., Klotz, M., Wurm, M., Schmieder, J., Wagner, B., Wooster, M., Esch, T., Dech, S. (2013). Delineation of Central Business Districts in mega city regions using remotely sensed data. *Remote Sensing of Environment*, 136, 386–401.

Toušek, V. (2003). *Geografické aspekty transformace českého průmyslu po roce 1989*. Bratislava: Univerzita Komenského v Bratislave.

Toušek, V., Mulíček, O. (2003). Brno - important industrial centre? *Acta Universitatis Carolinae Geographica*, 1, 437–444.

Toušek, V., Vašková, L. (2003). Trh práce ve městě Brně - změny po roce 1989. In *Sborník referátů z VI. mezinárodního kolokvia o regionálních vědách*. Brno: Masarykova univerzita, 229–240.

Turok, I., Mykhnenko, V. (2007). The trajectories of European cities, 1960–2005. *Cities*, 24 (3), 165–182.

Wolff, M., Haase, D., Haase, A. (2018). Compact or spread? A quantitative spatial model of urban areas in Europe since 1990. *PLoS ONE*, 13 (2). [on-line]. Available at: https://doi.org/10.1371/journal.pone.0192326

Wiest, K. (2012). Comparative Debates in Post-Socialist Urban Studies. *Urban Geography*, 33 (6), 829–849.

Wirth, P. and Lintz, G. (2006). Rehabilitation and Development of Mining Regions in Eastern Germany – Strategies and Outcomes. *Moravian Geographical Reports*, 14 (2), 69–82.

Žídek, L. (2017). *From Central Planning to the Market: Transformation of the Czech Economy 1989–2004*. Budapest: Central European University Press.

Žídek, L. (2019). *Centrally Planned Economies: Theory and Practice in Socialist Czechoslovakia*. London, New York: Routledge.

9 Challenges and problems of re-growth

The case of Leipzig (Eastern Germany)

Dieter Rink, Marco Bontje, Annegret Haase,
Sigrun Kabisch and Manuel Wolff

Introduction

As an East German city, Leipzig underwent a specific transformation after the end of the Cold War in 1989/90. This transformation was unlike the changes that took place in other post-socialist cities in Poland, the Czech Republic and Hungary. East Germany joined a "ready-made state", whereby West Germany became the blueprint for the targets of post-socialist transformation. The transformation was designed as "catch-up modernisation". For the former German Democratic Republic (GDR), the unification of the two German states meant immediate accession to the European Union, without the usual transitional rules and adjustment processes. Re-integration into the capitalist world market had already been formally completed with the economic and monetary union by July 1990. This led to a shock transformation after 1990; within a short period of a few years, profound economic, political, social, and cultural changes took place. The transformation was associated with extensive and widespread de-industrialisation, high and persistent unemployment, and a massive decline in population. However, unlike other East German cities, Leipzig returned to a path of gradual re-growth after 2000, not least thanks to major public investments and subsidies. From 2010 onwards, Leipzig has experienced dynamic re-growth and has been described as a "comeback city" (Bernt 2009) or "Phoenix city" that has "risen from its industrial ashes" (Power et al. 2010). Leipzig's shrinkage experience has already been studied in detail (see Bontje 2004; Rink et al. 2012); however, the re-growth process has so far received much less analytical attention (Haase et al. 2020, Power et al. 2010; Rink et al. 2012). We define re-growth as a particular urban development process affecting cities after a shrinking experience. It is a process towards economic recovery or revival accompanied by stabilisation and an increase in population. Leipzig's re-growth was welcomed by many stakeholders, including urban policy-makers, who looked for and supported projects and strategies for successful re-growth. However, re-growth has also led to serious challenges and problems in a number of policy fields. The aim of this chapter is to analyze the specific problems caused by strong re-growth after a phase of (severe) shrinkage.

This chapter focuses on the following questions: What specific challenges and problems arise from this constellation of re-growth after a period of intense shrinkage? How are they reflected and dealt with in relation to different policy fields? To

DOI: 10.4324/9781003039792-9

answer these questions, three policy fields have been selected: housing provision, schools as social infrastructure, and public transport as technical infrastructure. The selection was based on the following criteria: 1. The policy fields are affected by re-growth, but to a distinct degree and on different scales, 2. they face political constraints and opportunities for action, and 3. they are currently a focus of municipal policy. These three policy fields are currently at the top of the public agenda in Leipzig. Major problems have emerged in all three fields with respect to the re-growth context since the 2010s, although the nature of the problems vary. In the following, the case of Leipzig will be used to illustrate the problems caused by re-growth after a transformation crisis and shrinkage, and how these new challenges can be met by urban actors.

Background, materials and methods

Urbanists or urban planners in many countries have been discussing the post-decline revival of inner or core cities since the 1960s and 1970s. In this debate re-growth not only refers to an increase in population, but also a qualitative change including new urban forms, mixed-use areas, and the idea of urban conservation. Different terminologies have been used for this phenomenon, such as re-growth, reurbanisation, resurgence, and revival and others. Apart from the urban life cycle models, another strand of discussion that has some anchor points in cyclical models deals with the reconcentration of populations in large cities, set against an overall regional context of shrinkage: Core cities are referred to as "islands of growth or stabilization" (Herfert 2007) or as winners in a continuous context of decline (Couch et al. 2009). According to these studies, large cities either remain the only places that do not decline, or recover first, or become destinations for inward migration because of their amenities and infrastructure (Rink et al. 2012).

Some studies deal explicitly with the factors that drive re-growth and this strand of the debate is especially relevant for our chapter's rationale. Based on a comparative study of various "Phoenix cities", Power et al. (2010) and Power and Katz (2016) point out the following key factors: land reclamation and environmental upgrading; sprawl containment; transport infrastructure improvements; physical redesign and restoration; neighbourhood renewal; job creation; the development of new skills in the population; civic leadership and increased participation; social inclusion; and new publicly sponsored agencies that help to deliver change. In another comparative work on the "remaking of postindustrial cities", Carter (2016) suggests that the most important factors are the consideration of the metropolitan (not the urban) scale; the need for a long-term vision; the development of a sustainable planning strategy; the need for alliances and partnerships, strong leadership and citizen engagement; diversification of the economy; a strengthening of the central city; investment in education, culture, quality of life, heritage and urban design; and a readiness to take risks. Rink et al. (2012) highlight the existing ambivalences of re-growth and discuss related risks for sustained new growth after shrinkage. Particularly, this study mentions the economic fragility of re-growing cities and their ongoing dependency on both external decisions (e.g. by large-scale investors and the political choices of national or regional governments) and external factors such as national or regional economic circumstances. The authors also underline the ambivalence of those success factors that were identified at the time of research: What might support re-growth today, may tomorrow lead to new problems and hinder re-growth. Aside from this debate about the factors that drive re-growth, it should briefly be mentioned that other scholarly work looks at the relationship between the debate on "resurgent cities" and changes affecting demographics, households or housing. Not least, Siedentop (2018) provides a summary and assessment of the hitherto mentioned reurbanisation (re-growth) debate.

This contribution is based on long-term research on Leipzig and the Leipzig-Halle region that was conducted in the context of various projects dealing with segregation, reurbanisation, urban sprawl, shrinkage, and re-growth. In the 1990s we produced the *Social Atlas of Leipzig*, which analyzed and examined the socio-spatial differentiation of the population at the district level (Kabisch et al. 1997). The increasing social inequality among the population was investigated in the *Life Situation Report* (City of Leipzig 1999), and a further project examined urban poverty in Leipzig (Richter 2000). In the 2000s, the European Union (EU) project URBS PANDENS analyzed suburbanisation in the Leipzig-Halle region, in particular the connection between shrinkage and urban sprawl (Nuissl and Rink 2005). In the same decade, reurbanisation was addressed from a comparative perspective by the EU project 'Reurban Mobil in Leipzig' (Haase et al., 2012). The EU project 'Shrink Smart' in the late 2000s/early 2010s then focused on the trajectory, consequences, and political and planning response to shrinkage in the two cities of Leipzig and Halle (Rink et al., 2012, 2014). To better understand the population growth of the 2010s, a survey was conducted among newcomers and immigrants in Leipzig (Welz et al. 2014, 2017). Studies from the 2010s on Leipzig's housing market, gentrification and housing policy provide specific insights regarding the issue of housing supply (Rink 2015, 2020; Haase and Rink 2015). For the topic of schools, Anne Walde's PhD thesis about how to deal with declining school student numbers in Leipzig and Timisoara (Romania) was very useful (Walde 2019).

Our methodological approach includes both quantitative and qualitative methods of primary data collection. Several quantitative household surveys addressing several groups of inhabitants were conducted in a number of Leipzig districts. The migration survey was a large postal survey. Our governance analyses are based on expert interviews and the analysis of policy and planning documents. Furthermore, since the beginning of the 1990s, secondary data on demographic, social, economic, and political developments have been systematically analyzed for different levels, including the city of Leipzig, the Leipzig-Halle region and the (post-)industrial region south of Leipzig. Furthermore, some policy areas have been systematically researched over a long period of time, such as urban development and redevelopment, housing, environment, and education.

The advantage of this long-term perspective is that despite the changing focus of the projects mentioned above, we can follow the development of the city and the region with our own empirical material. It includes data on population and household development, socio-economic and income data, the housing market situation and housing conditions, as well as expert and household interviews, surveys, research formats, workshops, in-situ observations and mapping, and the evaluation of municipal policy and planning documents. We collaborated closely with a variety of urban practitioners during all the projects. The authors are engaged in expert panels, invited advisory boards, and specific working groups. This chapter provides new insights in the discussion on re-growth by focusing on selected policy fields that have so far not gained much attention.

Leipzig's development as a shrinking and re-growing city

Leipzig was founded in the Middle Ages around the year 1000 and was conveniently located at the intersection of two European trade routes. In 1394 it was granted the privilege of a trade fair and subsequently enjoyed development benefits. In 1409 the University of Leipzig was founded, making it one of the oldest in the German-speaking world. At the end of the 18th/beginning of the 19th century, the book trade in Leipzig experienced an upswing and the city's first commercial area was created east of the city wall between 1800 and 1830.

Leipzig as a growing city during industrialisation

In the course of industrialisation and urbanisation, Leipzig experienced dynamic growth and became the sixth-largest German city in 1871 in terms of population. By the end of the 19th century, Leipzig had developed into an industrial city with a focus on book and textile production and mechanical engineering. Leipzig became a "city of books" during the 19th century due to the major role played by book production and book trade; before the Second World War almost 50% of all books produced in Germany were printed and published in Leipzig (von Hehl 2019). The population grew exponentially in the second half of the 19th century; the population of Leipzig grew 4.2 times between 1871 and 1899, from 107,000 to 450,000 people. Shortly after 1900 the city had more than half a million inhabitants. In connection with this, the number of rail passengers quadrupled and Leipzig became the "heart of German rail transport". In anticipation of further growth, construction began on Europe's largest terminus station in 1900 and the station opened 15 years later. The First World War interrupted this increase in population, yet it continued unabated in the 1920s. At that time, the industrialisation of the area around Leipzig and Halle saw the addition of lignite mining and the chemical industry, and the so-called Leipzig-Halle-Bitterfeld conurbation was created. Leipzig developed central functions for this industrial region, including banks, insurance companies, wholesale trade and the trade fair. During the Great Depression, the population peaked in 1931 at 719,000, making Leipzig the fifth-largest city in Germany. The global economic crisis was accompanied by a halting of dynamic in-migration and a decline in the birth rate. When the Nazi dictatorship came to power in 1933, Jews and politically oppressed people began to emigrate or were deported. The Second World War was the most severe cut in the city's development: The population losses amounted to more than 100,000 and the number of inhabitants fell to less than 600,000 by the end of the war. As a result of the Anglo-American bombardments, the historic buildings in the city center and the inner city were badly damaged (von Hehl 2019).

The GDR era: Towards long-term shrinkage

Shortly after the end of the war many companies emigrated west with the American troops, which meant a deep cut, especially in the publishing industry, as many publishers moved to West German cities such as Frankfurt, Munich and Stuttgart (von Hehl 2019). Former armament and heavy industry factories were expropriated or transferred to Soviet stock corporations, and further economic losses were linked to reparations to the Soviet Union. It took a correspondingly long time to restore the industries and the economy in the 1950s and 1960s. At the beginning of the 1950s, the population reached the 600,000 mark once again due to return migration and the immigration of German refugees from Germany's former eastern territories. After this, Leipzig's population shrank almost continuously until the end of the 1980s, although it was a slow decline. By this time, other cities in the region were more strongly developed than Leipzig, e.g. Halle, Dessau and Bitterfeld-Wolfen. New industries were established in those cities and they were therefore the object of state investments (also in relation to infrastructure and housing). The structure of the industrial sector in Leipzig did not change after the Second World War; in the adjacent regions to Leipzig's north and south, large areas of lignite coal were exploited from the 1970s onwards. Although the surrounding region continued to grow, Leipzig became one of the few major European cities with long-term shrinkage. As a result of the economic weakness of the GDR, too little investment was made in a number of industrial sectors as well as in infrastructure and the housing sector. The consequences were gradual decline and decay, poor working and living conditions, and a polluted environment. All of this then formed part of the background to the famous Leipzig Monday demonstrations, which in 1989 ushered in the peaceful revolution and then led to German Reunification in 1990 (Doehler and Rink 1996).

Massive shrinkage after 1989

After German Reunification, there was great hope that the city would be saved from imminent collapse, e.g. in relation to its technical and housing infrastructure. But after the fall of the Berlin Wall in November 1989, the city lost more than 30,000 inhabitants in just 14 months. German Reunification was not associated with a rapid and sweeping economic upswing that many had hoped for; some had even expected a second "economic miracle", like the one that took place in West Germany after the Second World War. The city therefore relied on optimistic forecasts and assumed its economy and population would grow and catch up to the level of West German development in about 10–15 years. Instead, as a result of the rapid and seamless integration into the EU and the world market, as well as shock therapy, there was a widespread de-industrialisation of the city and the industrial conurbation by about 80% (Doehler and Rink 1996). Only a few industrial cores remained in the region (petrochemical industry, coal mining and energy generation) and in the core cities of Leipzig and Halle (mechanical engineering). The reasons for this massive and rapid de-industrialisation were 1. the age and poor state of the industrial plants, 2. the upward revaluation shock associated with the exchange rate between the two German currencies (exchange rate of 2:1 from the perspective of former GDR) and 3. the hasty privatisations. The combination of those factors led to many insolvencies within a very short time. The conurbation fragmented, the former production and economic relationships were almost completely discontinued. Only massive state intervention was able to cushion the crisis socially during this phase. The federal government provided billions of euros for re-qualification, further training and subsidised job creation measures; for Leipzig, this was estimated at around one billion euros in the 1990s (Rink and Kabisch 2019). The transformation crisis was associated with a huge drop in the birth rate, suburbanisation and migration to West Germany. Young and well-educated people migrated to West Germany because they found jobs there with better salaries and, furthermore, benefitted from better living conditions. The city lost a large amount of its younger inhabitants, including young families. In addition to the shrinkage in the city, the entire region also shrank. Only communities in the suburban zones of Leipzig and Halle were able to generate population gains by offering detached housing construction. With a population loss of about 100,000 people (about 20%), the loss of its industry and an unemployment rate of 20% (Rink and Kabisch 2019), the city became the paradigmatic example or extreme case of post-socialist shrinkage in East Germany.

Moderate re-growth: The 2000s

As was the case throughout all of eastern Germany, investments in Leipzig were made in the opposite direction to the shrinking trend. Around 1.5 billion euros were spent on renovating and modernising the housing stock. For the New Trade Fair Centre, around 680 million euros were provided by public funds from the federal government and the Free State of Saxony. During this period, Leipzig attracted attention thanks to other major projects, such as the Medical Heart Centre, the Media City and the Quelle mail-order centre, although this did not initially translate into significant growth. At the time, Leipzig proclaimed itself to be the "boom city of the East" and residents took heart from the city's image campaign "Leipzig is coming". In addition, substantial subsidies were granted to businesses and practically all infrastructure was improved, such as transport, communications, water/wastewater, schools and kindergartens. Investments were also made in the environmental sector, for example in the rehabilitation of contaminated facilities, rivers and canals. Hundreds of millions of euros were spent by the EU, the federal government and the Free State of Saxony to rehabilitate the open-cast mines and turn them into recreational areas, thereby completely transforming the character of the entire region. The first successes of these massive investments were evident at the end of the 1990s/beginning of the 2000s (Heinig and Herfert 2012). In Leipzig, Porsche and BMW decided to build car factories. In Halle, Dell decided to set up a new production centre, and in Bitterfeld-Wolfen, solar industry companies received investments. In the north of Leipzig along the A14 motorway in the direction of Halle and in the vicinity of the airport, companies in the logistics and service sector also established new production and distribution centers. The airport itself was gradually developed into an intercontinental airport with public investments from the Free State of Saxony and the State of Saxony-Anhalt. In 2008, DHL set up a branch at the airport, which developed into a job engine. In 2019, almost 10,000 people were employed at Leipzig-Halle Airport (Rink and Kabisch 2019).

A political and legal reform that altered municipal boundaries led to a change in the population figures in 1999 and 2000. By incorporating a number of adjacent municipalities, Leipzig was able to statistically increase its population by 50,000 and thus approached the 500,000 mark (Figure 9.1). The 500,000-inhabitant threshold was important for the allocation of additional financial support from the federal government. The city of Leipzig experienced moderate growth in the 2000s, with annual growth rates of between 0.5% and 1%, but there was further shrinkage in Halle and across the entire region. In Leipzig, it was mainly the extensively rede-veloped Wilhelminian-style inner-city districts that experienced an influx, with growth taking the form of reurbanisation (Haase et al. 2012). This influx came primarily from the southern part of eastern Germany (Saxony, Thuringia and Saxony-Anhalt). This influx became more permanent, mainly due to economic re-growth. Since the mid-2000s, Leipzig has been experiencing a sustained employment dynamic; from 2005 onwards, the number of unemployed began to fall and the number of jobs started to rise.

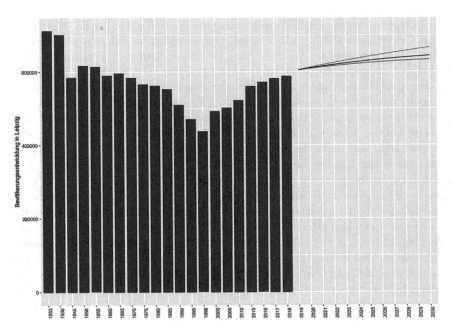

Figure 9.1 Population development 1933–2018 and forecast until 2030.
Source: Own elaboration on the basis of data from Stadt Leipzig 2019.

Challenges and problems of re-growth

At the beginning of the 2010s, immigration doubled compared to the yearly growth rates in the 2000s. Leipzig experienced a very dynamic population growth of over 2%, in some years even 3%. It was now no longer only young people from eastern Germany who immigrated to Leipzig, but also people from western Germany and from abroad (Welz et al. 2017). In addition, young people came from the southern EU countries that were deeply affected by the 2007/08 Global Financial Crisis, as well as from Eastern Europe. In 2015/16, Leipzig experienced strong refugee immigration, particularly from Syria, Afghanistan, and Iraq. In the middle of the 2010s, Leipzig became the fastest-growing metropolis in Germany. Due to the strong economic boom and availability of newly created jobs, the attractiveness of universities and colleges, as well as its positive image as a good place to live, Leipzig has experienced a significantly higher level of in-migration than other eastern German cities such as Chemnitz, Dresden or Halle. The growth now spread to almost all parts of the city; only a few districts experienced stagnation or even shrinkage. Towards the end of the decade growth slowed down, but still stood at around 1%. Leipzig is thus still an island of growth in a region that is shrinking (Herfert 2007) (see Figure 9.2). Since 2014, however, suburbanisation has also been taking place again, especially in the directly adjacent municipalities; every year about 1,000–1,500 people moved into the surrounding area (just under 0.25%)—an overall weak suburbanisation (see Figure 9.2). The main source of this new suburbanisation has been young families building or buying their home outside the city borders. Halle, the second-largest major city in the region, has been experiencing weak growth for some years, but

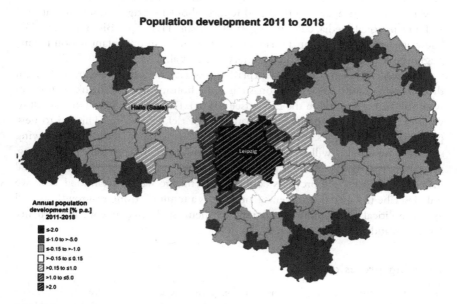

Figure 9.2 Population development in the Leipzig-Halle region 2011–2018.
Source: Statistical Office of the Laender 2019; GeoBasis-DE/BKG 2019.

the rest of the region is shrinking to a greater or lesser extent. In this respect, one can already speak of a polarised development.

At the peak of immigration in 2016, a very optimistic expert forecast was made for the population development. The forecast expected the population to grow to approximately 700,000 inhabitants by 2030 (Stadt Leipzig 2016). Against the backdrop of declining immigration in the subsequent years and the 2019 population figure of 600,000, the new population forecast from 2019 had been adapted to approximately 640–650,000 by 2030 (Stadt Leipzig 2019, see: Figure 9.1). This equates to slightly less than 1% growth per year, or 8% by 2030. National population forecasts continue to see Leipzig as one of the fastest-growing cities in Germany. In its latest study, the renowned Berlin Institute forecasts that Leipzig will be the front-runner in Germany with 16% growth by 2035 (Berlin Institut 2019), which seems to be too optimistic.

The population development is based on economic recovery and the provision of jobs; between 2005 and 2018 more than 70,000 new jobs were created, although many of them are precarious and poorly paid. A radical revival of the labour market in the city and, to some extent, in the region could be observed. While in 2005 unemployment in Leipzig was around 20%, it fell to 5.3% by 2019—its lowest level since German Reunification. This background makes it possible to explain the continuous immigration, especially of young people. The influx was additionally promoted by the positive image that Leipzig had gained in the 2010s. This image was largely due to the German public's growing awareness of the city's economic success, population growth, and attractive housing market (good housing available for reasonable costs). Successful PR work also played a major role, e.g. the marketing campaign "Leipziger Freiheit" and other activities. This led to a real boom in mass media reporting: More than 200 reports about Leipzig appeared in national and international media, culminating in the slogan "Hypezig" (Bischof 2015). The "new attractiveness" of Leipzig had been expressed as an important reason for in-migration in a survey among newcomers (Welz et al. 2017).

This robust growth came as a surprise to the city, nobody had expected it. On the contrary, in some areas of action, such as the housing market, shrinkage policies were still being pursued in the 2010s (see below). The strong re-growth was felt to varying degrees in the individual urban areas or fields of action. This led to new problems and challenges, to which different responses were made. In the following sections, three fields will be presented in more detail: the provision of housing, schools as part of social infrastructure, and public transport as part of technical infrastructure. First, the previous context of shrinkage will be examined, then we will describe the specific problems that arose as a result of strong re-growth, as well as the political and planning reactions to them, such as concepts, instruments, resources, and specific governance.

Housing provision

In Germany, housing provision is not completely left to the market. Instead, to a certain extent it is seen as a socio-political task. In addition to the federal and state governments, local authorities also have a responsibility here. Immediately after

German Reunification in the beginning of the 1990s, the housing market in Leipzig was tense because on the one hand about 70,000 people were looking for a suitable apartment, while on the other hand about 25,000 apartments (about 10% of the stock) were empty due to lack of renovation and repairs (Doehler and Rink 1996). In the 1990s a capitalist housing market was gradually re-established, the public ownership of apartments was restituted to former private owners or their descendants, and rents and prices were no longer determined by the state. With urban development subsidy programmes and special tax deductions, much public money was channelled into housing stock and private capital was mobilised (Doehler and Rink 1996). Thus, in the 1990s, about two-thirds of the 110,000 pre-1918 (so-called Wilhelminian-style) apartments were renovated and modernised. In addition, about 47,000 new housing units were built during this phase. The combination of the above-mentioned population shrinkage with the expansion of the housing supply led to a massive oversupply on the housing market at the end of the 1990s/beginning of the 2000s (Wolff et al. 2017). In 2000, approximately 68–69,000 housing units were vacant, which corresponded to a vacancy rate of 21%. Just under half of these were not renovated or modernised, but the other ample half were either freshly renovated/modernised or newly built. This extremely high vacancy rate was the highest of any major German city and one of the highest in eastern Germany, making Leipzig the "metropolis of vacancy" (Rink 2015).

Although the population forecasts at that time predicted a certain amount of growth for the 2000s, for the longer-term perspective after 2010 experts expected stagnation or a return to shrinkage. Given the long-term lack of demand in the housing market, the demolition or dismantling of apartments seemed inevitable. Starting in 2002, public funds were once again made available for both demolition of apartments and upgrading of the remaining stock as part of a joint federal and state programme called Stadtumbau Ost (Urban Restructuring East). Between 2001 and 2012, almost 13,000 residential units were thus demolished in Leipzig; the majority of them were prefabricated apartments in large housing estates at the urban fringe that belonged to the municipal housing company's and the cooperatives' portfolio. In conjunction with the renewal of the Wilhelminian-style inner-city areas, the vacancy rate fell significantly and in 2011 amounted to about 39,000 residential units, which was still quite high compared to other major German cities (Rink 2020).

By the end of the 2000s, the real estate market had already resumed its course of moderate growth. The subsequent dynamic growth led to a significant increase in investment in the renovation of old buildings and new construction in the city centre. Due to strong demand in the 2010s, the vacancy rate continued to decline sharply. The vacancy rate in 2018 was approximately 4.5%, but the market-active vacancy rate was only 1.8% (Rink 2020). Together with the significant increase in property prices since the mid-2010s and rising rents, the city administration indicated that the housing market had now become tight. The city council began reacting to the new housing market situation in the mid-2010s and adopted a new housing policy concept in 2015 (Stadt Leipzig 2015), which has been implemented since 2016. With this new concept, the general orientation is towards medium- and long-term population growth in the city, thereby increasing demand on the housing market without any demolitions. However, the new housing policy and associated instruments can only be implemented on the proviso that there are indications of a tight housing market. A set of indicators was developed and used specifically for this purpose. The City of Leipzig has thus reacted cautiously to the new situation on the housing market, which is understandable for a city coming out of a deep and long vacancy depression. Although the city administration claims that Leipzig has a tight housing market, this is still a matter of dispute. The housing cooperatives doubt this in view of the vacancy rates in their housing stocks (Rink 2020). Nevertheless, housing policy instruments have been successively implemented since 2016. On the one hand, the aim is to influence new construction and renovation in such a way that affordable housing comes onto the market, for example through the promotion of social housing and urban development contracts with investors. In addition, the municipal housing association (LWB) intends to build new housing and significantly expand its stock in the affordable segment by around 5,000 units by the mid-2020s. On the other hand, price increases in the existing stock are to be curbed, for example by limiting rent increases to a total of 15% within three years or by designating conservation areas to prevent expensive renovations.

In the past the state has repeatedly intervened in the housing market with regulations and financial interventions. In the 1990s, for example, the supply of housing was quantitatively ensured by new construction and improved by renovation—through both public investment and subsidies. In the 2000s, on the other hand, housing demolition was largely supported by state-led programmes. The housing demolitions subsequently came under criticism in recent years due to developments in the housing market. Civil society actors have also criticised the public sector, claiming that it reacted too late and too little to the new growth in recent years. Private and cooperative housing companies, however, reject further or stronger regulation. In the years 2018–2020 there has been a significant increase in investment in Leipzig's housing market, with more than 2,000 new or refurbished apartments coming onto the market each year. The current and planned investment volume in the housing market for the years 2021–2025 amounts to over 4 billion euros of private investments, and the 20 million that the Federal State of Saxony spends on this each year is quite modest (Rink 2020). As can be seen, the city has had to intervene more strongly in the market again in the field of housing supply. A different approach can been witnessed concerning schools, where the municipality is fully responsible.

Schools as an example of social infrastructure

School infrastructure is the area where the greatest changes and problems have been caused by Leipzig's shrinkage and re-growth. As described above, there was a decrease in the birth rate in the early 1990s and the total fertility rate in eastern Germany fell to below one—the lowest level ever measured in Germany. At the same time, young people in particular and, to some extent, families with children emigrated to western Germany and the greater Leipzig area (suburbanisation). As a result, the number of school students fell rapidly from the second half of the

1990s onwards. The size of the school population almost halved, decreasing from just under 60,000 in the 1989/90 school year to about 35,000 in 2008 (Walde 2019). This means that the number of school students fell far more sharply than the average population decline, so the school sector was therefore particularly affected by the shrinkage. It was only through the incorporation of numerous surrounding municipalities that the number of school students rose again for a short time between 1998 and 2000, after which it continued to decline. The political response to the declining number of school students was to reduce the number of schools and to thin out the school network, not only in order to adapt the number of school places to the reduced demand, but also to save costs. Following reunification and the conversion of the school system, the City of Leipzig established a total of 161 primary and secondary schools in the 1992/93 school year. In the course of the following two decades, 75 of the schools were closed (Walde 2019). Some of the closed schools could not be used for other purposes and decayed over the years, becoming ruins within residential housing areas. The school closures were sequential, starting with the primary schools and ending with the secondary schools. The driving force behind the school closures was the city council, which made the relevant decisions. This process took a very conflictual course compared to other fields of action. Practically every school closure provoked resistance from the affected school students, their parents and teachers, who carried out numerous protest actions (Walde 2019).

The renewed increase in the number of school students made itself felt in primary schools from the 2003/04 school year onwards, and in the first grade of secondary schools from the 2009/10 school year onwards (Stadt Leipzig 2018). Since the beginning of the 2010s, there has been dynamic growth in the number of school students across all grades. This growth is also fed by higher birth rates. Birth rates already began rising in the mid-2000s and in 2014 the balance of births and deaths was positive for the first time since the mid-1960s. Since then, Leipzig has been one of the few cities in eastern Germany to record natural population growth, even if it is quite low. The high level of immigration in the 2010s meant that (young) families with school-age children also came to Leipzig at that time. Many of them arrived during the peak of the wave of refugees in the years 2015/16, as a result of which the number of school students increased by 50% between 2008 and 2018, from 35,000 to 52,000. In the years 2015 to 2017 alone, the number of inhabitants in the 6- to 15-year-old age cohort rose by 5,620 (Stadt Leipzig 2020). This quantitative development was closely connected with qualitative efforts, because of the large number of children without sufficient German language skills. As a consequence, new student groups were formed to provide particular support for those learning the German language and entering the German school system.

The disproportionate decline in the number of school students during the shrinking phase was followed by a disproportionate increase in the number of students. This strong growth in the number of children and school students is fed by both immigration and natural growth. As a result, the demand for schools in particular has grown enormously in the short term. The city or public authorities are legally obliged to provide school places for all school-age children. A new school development plan was drawn up in 2016 in response to this growth, which featured up to 40 construction measures across all school types, including new buildings, reactivation of closed schools as well as renovation and capacitiy expansion of existing schools. Over the following years, this plan was revised and adapted to the requirements, and in 2018 the 'Immediate Construction Program for Schools' was launched and an independent task force responsible for school construction was set up within the city administration. The number of grade one enrolments at primary schools will increase by approximately 50% by 2030, by which time there will be around 21,000 more students in total. This means that by 2030, in purely mathematical terms, Leipzig will require more than 70 new schools or 120 capacity-enhancing school construction measures. Between 2016 and 2023 the city will invest a total of about half a billion euros in school construction (Orbeck and Dunte 2019).

A substantial part of this will come from the Free State of Saxony, but the city will also have to make its own contributions. In order to be able to raise these funds, in 2019 the mayor of finance announced a budget freeze on current expenditure to save money for the construction programme. In addition, the municipality of Leipzig wants to raise up to 500 million euros in new debt to finance the school construction programme (Orbeck and Dunte 2019). However, the implementation of many construction measures is not progressing fast enough and is not keeping pace with the growth of the school student population. For example, finding suitable sites for the construction of schools is a major problem and there is also a lack of building capacity, because housing construction is booming and competing with public infrastructure such as schools (see above). Bottlenecks had already started appearing in the 2019/20 school year, especially in secondary schools, and these will not ease significantly in the coming years despite considerable additional investment measures. For example, ten classes are already missing from secondary schools in the 2019/20 school year (Orbeck 2019). Due to this dynamic growth in the number of school students, the existing bottlenecks and the statutory mandate, schools are currently right at the top of the city's political agenda, and the school construction programme has top priority. The city had to do a 180-degree turn from school closures to an extensive programme of new construction in this field. Not quite as dramatic is the development in the area of public transport, which we will look at next.

Public transport as an example of technical infrastructure

Public transport is another area where the problem of re-growth can be observed. In the GDR, local public transport was given priority, while motorised private transport played a subordinate role. Cars were expensive luxury products, which explains the GDR's low level of motorisation, especially in cities. The share of public transport in mobility was quite high and walking also played a major role. Thus, shortly before German reunification, 35% of all trips in Leipzig were made by public transport and 36% on foot, while motorised private transport accounted for only 24% (Jana 2018). In 1987, about 200 million passengers travelled on Leipzig's public transport system. Admittedly, the system also suffered from the investment weakness of the GDR, as a result of which the vehicle fleet of the Leipzig transport company was completely outdated and the technical equipment was largely worn out. Nevertheless, the inhabitants were dependent on the public transport system consisting of trains, urban railways, buses and trams. There was a surge in motorisation immediately after the economic and monetary union in summer 1990. At last, long sought-after Western cars were available to buy, although mostly older models were sold. Despite the transformation crisis and high unemployment, the strong growth in motorisation continued in the 1990s. The modal split worsened to the disadvantage of public transport: The share of public transport halved between 1987 and 2003 from 35% to 17.3%, while the share of motorised private transport almost doubled in the same period from 24% to 44% (Jana 2018). This was associated with a sharp decline in the number of passengers using public transport (tram, urban railway, and bus). All in all, there was a strong increase in overall traffic, fueled by suburbanisation. In the 1990s about 30,000 inhabitants moved from the city to the surrounding area and many started commuting to Leipzig for work. The 1990s also saw the emergence of numerous large shopping centres in the suburban area, which led to a considerable increase in shopping traffic. Finally, there was also a partial relocation of trade and services to the suburban area, which similarly led to an increase in traffic. We are dealing here with opposing developments; the shrinkage did not in any way result in a decrease in traffic. The decline in the number of public transport passengers could not be stopped by the massive investments either. In the 1990s, new trams and buses replaced the outdated models, and the rail network and bus shelters were renewed. The attractiveness of public transport has increased enormously since the 1990s. However, public transport still couldn't compete with the convenience and attractiveness of the car.

At the end of the 1990s, the number of public transport passengers in Leipzig appeared to reach a low point, after which the number of passengers rose again, reaching 156.4 million in 2018 (www.l.de 2019). This is the highest figure since the German Reunification, but it is still below the 1987 figure at the end of the GDR. Since reunification, roads and transport infrastructure have been renewed or modernised in parallel with investments in public transport. Motorway connections in the region have been expanded and modernised, and the A14 motorway has been completely modernised and extended to the A2 to Magdeburg. With the new construction of the A38 motorway, the region now has a southern bypass from Leipzig and Halle and an additional connection to western Germany. Motorised private transport has increased strongly in the course of the economic situation since the mid-2010s, which can also be seen in the growing number of motor vehicle registrations, not only in the city but across the entire region. At the end of the 1990s, construction began on the City Tunnel from the central railway station through the inner city. This was Leipzig's major public infrastructure project, which the federal government financed with one billion euros. Its completion was originally planned for the 2006 Football World Cup, but it was completed much later and began operating at the end of 2013. It remarkably improved the urban railway connections within the city of Leipzig. For instance, the project made it possible to reopen a railway line to a large housing estate that had been closed in 2011 despite massive protests by residents. Furthermore, the regional railway connections between the city and the central German region were substantially improved. Thus, the completion of the City Tunnel fitted perfectly into the phase of dynamic growth and has since provided relief for some of the problems associated with shrinkage and re-growth.

Current plans envisage a significant expansion of the public transport system, with passenger numbers expected to double by 2030 and the share of public transport in the modal split estimated to rise to 23% (Jana 2018). This increase is the result of the city's expected growth and its sustainability goals: Leipzig aims to gradually reduce motorised private transport to 30% by 2025 (Jana 2018). At the same time, foot traffic is set to increase to 27% and cycling to 20% (Jana 2018). The problem here is how to implement and finance the politically desired and planned disproportionate growth in public transport. In the mayor's election campaign in 2020, for example, a 365-euro ticket for one year was proposed, but this would deprive the municipal transport company (LVB) of millions in revenue and place an enormous burden on the municipal budget. Here in the transport sector, a cut-back in local public transport usage clearly took place in the 1990s, i.e. in the phase of de-industrialisation and shrinkage. Due to motorisation and suburbanisation processes, however, traffic continued to increase during this phase instead of shrinking. In the 2000s, public transport recorded growth in passenger numbers again, but the share of the modal split has stagnated at a low level since the early 2000s. Given that the share of motorised individual transport stagnated at a high level during this period, there has also been a growth in traffic volume.

Discussion

What did we learn about the re-growth of Leipzig with our particular focus on the commonalities and differences between these three policy fields? What challenges and problems arose and how are they reflected by the development in the selected policy fields?

First of all, a *number of a) parallels became apparent*: In all three fields, the consequences of the population shrinkage in the 1990s were drastic, especially with respect to schools and the transport sector (although the dramatic changes in transport were mainly due to increased car usage). The same is true for the dynamic re-growth of the 2010s, which is also causing massive problems, particularly in the school sector. However, we are *also observing certain b) counter-trends*: For example, in the 1990s, new housing construction and renewal was promoted and supply was massively expanded, although demand fell sharply. In the 2000s, housing stock was then torn down, although demand rose again. In the school sector, schools were still being closed as late as 2012, even though the number of students has been rising again since 2008. These counter-trends are mainly caused by the *c) inertia of policies*. This is particularly visible in the housing sector: The housing shortage at the beginning of the 1990s triggered a decade of new construction and renewal that took place alongside a simultaneous decline in housing demand. It was not until around 2000 that the massive problem of housing vacancy was recognised and then dealt with by large-scale demolition. After 2010 this inertia or mismatch between new developments and policies is still evident, and although demand grew rapidly, demolition continued until a new housing policy was introduced in 2015. In the school sector, the reaction to the sharp decline in the number of students in the

1990s was much faster and schools were closed. However, schools were still being closed until 2012, even though the number of students had begun rising again in 2008. It took several years for school policy to change completely. But there are also *d) learning effects* from the experience of extreme shrinkage and dynamic re-growth. After only sporadic population forecasts were made in the 1990s, this instrument has been systematically developed since 2001. Every three to four years, comprehensive and differentiated population projections are prepared in different variants. This is done by including many local stakeholders and experts from administrative and scientific bodies, and has formed the basis for planning in fields such as housing provision, schools, and public transport. In the 1990s, there had been divergent developments and counter-trends in various policy fields, and undesirable processes such as suburbanisation. Since the early 2000s, master plans have been regularly drawn up to enable coordinated and integrated action for a number of policy fields. In addition, all three policy fields considered here are still subject to specific monitoring, planning concepts and action programs that are constantly being updated.

As we can see, shrinkage and re-growth are each associated with specific problems that are quite different in the researched policy fields. Leipzig, in a way, represents a "city of extremes" (Rink 2015) with its experience of massive shrinkage which was followed by initially moderate and then dynamic re-growth. While re-growth "solved" some problems such as weak demand in the housing sector, it also created new problems such as increased demand in the school sector. The political and planning responses also depended on the responsibilities and possibilities for action of local policy-makers in each individual policy field. For instance, the municipality is legally obliged to facilitate compulsory school attendance. Housing provision is largely seen as a task to be handled by the market; the municipality has only limited responsibilities and opportunities in this sector. Regardless of shrinkage and re-growth, the city has consistently pursued the goal of strengthening public transport. Due to the strong increase in traffic during the re-growth period, the expansion of public transport is imperative in the future. Table 9.1 summarises the processes within the policy fields under investigation for the periods of shrinkage, moderate and dynamic re-growth in Leipzig.

Conclusion

The current re-growth in Leipzig must be interpreted primarily as the result of welfare state policy, which had (and has) the goal of equalising living conditions. Without the permanent and substantial subsidies and investments from the federal government, the European Union and the State of Saxony, this development would not be possible. Leipzig has thus caught up to the level of urban development evident elsewhere in Europe; its re-growth is comparable to that of other European cities of this size, such as Gdansk, Liverpool or Lyon. The further prospects for Leipzig's population development foresee further re-growth, but it remains to be seen whether the high level of in-migration of recent years will continue. The city has consistently pursued classic growth objectives. In the 1990s, massive state support from the federal government, the EU and the Free State of Saxony was used

Table 9.1 Processes in different policy fields during shrinkage, moderate and dynamic re-growth

Policy Field	Housing Provision	Public Schools	Public Transport
Shrinkage 1990s	Massive new construction, renewal and modernisation	New structure, start of school closures	Decline in public transport, growth in motorised private transport
Moderate re-growth 2000s	High vacancy rates, intense demolition and some upgrading	School closures, some demolition	Recovery of public transport, growth in motorised private transport
Dynamic re-growth 2010s	New construction and renewal, some demolition	First closures, then re-opening, renovation and new construction of schools	Growth in both public and motorised private transport

Source: Authors' elaboration

to pursue growth policies and counteract shrinking processes. In the 2000s, investments then began to take effect with the establishment of new businesses, which boosted re-growth. Despite the emergence of re-growth the management of shrinkage continued to be pursued in order to deal with the consequences of shrinkage exemplified by provision of school infrastructure and housing. In the 2010s, the city then experienced dynamic re-growth, which confronted the city with a number of new problems, particularly in the areas under investigation. The changing requirements meant that the city constantly had to react and, as we have seen, the reactions differed greatly between the various policy fields. Both the shrinkage and the re-growth revealed that the city cannot cope with the associated problems on its own. In all three fields of action considered here, Leipzig is absolutely dependent on public funding from the federal government, the Free State of Saxony and the European Union. Concerning the present problems and challenges caused by the coronavirus crisis, several future scenarios are more or less likely to happen: (a) resumption of growth after a brief slump, (b) stagnation, or (c) renewed shrinkage. At present, all levels of government are in crisis mode and it remains to be seen what policy response patterns will develop as a result. What can the urbanistic discourse learn from Leipzig's re-growth experience? As we've shown, shrinkage and re-growth affect the policy areas differently and trigger different government responses. Re-growth solves some shrinkage-related problems, but brings with it new problems and challenges. Policy-makers should generally be more open to varying future developments and react more quickly to both shrinkage and re-growth. While some learning effects can be identified, the scope and ability of policy-makers to act promptly and flexibly is often limited.

References

Berlin-Institut (2019). Die demographische Lage der Nation, Berlin.

Bernt, M. (2009). Renaissance through demolition in Leipzig. In: Porter, L., Shaw, K. (Eds.): Whose urban renaissance. In *International Comparison of Urban Regeneration Strategies*. Routledge Studies in Human Geography 26: 75–83.

Bischof, A. (2015). #Hypezig. Die Verkleinbürgerlichung des Alternativen. In: Eckardt, F., Seyfarth, R., Werner, F. (Hg.): *LEIPZIG. Die neue urbane Ordnung der unsichtbaren Stadt*, Münster, 72–87.

Bontje, M. (2004). Facing the challenge of shrinking cities in East Germany: the case of Leipzig. *GeoJournal* 61: 13–24.

Carter, D.K. (2016). *Remaking Post-Industrial Cities*. London: Routledge.

Couch, C., Fowles, S., Karecha, J. (2009). Reurbanization and housing markets in the central and inner urban areas of Liverpool. *Planning Practice and Research*, 24: 321–334.

Couch, C., Cocks, M., Bernt, M., Grossmann, K., Haase, A., Rink, D. (2012). Shrinking cities in Europe. *Town & Country Planning*, 264–270.

Doehler, M., Rink, D. (1996). Stadtentwicklung in Leipzig: Zwischen Verfall und Deindustrialisierung, Sanierung und tertiären Großprojekten. In: Häußermann, Hartmut, Neef, Rainer (eds.): *Stadtentwicklung in Ostdeutschland*. Soziale und räumliche Tendenzen, Opladen, 263–286.

Haase, A., Rink, D. (2015). Inner-city transformation between reurbanization and gentrification: Leipzig, eastern Germany. *Geografie*, 120(2): 226–250.

Haase, A., Herfert, G., Kabisch, S., Steinführer, A. (2012). Reurbanizing Leipzig (Germany): Context Conditions and Residential Actors (2000–2007). *European Planning Studies*, 20(7): 1173–1196.

Haase, A., Bontje, M., Couch, C. Marcinczak, S., Rink, D., Rumpel, P., Wolff, M. (2020). Factors driving the regrowth of European cities and the role of local and contextual impacts: a contrasting analysis of regrowing and shrinking cities, re-submitted to: Cities: The International Journal of Urban Policy and Planning, 108, [102942]. https://doi.org/10.1016/j.cities.2020.102942

Hehl, U.V. (Hg.) (2019). Geschichte der Stadt Leipzig Band 4: Vom ersten Weltkrieg bis zur Gegenwart, Leipzig.

Heinig, St., Herfert, G. (2012). Leipzig – intraregionale und innerstädtische Reurbanisierungspfade. In: Brake, L., Herfert, G. *Reurbanisierung, Materialität und Diskurs in Deutschland*. Springer VS Wiesbaden, 323–342.

Herfert G. (2007). Regionale Polarisierung der demographischen Entwicklung in Ostdeutschland – Gleichwertigkeit der Lebensverhältnisse? *Raumforschung und Raumordnung* 65: 435–455.

Jana, M. (2018). Überblick über die Finanzierung des ÖPNV durch die Stadt Leipzig, unter: https://static.leipzig.de/fileadmin/mediendatenbank/leipzig-de/Stadt/02.6_Dez6_Stadtentwicklung_Bau/61_Stadtplanungsamt/Stadtentwicklung/Leipzig_weiter_denken/Nachhaltiger_Finanzhaushalt/Praesentation_Herr_Jana.pdf; Zugriff: 16 April 2020.

Kabisch, S., Kindler, A., Rink, D. (1997). Sozial-Atlas der Stadt Leipzig, Leipzig, UFZ Leipzig-Halle, Leipzig 1997.

Nelle, A., Großmann, K., Haase, D., Kabisch, S., Rink, D., Wolff, M. (2017). Urban shrinkage in Germany: An entangled web of reality, discourse and policy. *Cities*, 69, 116–123.

Nuissl, H., Rink, D. (2005). The "production" of urban sprawl in eastern Germany as a phenomenon of post-socialist transformation. *Cities*, 22(2), 123–134.

Orbeck, M. (2019). Leipzig braucht bis 2030 etwa 21.000 Schulplätze mehr. In: Leipziger Volkszeitung 16.6.2019 (www.lvz.de/Leipzig/Stadtpolitik/Leipzig-braucht-bis-2030-etwa-21-000-Schulplaetze-mehr; accessed 16 April 2020).

Orbeck, M., Dunte, A. (2019). Städte investieren Millionen in den Schulneubau. In: Leipziger Volkszeitung, 13.7.2019 (www.lvz.de/Region/Mitteldeutschland/Staedte-investieren-Millionen-in-den-Schulneubau; accessed: 16 April 2020).

Power, A., Plöger, J., Winkler, A. (2010). *Phoenix Cities: The Fall and Rise of great industrial Cities.* Bristol, Portland: The Policy Press.

Power, A., Katz, B. (2016). *Cities for a Small Continent.* Bristol, Portland: The Policy Press.

Richter, J. (2000). Städtische Armut in Leipzig, UFZ-Bericht 27/2000, Leipzig.

Rink, D., Haase, A., Grossmann, K., Couch, C., Cocks, M. (2012). From long-term shrinkage to re-growth? A comparative study of urban development trajectories of Liverpool and Leipzig. *Built Environment*, 38(2), 162–178.

Rink, D., Bernt, M., Großmann, K., Haase, A. (2014). Governance des Stadtumbaus in Ostdeutschland – Großwohnsiedlung und Altbaugebiet im Vergleich. In: Jahrbuch StadtRegion, Berlin, 132–147.

Rink, D. (2015). Zwischen Leerstand und Bauboom: Gentrification in Leipzig. In: Eckardt, F., Seyfarth, R., Werner, F. (Hg.): LEIPZIG. Die neue urbane Ordnung der unsichtbaren Stadt, Münster, 88–107.

Rink, D. (2020). Wohnungspolitik in einem Wohnungsmarkt mit Extremen: Leipzig. In: Rink, D., Egner, B. (Hg.): *Kommunale Wohnungspolitik. Beispiele aus deutschen Städten*, Baden-Baden: Nomos, 177–195.

Rink, D., Kabisch, S. (2019). Von der Schrumpfung zum Wachstum. Demographische und ökonomisch-soziale Entwicklungen. In: von Hehl, U. (Hg.): *Geschichte der Stadt Leipzig Bd. 4. Vom Ersten Weltkrieg bis zur Gegenwart*, Leipzig, 842–861.

Siedentop, S. (2018). Reurbanisierung. In: Rink, D., Haase, A. (eds.): *Handbuch Stadtkonzepte*, Opladen, Toronto: Barbara Budrich, 381–403

Stadt Leipzig (1999). *Lebenslagenreport Leipzig*. Leipzig: Bericht zur Entwicklung sozialer Strukturen und Lebenslagen in Leipzig.

Stadt Leipzig (2015). *Wohnungspolitisches Konzept*. Leipzig: Fortschreibung.

Stadt Leipzig (2016). *Bevölkerungsvorausschätzung 2016*. Leipzig: Methoden-und Ergebnisbericht.

Stadt Leipzig (2019). *Methoden und Ergebnisse der Bevölkerungsvorausschätzung 2019*. Leipzig.

Stadt Leipzig (2018). INSEK Integrated Urban Development Concept for Leipzig 2030, accessed: 6 May 2020.

Stadt Leipzig (2020). Schulbauprogramm – Ausbau von Schulkapazitäten, unter: https://www.leipzig.de/jugend-familie-und-soziales/schulen-und-bildung/schulen/schulbauprogramm/; Zugriff: 16 April 2020).

Turok, I., Mykhnenko, V. (2007). The trajectories of European cities, 1960–2005. *Cities*, 24: 165–182.

van den Berg, L., Drewett, R., Klaassen, L.H., Rossi, A., Vijverberg, C.H.T. (1982). *Urban Europe: A Study of Growth and Decline*, Vol. 1. Oxford: Pergamon Press.

Verkehrs- und Tiefbauamt Stadt Leipzig (2015). Kennziffern der Mobilität für die Stadt Leipzig, https://static.leipzig.de/fileadmin/mediendatenbank/leipzig-de/Stadt/02.6_Dez6_Stadtentwicklung_Bau/66_Verkehrs_und_Tiefbauamt/SrV-2015-Information-zu-Kennziffern-der-Mobilitat-fur-die-Stadt-Leipzig.pdf, accessed: 6 May 2020.

von Hehl, U. (ed.): (2019). *Geschichte der Stadt Leipzig Bd. 4. Vom Ersten Weltkrieg bis zur Gegenwart*. Leipzig: Universitätsverlag.

Walde, A. (2019). *Schulpolitik in Städten mit Schülerrückgang.* Berlin: Eine Governance-Analyse am Beispiel von Leipzig und Timişoara.

Welz, J., Kabisch, S., Haase, A. (2014). Meine Entscheidung für Leipzig. Ergebnisse der Wanderungsbefragung 2014. Statistischer Quartalsbericht II/2014, Amt für Statistik und Wahlen, Stadt Leipzig, 19–24.

Welz, J., Haase, A., Kabisch, S. (2017). Zuzugsmagnet Grossstadt – Profile aktueller Zuwanderer. Das Beispiel Leipzig. Major cities as influx magnets - A profile of current immigrant groups. *The Example of Leipzig, disP - The Planning Review*, 53(3), 18–32.

Wolff, M., Haase, A., Haase, D., Kabisch, N. (2017). The impact of urban regrowth on the built environment. *Urban Studies*, 54(12), 2683–2700.

10 Łódź

A multidimensional transition from an industrial center to a post-socialist city

Jolanta Jakóbczyk-Gryszkiewicz

Introduction

The chapter is devoted to Łódź—a large city in central Poland. Its aim is to present the historical conditions of the city's development and conduct an analysis of the socio-economic transformations following the fall of communism in 1989. Poland entered the period of post-socialist changes after the political negotiations described as the Round Table Talks. These were negotiations between Polish opposition, from the Solidarity movement, and the communist party. They were ongoing between 6 February and 5 April 1989. Round Table Talks resulted in the change of the Polish political system. During the talks, the Polish communist party agreed to organise democratic elections and introduce the rules of the free-market economy into the country (Cudny et al. 2012).

The post-1989 system transformation was a long process, often referred to as post-socialist transition, which strongly affected Polish cities and towns (see Słodczyk 2001; Parysek 2004; Young and Kaczmarek 2008; Marcińczak et al. 2013; Kazimierczak and Szafrańska 2019). The first years of transformation were the time of so-called shock therapy, resulting, among others, in the collapse of traditional branches of industry (including textile manufacturing), rapid privatisation, rising unemployment, and social problems. This initial phase of post-socialist transformation was followed by the development of the private sector, the rise of foreign investments, and economic restructuring. The economic recovery was strengthened by the enlargement of the European Union in 2004, when Poland, together with nine other countries, entered this organisation (Kołodko 2005).

In 1989 Łódź was the second-largest Polish city with ca. 850,000 inhabitants. It was one of the biggest industrial cities and an important regional urban centre comparable to other largest cities in Poland, i.e. Wrocław, Poznań, Gdańsk, and Kraków. The disadvantage of Łódź's economic structure was its monofunctional structure based on the domination of the old-fashioned textile industry and outdated infrastructure. This traditional specialisation of Łódź was the result of its 19th-century history, when the decision was taken to locate a small cloth-making settlement in Łódź, and the resulting development of the textile industry led to the rapid growth of the town. As early as the second half of the 19th century, Łódź became the second-largest city and the most important textile production centre on Polish lands.

DOI: 10.4324/9781003039792-10

The analysis presented in this chapter includes the issues of post-socialist cities in Central Europe, which emerged as a result of the fall of the communist system (Sailer-Fliege 1999; Stanilov 2007; Smith et al. 2008; Sýkora and Bouzarovski 2012). There are numerous examples of such cities in this region, where many countries were dominated by communism in the 20th century (see: Jacobs 2013; Kovács et al. 2019; Špačková et al. 2016). The transformations in post-socialist cities are a long and complex process, therefore we distinguish a number of stages of these transformations and their different types. According to Korec and Ondoš (2006), a transformation of a post-socialist city consists of the following stages: the early post-socialist city, the mature post-socialist city, the late post-socialist city, followed by the early capitalist city.

Contemporary Łódź is an example of a post—socialist city which, according to Liszewski (2001 cited in Cudny et al. 2012 p. 17–18), is defined as

> any city which has been functioning in new political and economic conditions for over 10 years, and earlier (before 1990) it had been functioning in the conditions of real socialism for 45 years, regardless the fact whether it was built in the system or much earlier, subject to its ideology and laws (centralization of power, lack of market economy, social and spatial egalitarianism, ideologization of life, etc.).

> Sýkora (2009, p. 394)

defined post-socialist cities as

> cities at the transition stage, characterized by dynamic processes of change rather than static patterns. The urban environment formed under the previous system is being adapted and remodelled to match the new conditions of the political, economic, and cultural transition towards the capitalist society. Many features of a socialist city suddenly stood in opposition to the capitalist principles. The contradictions between the capitalist rules and the socialist urban environment led to the restructuring of the existing urban areas. With time, new capitalist urban developments are having more and more influence on the overall urban organization. The post-socialist developments bring the re-emergence of some pre-socialist patterns, transformations in some areas from the socialist times, and creation of new post-socialist urban landscapes.

Matlovič et al. (2001, p. 11) presented several basic groups of transformations which affected the socio-economic and spatial structure of post-socialist cities after 1989. The processes observed in the morphological-spatial structure included suburbanisation, gentrification, revitalisation, intensification, recession and urban fallows. Those which took place in the functional structure included suburbanisation, commercialisation, deindustrialisation, demilitarisation, sacralisation and functional fragmentation. The changes that occurred in the demographic-social structure included suburbanisation, gentrification, segregation, separation and regression of the socio-economic status.

Research into post-socialist cities was commonly conducted in Polish urban studies and human geography. There are numerous examples of publications encompassing the complex analysis of post-socialist urban transformations (see: Słodczyk 2001; Liszewski 2001; Parysek 2004; Cudny 2012). Some works focused on the functional transformations of cities (e.g. Suliborski 2001; Suliborski et al. 2011) while others analysed the demographic and social transition (see: Jakóbczyk-Gryszkiewicz 2010; Dzieciuchowicz 2009; Marcińczak 2009; Stryjakiewicz 2014).

In Łódź the period after 1989 was a time of vast socio-economic changes. They are often referred to as the post-communist transformation, which led to a transformation of the city into a post-socialist urban centre (see: Liszewski 2001; Cudny 2012). The socio-economic transformation after 1989 limited textile production to the clothing industry alone and caused a vast socio-economic crisis during the 1990s. Due to the later diversification of functions and restructuring, other industries and services started to develop. The recent economic growth of the city was enhanced by its central location, convenient road links, economic traditions, access to EU funds and low prices of land and commercial spaces, compared to other large cities in Poland. Nevertheless, the good economic situation of recent years has not slowed down the process of depopulation and Łódź is currently the fastest-shrinking large city in Poland.

Łódź is currently the third-largest city in Poland, according to the number of inhabitants, and is the capital of a centrally located large voivodeship. The chapter points to the contemporary problems of the city of Łódź, related to the change of functions, progressing depopulation and the revitalisation of the urban space after 1989. The analysis was based on a literature review and statistical analysis. The study also regards Łódź development opportunities and future directions.

Presentation of the research area

After the change of the Polish borders after the Second World War, Łódź became the geographical centre of the country (Figure 10.1). Having received municipal rights in the Middle Ages (1423), Łódź remained a small, farming town, lying away from major roads, for the next 400 years.

In the early 19th century Łódź was populated by a mere 800 inhabitants. In 1823, as a government town (a part of the Kingdom of Poland, dependent on Russia), Łódź was selected for development as a cloth-making settlement. Due to the access to the huge Russian market, the cloth and later cotton-wool industry flourished. From a small town, Łódź grew to become a huge city and the largest textile centre on Polish lands.

After the First World War, and after Poland had regained independence in 1918, Łódź became the capital of a newly established province (*voivodeship*). At the threshold of the 20th century, the population of Łódź increased to over 300,000, reaching nearly 799,000 before 1939. The Second World War significantly decreased the population of two major ethnic groups: Jews and Germans. As a result, the number of the city's population dropped to 250,000 in 1945.

The Second World War also brought a change of Polish borders and Łódź became a central city after 1945. It developed new service functions and became a

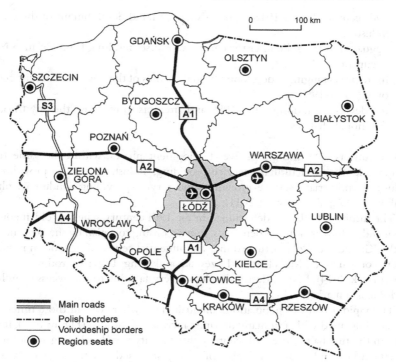

Figure 10.1 Contemporary location of Łódź in Poland, against the provinces and motorway network.

Source: Author's elaboration.

major university centre. However, the traditional textile industry continued to play the main role until the fall of communism in 1989. The post-war population boom and migrations caused a continuous increase in the number of Łódź inhabitants, up to 855,000 in 1988. Towards the end of 2018, the population of Łódź went down to 685,000 due to the negative natural increase rate and negative migration balance (Central Statistical Office, based on the Local Data Bank information).

The textile industry in Łódź always depended on the access to the Russian market, Therefore, during the interwar period as well as after 1989, it declined and currently it is represented by the clothing industry alone, due to the favourable location of the city, convenient transport connections, economic traditions, as well as low prices of land and commercial spaces.

Research results

Before 1989

The development of Łódź has been divided into several stages (Liszewski 2009, 2015). The longest period of Łódź as a small town lasted 400 years (1423–1823). The next 200 years can be divided into four further periods:

- 19th century (up to 1918)—intensive, even rapid development of the textile industry;
- stagnation during the interwar period, war losses resulting from German Nazi occupation;
- more wide-ranging redevelopment in the times of the post-war socialist economy (1945–1989);
- the decline of industry after 1989 and attempts to revitalise the city through functional changes.

Other authors also distinguish five development phases, within the same time framework: the agricultural—service-providing, industrial, service-providing—industrial, industrial—service-providing and service-providing—industrial phase (Suliborski, Wójcik, Kulawiak 2011).

The future of Łódź was determined in the 19th century, when an old, medieval town and a cloth-making settlement, set up according to a plan by the government of the Kingdom of Poland, turned into an influential centre of the cotton industry, often compared to Manchester (Liszewski and Yong 1997; Ogrodowczyk and Marcińczak 2014). Łódź industry developed very fast due to the access to unlimited Russian markets.

The rapid growth of population, unequalled anywhere else in those times, was not accompanied with a similarly dynamic increase of the area (Figure 10.2). It was only in the first half of the 19th century that the city was developing according to an established and clear spatial plan of clearly specified functions for each of its parts. What happened next was a chaotic and spontaneous growth of Łódź (Koter 1979). The number of the population increased from 800 inhabitants in 1821 to 314,000 in 1900, while the city area only doubled. The living conditions in a densely populated city (12,000 people/km^2) were unfavourable; this was exacerbated by the poor infrastructure, e.g. the lack of a water and sewage system. At that time, Łódź became a monofunctional industrial city with very poorly developed services and scarcely represented intelligentsia (Liszewski 2015).

The city was inhabited by four ethnic groups. The most numerous was the group of Poles (mainly factory workers), followed by Germans (engineering staff in factories, many factory owners) and Jews (workers, merchants, industrialists). The least numerous was the group of Russians, who represented the authorities of the Russian Empire (office workers, police, soldiers). The Kingdom of Poland, which included Łódź after the partitions of the country in the 18th century, was a part of the Russian Empire.

The most powerful Łódź industrialists—Karol Scheibler (a German) and Izrael Poznański (a Jew) created huge industrial complexes performing industrial-housing-residential functions. Apart from the factory buildings, they included the industrialist's palace, family houses for the factory workers, a school, hospital, or fire station. The most impressive complex of this type, built by K. Scheibler, covered one-seventh of the city area (Liszewski 2015). Cotton-wool factories belonging to both industrialists, employing 6000 workers each, were among the largest in Europe. There were no industrialists among Poles, because the most affluent of

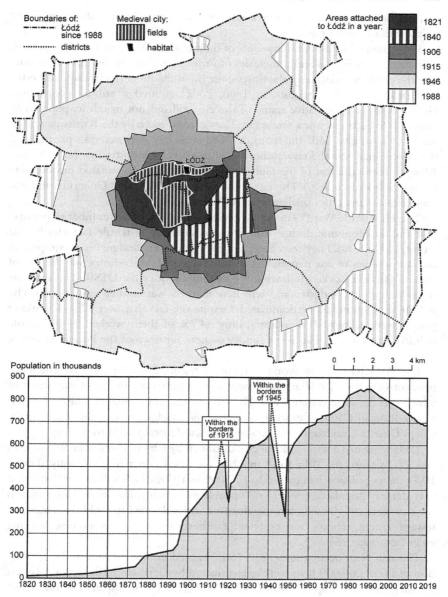

Figure 10.2 Spatial and population development of Łódź.

Source: Author's elaboration.

them (aristocracy, noblemen) were not interested in investing in industry and lived on their vast landed properties.

19ᵗʰ-century historical monuments of industrial architecture as well as palaces, churches, prayer houses and cemeteries of various religious denominations constitute the precious heritage of the times gone by, in the modern cityscape of Łódź.

Thanks to its industrialists, in 1866, Łódź acquired a railway connection. However, the administrative status of the city still did not match its population. From 1876, Łódź was just a *powiat* city. The decision taken by the Russian authorities in 1841 to give Łódź the status of a Gubernia city only raised its prestige and the efforts made by the city authorities supported by industrialists (in 1893 and 1914) to move the seat of the Gubernia from Piotrków Trybunalski proved ineffective (Sobczyński 2000). The city still belonged to Piotrków Gubernia, and was subject to the ten times smaller town of Piotrków.

The First World War (1914–1918) caused a decrease in the number of population and an economic decline due to the devastation of textile factories by the retreating Germans. They took away not only machinery and pieces of equipment, but also reserves of raw materials. A serious blow to Łódź industry was the loss of eastern, Russian markets following the establishment of the USSR. The attempts to restructure the industry and win new markets were rather unsuccessful. The industrial function still predominated. Despite the fact that over 62% of working people were employed in industry, only 34.7% of them worked in the textile industry (Liszewski 2015). Some factory owners, for instance the Poznański family, went bankrupt.

In 1918, after the First World War, Poland regained freedom and became independent of Russia. Łódź acquired new service functions, the most important of which was the administrative one. The city became the capital of a new province, a new Roman Catholic diocese as well as the headquarters of the Polish Army Corps District Command. All the above attracted intelligentsia to the city.

Łódź remained a multinational city, inhabited by 62% of Poles, over 30% of Jews and 7% of Germans. While the area had increased rather insignificantly (Figure 10.2), the population had grown to 672,000 in 1939, which still meant that it was very densely populated (over 11,000 people per 1 km²). The living conditions in the city were still difficult, but the city authorities decided to build the water and sewage system, schools and hospitals. Furthermore, a number of new housing estates also appeared.

German Nazi occupation (1939–1945) caused immense losses in the population of Łódź. 200,000 Jews were exterminated by Germans in the second-largest Jewish ghetto in occupied Poland. Łódź intelligentsia were displaced or murdered, and many Germans (e.g. those coming from the Baltic countries) settled in the city taking their place. The textile industry worked to satisfy the war needs (Liszewski 2015).

Right after the war, Łódź was inhabited by only ca. 250,000 people. The city's Jewish population was exterminated in the Holocaust and the vast group of Germans escaped from the approaching Russian Army in the early 1945, therefore Łódź population had drastically decreased. The city had not suffered substantial material losses, so it was flooded by migrants from Warsaw (capital city of Poland), destroyed during the war, Eastern borderlands (Vilnius, Lviv) and other Polish cities. For a few years, Łódź informally took over the functions of the ruined capital city of Warsaw. The inflowing population settled in numerous former Jewish and German tenement houses. In 1946, the number of population increased to 500,000 and the area of Łódź grew considerably as well (Figure 10.2).

The introduction of a new socialist political-economic system in Poland, in 1945, brought a change from the capitalist system to the centrally planned economy. Private property was abolished and, as a consequence, factories were nationalised. Textile production with goods exported to the traditional eastern markets (USSR) regained its former significance. It was revitalised and attracted 70% of the total number of people employed in Łódź industry (Suliborski et al., 2011). It was modernised and restructured, new, huge factories were built on the city outskirts and seven industrial-warehouse districts were created. Those districts were often built next to housing estates, inhabited by thousands of people. The city authorities invested in technical infrastructure by developing the water and sewage, gas and district heating systems. The living conditions in the city visibly improved. The number of the population continued to grow, reaching 855,000 in 1988, which was the highest number ever.

After the war, the new service functions of Łódź confirmed and at the same time raised its status as the second-largest city in the country. Łódź became a new Polish academic centre, with seven new universities: the University of Łódź, the Technical University of Łódź, Music Academy, Academy of Fine Arts, Film School, Medical Academy and Military Medical Academy. The most famous of them was the Film School. Having no infrastructure of their own, until the 1990s, the universities had been accommodated in various buildings, adapted to research and didactic purposes. Thanks to them, the percentage of people with higher education increased considerably.

In 1975, as a result of the administrative reform of the country, the city became the capital of a greatly reduced Łódź urban province—several times smaller than previously and at the same time the smallest one in Poland. Its range of influence drastically decreased.

Throughout the whole period discussed above, the predominant function of the city was the industrial one. Industry, mostly the textile industry, employed nearly 54% of the population in 1970 and 46% in 1983. However, the earnings were among the lowest in the country (Liszewski, 2015).

After 1989

The year 1989 was a breakthrough in the latest history of Poland and Central Europe, bringing some radical changes. Poland regained full independence (from the Soviet Union) and underwent a political and economic transformation, changing the socialist system to the capitalist system. Unfortunately, privatisation led to the decline of industry in Łódź, followed by the economic crisis. During the period 1989–1997, most large state-owned factories were closed down, or partly sold out, laying off nearly 100,000 employees without any compensation. The unemployment affected 19% of the city inhabitants being at the productive age, and young, educated people migrated from the city as they were unable to find jobs (Liszewski 2015).

Łódź became a demographically old and fast-shrinking city. The number of the population was falling and in 2018, it dropped to only 685,000 inhabitants. Compared to 1988, the city had lost nearly 170,000 people (Table 10.1). Łódź dropped in the hierarchy of the settlement network of Poland to the third position, following Warsaw and Krakow. The decreasing population was first of all connected with a dramatically low natural increase rate. The difference between the index of births and the index of deaths was the lowest in 1999—about −7.6‰. It is predicted that in 2030, Łódź will be inhabited by 606,000 people, and in 2050— by only 484,000 (Dzieciuchowicz 2009).

Compared to other large cities in Poland, Łódź presents considerably worse demographic qualities, such as the fastest-decreasing number of population (Table 10.2), the lowest negative natural increase rate, caused by the low number of births and a very high number of deaths at the same time, resulting from a poor health condition of Łódź inhabitants, strong feminisation, negative migration balance, fast ageing of the society (Dzieciuchowicz 2009; Jakóbczyk-Gryszkiewicz, 2015b).

Persistent unemployment, pauperisation of the society and the decline of industrial plants that followed after 1989 were typical mostly of the central part of the

Table 10.1 Changes in the number of population and demographic structures of Łódź, in 1988–2018

Year	Population (in thousands)	Women (%)	Pre-productive age. (%)	Productive age (%)	Post-productive age (%)	Natural increase per 1000 people	Migration balance No of people
1988	854.3	54.1	22.5	61.0	16.5	−2.9	+1265
1995	816.7	54.3	20.2	61.3	18.5	−6.6	+146
2002	789.3	54.4	16.1	64.8	19.1	−6.2	−743
2011	728.9	54.5	14.1	59.9	26.0	−4.9	−1427
2018	685.3	54.8	14.7	58.0	27.3	−5.5	−1189

Source: Central Statistical Office data, based on Local Data Bank.

Table 10.2 Changes in the number of population in the largest cities of Poland, in 1985–2015, in thousands

Year	Warsaw	Krakow	Łódź	Wrocław	Poznań
1985	1659.4	740.1	847.8	637.2	575.1
1995	1635.1	745.0	823.2	641.9	581.1
2005	1697.5	756.6	767.6	635.9	567.8
2015	1744.5	761.0	718.9	631.1	550.7
Decrease/increase	**+85.1**	**+20.9**	**−128.9**	**−6.1**	**−24.4**

Source: Central Statistical Office data, based on Local Data Bank

city. They resulted in the appearance of urban fallows and poverty enclaves (Marcińczak, 2009). Łódź became comparable to the declining city of Detroit (Ogrodowczyk and Marcińczak 2014).

Employment in Łódź industry rapidly decreased, from 171,000 in 1990, through 91,000 in 1996, to 50,000 in 2007, causing serious unemployment (19% in the 1990s) (Jakóbczyk-Gryszkiewicz 2010). Currently, in 2018, industry employs about 23% of all inhabitants, only one-fifth of whom work in the textile industry.

Another change took place in the ownership structure of industry, which was being gradually privatised. Private companies often took over the post-industrial premises of former state enterprises, which accommodated not only production but also services. Huge state textile factories were replaced with dozens of small, private companies, employing up to a few dozen people. The majority of firms (94.8%) were small, employing up to ten workers (Jewtuchowicz, Suliborski 2009, p.406) and often specializing in clothes making (a traditional Łódź speciality). The entrepreneurs sold their products at huge trading centres, created in the 1990s on the outskirts of Łódź. They had been very successful until Poland joined the European Union, mainly thanks to the wholesale buyers from across our eastern border. Currently, most clients are Polish (Jakóbczyk-Gryszkiewicz 2009).

The most vital economic problem was choosing the direction of changes which could prevent the economic collapse of Łódź. The city authorities commissioned constructing a development strategy with a renowned consulting company, McKinsey & Company. The strategy, which cost the city one million Euro, (mostly paid from the EU funds) continued to focus on the development of industry in Łódź, but not the production of textiles any more, but of home appliances, as well as on Business Process Outsourcing (BPO), developing a logistics centre, and IT (data processing) (Jakóczyk-Gryszkiewicz 2010).

Foreign investors favoured new industries, such as the production of home appliances, computers, shavers, etc. New, world-known brands settled in Łódź, including Gillette, Dell, Indesit, BSH, Infosys, Fujitsu, Ericpol, Ceri, Comarch and Accenture (Jakóbczyk-Gryszkiewicz 2015a).

A symbol of the transformations in Łódź are special economic zones, created in post-industrial areas (brownfields), as well as over non-built-up spaces (greenfields).[1] The Łódź Special Economic Zone, covering 383 ha. in Łódź and other towns of the Province, was created in 1997. It has become the fifth-largest zone in Poland, with regard to size and employment, and it gathers the best opinions.

The economic zone in Łódź consists of nine sub-zones, covering the total area of 100 hectares, mostly in former industrial areas (Marcińczak and Jakóbczyk-Gryszkiewicz 2006). The biggest western investor is Dell—an American company producing computers and planning to eventually employ over 10,000 people.

Since 2004, we have been observing an increase in the value of sold industrial products, due to the large consumption and investment demand. Foreign capital companies make up over 20% of trading companies and 3% of all economic entities in Łódź. The Łódź economy is becoming increasingly competitive and the city is considered to be one of the best-developing cities in Poland (Jewtuchowicz and Suliborski 2009).

Łódź attracts investors with its cheap, but professional workforce. As a major university and research centre, it the city has adequately educated human resources. On the other hand, the tradition of being an industrial city is equally significant. Lower salaries, low prices of land and commercial space, as well as convenient in terms of transport and the close proximity of Warsaw additionally increase the attractiveness of Łódź.

It is worth mentioning the changing significance of the city as regards transport. Despite its central location, Łódź has only recently been included in the system of major roads. Since 2016, there has been a junction of two important European motorways, A1 and A2, near Łódź (Figure 10.1). The situation regarding railway connections is worse. Major train routes skirt Łódź, but the city authorities' investment in connecting the two most important railway stations by means of an underground tunnel will raise its status as a railroad node. Łódź airport, functioning since the interwar period (at that time, it was located outside the city limits), is of minor significance compared to the Warsaw airport, but the A2 motorway considerably shortens the journey to the capital (Figure 10.1).

Due to revitalisation, the historical buildings changed their functions. The 19th-century industrial buildings which used to belong to Łódź industrialists were taken over by modern apartments (lofts at Scheibler's arranged in Księży Młyn—the largest 19th-century cotton-wool spinning factory in Europe, or the Manufaktura shopping-cultural-entertainment centre, on the premises of the former Poznański industrial complex (Figure 10.3)). Manufaktura is becoming a symbol of the city, being an example of a successful combination of the old with the new. The enormous challenge of transforming the dilapidating industrial complex, once owned by I. Poznański, into a modern centre was taken on by a French company, Apsys. Old factory buildings were renovated, with their overall shape preserved and interiors partly changed. Nowadays, they perform a wide range of functions—as museums (a branch of the Modern Art Museum and the Factory Museum), restaurants, entertainment facilities or a hotel. Also universities have their part in the revitalisation process—e.g. the Technical University of Łódź adapts former Łódź factories to suit its purposes (Jakóbczyk-Gryszkiewicz 2010).

Figure 10.3 Examples of revitalising old, 19th-century industrial architecture.
Lofts at Scheibler's and the Manufaktura centre. Photo: Author's photo.

The beginning of the 21st century brought further improvement of the economic situation of Łódź, due to Poland joining the EU in 2004, among other things. In 2007–2013 alone, within the framework of the European Regional Development Fund and the Cohesion Fund, 1418 projects were implemented. The value of the projects put into life in 2014–2020 comes to 4.3 billion PLN (ca. 970 million euros) (Jakóbczyk-Gryszkiewicz 2015a). The local authorities use EU funds to revitalise old, historical tenement houses in the city centre, particularly those standing along the most prestigious street in Łódź—Piotrkowska Street. Creating the New Centre of Łódź is also based on EU funds (Jakóbczyk-Gryszkiewicz 2015a).

The aim of the New Centre of Łódź (NCŁ) programme is to create a new functional centre of the city, connected with the historical centre situated along the Piotrkowska Street axis. The city authorities regard this investment as one of the greatest urban planning challenges in the history of Łódź. The NCŁ involves reconstructing the city over the central area of 100 ha, by the City of Łódź, railway companies and private investors. As a result, a new district of creative industries is emerging. The NCŁ programme encompasses 51 projects, worth 4.4 billion PLN (ca. 1 billion euros) (Jakóbczyk-Gryszkiewicz 2014). The buildings that have been revitalized so far include the oldest Łódź power house from the early 20th century, currently accommodating a culture and art centre (EC1 East) and a science and technology centre (EC1 West), the most modern railway station in this part of Europe (Łódź Fabryczna), as well as several service institutions. At the moment, renowned developers are building modern multi-family houses and there are plans to build a modern market square.

The service function in the city started to gradually displace the industrial one. The changes in the administrative division of Poland in 1999 raised the status of Łódź by restoring the former, larger area of the province. The 1990s also witnessed rapid growth of higher education, when 18 private universities reinforced the seven state universities, which had developed since 1945, and three church universities. The number of students exceeded 100,000. The largest university is still the University of Łódź. Currently, it is also the largest employer in the city. Some private universities ceased to exist after 2010, and the number of students decreased below 80,000.

Next to home appliances, hardware and software production, the industries developing in Łódź include logistics and professional services for business (BPO). The city is turning into a major financial and clearing centre (Jewtuchowicz, Suliborski 2009). The low prices of modern commercial spaces, relatively low earnings and the proximity and good transport connection with Warsaw enhance the attractiveness of Łódź.

Tourism is yet another area which has been recently developing in Łódź in appreciation of the unquestionable value of the 19th century historical monuments—the factories and their owners' palaces and villas. One important asset of Łódź is the large amount of greenery—which is among the biggest among Polish cities. Łódź has 37 parks, a Botanical Garden, a zoo and a vast urban forest, which is unique in Europe.

Łódź is increasingly visited by tourists who are interested in the industrial and cultural heritage of the city. Many of them, mostly Jews and Germans—the descendants of the pre-war community of Łódź, take part in sentimental tourism. Łódź is famous for its unique museums, such as the Museum of Modern Art, the Museum of Cinematography, and the Museum of Textile Industry, as well as interesting international events, such as the Łódź of Four Cultures Festival or numerous theatre, film and ballet festivals (Jakóbczyk-Gryszkiewicz 2010). We should also mention the unique trail of murals, created by world-famous representatives of street art.

The tourist attraction of the city considerably increased after opening the Manufaktura service and shopping centre in 2006, as well as the facilities at the New Centre of Łódź. According to the Brief of Poland ranking, in 2011, Łódź occupied the second position among Polish creative cities, closely following Wrocław. In 2017, the city joined the UNESCO Creative Cities Network, receiving the title of the City of Film. Regrettably, the application filed by the authorities of Łódź to host EXPO 2022 did not manage to go through—we were beaten by Buenos Aires by merely six votes. However, the city was granted the privilege of organising Green EXPO in 2024.

In its publication Best in Travel 2019, the prestigious Lonely Planet tourism magazine pointed to Łódź as the second most interesting place in the world, offering tourist attractions such as the Manufaktura complex, the Łódź Fabryczna Railway Station at the New Centre of Łódź, or a rickshaw ride along Piotrkowska Street—the longest shopping promenade in Europe.

Summary

Over the years, Łódź has gone through periods of intensive growth as well as stagnation or even crisis. At present, it is an example of a post-socialist city, which displays the features mentioned in definitions formulated by Liszewski (2001) and Sýkora (2009). After the fall of communism in 1989, the city went through a deep socio-economic and spatial transformation. The changes included all the elements listed by Matlovič et al. (2001), including the morphological-spatial, functional and demographic structures.

The strengths of Łódź certainly include the traditions of an influential industrial city with qualified workers. The resourceful inhabitants of Łódź managed to cope with high unemployment largely by themselves, setting up small businesses, often involved in the traditional textile production.

Due to the increasing academic function after the First World War, Łódź gained well-educated specialists in a variety of domains, which allowed the city to develop new industries, mainly high technology (Dell) and home appliances production. This particular economic activity is being reinforced by special economic zones, situated in different parts of the city. Creating special economic zones is one of the most important strategies for regenerating the shrinking cities in Poland (Stryjakiewicz 2014).

Another advantage is the rapid development of higher-order functions, related to higher education, business service, logistics and finance. Experts in these domains are attracted to Łódź with the competitive prices of renting modern commercial spaces or apartments.

Łódź is more and more often perceived as an interesting tourist city. The current tourist offer includes historic architecture, unique museums, festivals and cultural events, and interesting tourist trails. A strong point for Łódź is being placed on the UNESCO list as the City of Film, as well as being indicated as one of the most interesting places in the world, from the tourism point of view.

The central location, convenient road connections, also with Warsaw, make way for wide-ranging contacts, encourage entrepreneurs to invest in Łódź and visit the city.

A great chance for the further development of Łódź is an efficient territorial self-government, successfully applying for EU funds and making rational use of them. The authorities' investments include revitalisation programmes, largely concerning the industrial architecture, the construction of the New Centre of Łódź, reconstruction of the city road network. It is also worth mentioning that rail connections are being improved by building a tunnel which will connect the two largest railway stations in Łódź. On the initiative of Łódź authorities, the city has been consistently applying to organise world exhibitions and it has been decided that Green EXPO 2024 will be held in Łódź.

Developing the IT and BPO sectors, creating the financial centre as well as developing tourism in Łódź contribute to the modernisation of the city's economy, which for decades was based on the monoculture of the textile industry.

As regards an improvement of the critical demographic situation, a certain chance lies in encouraging Łódź university students to stay in the city after graduating. The incentives include the interesting work market, the rich cultural offer of the city, and the low cost of living, which substantially compensates for earnings, which are lower than in other large Polish urban centres.

Despite the advantages presented above, Łódź still faces many socio-economic problems. For years, the city's main issue was the underdevelopment of the service sector. It is not long since the situation changed radically.

However, the most problematic issue is the demographic condition of the city. Since 1988, depopulation has caused the loss of almost 170,000 inhabitants, making Łódź the fastest-shrinking large city in Poland. The present situation may still deteriorate, considering the fast increasing number of people at the post-productive age.

Another serious problem in Łódź is the inefficient city transport system, due to the large number of vehicles and traffic congestion. The poor condition of street surfaces, despite continuous repairs, does not help the situation either. The rail and air transport systems are inadequate, considering the status of Łódź and its convenient location in the middle of the country.

The progressing depopulation combined with the ageing of the city pose the greatest threat to the development of the city. According to demographic forecasts, without a substantial migration inflow, the population of the city will fall below 500,000 in the middle of the 21st century.

The problem of urban shrinkage requires the adoption of a proper strategy. Jaroszewska and Sryjakiewicz (2014) divide these strategies into active and passive. The active ones include expansion, i.e. actions which aim at maintaining the population numbers at the same level or increasing them, creating new residential areas and maintaining the existing resources by making the existing urban structures attractive, developing the existing forms of land use and functions, as well as implementing programmes oriented towards selected target groups. A passive strategy means, on the one hand, planned decrease, controlled shrinkage, qualitative development, and—on the other hand—a lack of activity or specific aims, as well as waiting for intervention from outside.

Łódź is an example of implementing an active strategy, which involves maintaining the existing resources, and, at the same time, a passive strategy, which involves control over the shrinking process and qualitative development.

Another problem is the growing debt of the city, resulting from the huge investments made by the city authorities. Even though they are largely funded by the European Union, this does not cover 100% of the expenses. The city must contribute 30–50% to every activity, and because these are sums exceeding its budget, it has to take out loans.

The long dominance of the industrial function caused substantial damage to the natural environment, which the local government are slowly trying to repair. The rapid development of the automobile use since 1990 causes air pollution. After the cities of Upper Silesia, Łódź is the most frequently mentioned with reference to smog alert, especially in the autumn and winter.

New industries, successfully developed in Łódź, are also, in a way, dangerous as regards the economic condition of the city. The home appliances industry means in fact their assembly and not production. Locating factories in the city mostly resulted from the low cost of Łódź workers. It is possible that investors may move their businesses to cheaper countries.

One of the main instruments of the Polish regional policy in the first years following the political and economic system transformation was the creation of Special Economic Zones. The investors who created new jobs were offered special tax relief. The zones helped to restructure old industrial centres. It also concerned Łódź, whose special economic zone is among the best in the country.

Note

1 Special economic zones are administratively separated areas of the country (the first special economic zone was established in 1995 in Poland) where economic activity may be conducted on preferential terms, including tax exemptions.

References

Cudny W. 2012. Socio-Economic Changes in Lodz–Results of Twenty Years of System Transformation. *Geografický časopis*, 64(1), 3–27.

Cudny W., Michalski T., Rouba R. 2012. *Tourism and the transformation of large cities in the post-communist countries of Central and Eastern Europe*. Łódzkie Towarzystwo Naukowe: Wydawnictwo Uniwersytetu Łódzkiego.

Dzieciuchowicz J., 2009. Przemiany ludnościowe Łodzi na przełomie XX i XXI wieku, [w:] red. S. Liszewski, Łódź. Monografia miasta, ŁTN, Łódź, p. 381–400.

Jacobs A. 2013. The Bratislava metropolitan region. *Cities*, 31, 507–514.

Jakóbczyk-Gryszkiewicz J., 2009. Centrum targowo-bazarowe w strefie podmiejskiej Łodzi i jego funkcjonowanie po wejściu Polski do Unii Europejskiej, Czasopismo Geograficzne.

Jakóbczyk-Gryszkiewicz J., 2010. Zmiany w rozmieszczeniu ludności i wybranych struktur demograficznych w Łodzi w latach 1988–2005, [w:] I. Jażdżewska (red.), Duże i średnie miasta Polski w okresie transformacji, Konwersatorium Wiedzy o Mieście 22, Wydawnictwo UŁ, Łódź, p. 211–228.

Jakóbczyk-Gryszkiewicz J., 2014. Nowe Centrum Łodzi – od idei do realizacji, [w:] A. Wolaniuk (red.), Centra i peryferie w okresie transformacji ustrojowej, Konwersatorium Wiedzy o Mieście 27, Wydawnictwo UŁ, Łódź, p. 165–176.

Jakóbczyk-Gryszkiewicz J., 2015a. Fundusze Unii Europejskiej jako czynnik rozwoju miasta, [w:] A. Wolaniuk (red.), Współczesne czynniki i bariery rozwoju miast, Konwersatorium Wiedzy o Mieście 28, Wydawnictwo UŁ, Łódź, p. 119–127.

Jakóbczyk-Gryszkiewicz J., 2015b. Łódź miasto malejące. Porównanie z największymi miastami Polski, [w:] M. Soja, A. Zborowski (red.), Miasto w badaniach geografów, 2, IG i GP UJ, Kraków, p. 91–104.

Jaroszewska E., Sryjakiewicz T., 2014. *Kurczenie się miast w Polsce*, Poznań, Bogucki Wydawnictwo Naukowe, pp. 67–78.

Jewtuchowicz A., Suliborski A., 2009. Gospodarka Łodzi na przełomie XX i XXI wieku, [w:] red. S. Liszewski, Łódź. Monografia miasta, ŁTN, Łódź, p. 400–410.

Kazimierczak J., Szafrańska E. 2019. Demographic and morphological shrinkage of urban neighbourhoods in a post-socialist city: the case of Łódź, Poland. *Geografiska Annaler: Series B, Human Geography*, 101(2), 138–163.

Kołodko G. W. 2005. Introduction: the seven lessons the emerging markets can learn from Poland's transformation. In Kolodko, G. W., ed. *The Polish miracle: lessons for the emerging markets*. Aldershot, Ashgate Publishing Limited, pp. 15–25.

Korec P., Ondoš S. 2006. Súčasné dimenzie sociálno-demografickej priestorovej štruktúry Bratislavy. *Sociológia*, 38(1), 49–82.

Kovács Z., Farkas J.Z., Egedy T., Kondor A.C., Szabó B., Lennert J., Baka D., Kohán B. 2019. Urban sprawl and land conversion in post-socialist cities: The case of metropolitan Budapest. *Cities*, 92, 71–81.

Liszewski S., 2001. Model przemian przestrzeni miejskiej miasta post-socjalistycznego. In: Jażdżewska I. Ed. *XIV Konwersatorium Wiedzy o Mieście, Miasto postsocjalistyczne-organizacja przestrzeni miejskiej i jej przemiany cz. II*, pp. 303–310, Łódź, ŁTN.

Liszewski S., 2015. *Dylematy wielkiego miasta w okresie postsocjalistycznym – przykład Łodzi*, Konwersatorium Wiedzy o Mieście 28, Wydawnictwo UŁ, Łódź, p. 9–23.

Liszewski S. (red.), 2009. Łódź. Monografia miasta, ŁTN, Łódź, p. 500.

Liszewski S., Yong C., 1997. *A Comparative Study of Łódź and Manchester. Geographies of European Cities in Transformation*, Uniwersytet Łódzki, Łódź.

Koter M., 1979. Struktura mofogenetyczna wielkiego miasta na przykładzie Łodzi, Acta Unniversitatis Lodziensis, Zeszty Naukowe UŁ, Nauki Matematyczno-Przyrodnicze, Folia Geographica, ser. II, z. 21, pp. 25–52.

Marcińczak S., 2009. *Przemiany struktury społeczno-przestrzennej Łodzi w latach 1988–2005*, Wydawnictwo UŁ, Łódź.

Marcińczak S., Jakóbczyk-Gryszkiewicz J., 2006. Lokalizacja wewnątrzmiejskich bezpośrednich inwestycji zagranicznych w Łodzi, Przegląd Geograficzny, t. 78, z. 4., pp. 515–536.

Marcińczak S., Gentile M., Stępniak M. 2013. Paradoxes of (post) socialist segregation: Metropolitan sociospatial divisions under socialism and after in Poland. *Urban Geography*, 34(3), 327–352.

Matlovič R., Ira V., Sýkora L., Szczyrba Z. 2001. Procesy transformacyjne struktury przestrzennej miast postkomunistycznych (na przykładzie Pragi, Bratysławy, Ołomuńca oraz Preszowa). In: Jażdżewska I. Ed. *XIV Konwersatorium Wiedzy o Mieście. Miasto postsocjalistyczne–organizacja przestrzeni miejskiej i jej przemiany*, pp. 9–21, ŁTN, Łódź.

Ogrodowczyk A., Marcińczak S., 2014. Łódź – od polskiego Manchesteru do polskiego Detroit? [w:] T. Stryjakiewicz (red.), *Kurczenie się miast w Europie Środkowo-wschodniej, Bogucki Wyd*. Naukowe, Poznań, pp. 79–88.

Parysek J. 2004. The socio-economic and spatial transformation of Polish cities after 1989. *Dela*, 21, 15–32.

Sailer-Fliege U. 1999. Characteristics of post-socialist urban transformation in East Central Europe. *GeoJournal*, 49(1), 7–16.

Smith A., Stenning A., Rochovská A., Świątek D. 2008. The emergence of a working poor: labour markets, neoliberalisation and diverse economies in post-socialist cities, *Antipode*, 40(2), 283–311.

Špačková P., Pospíšilová L., Ouředníček M. 2016. The long-term development of socio-spatial differentiation in socialist and post-socialist Prague. *Czech Sociological Review*, 52(6), 821–860.

Stanilov K. (ed.) 2007. *The Post-Socialist City. Urban Form and Space Transformations in Central and Eastern Europe after Socialism*. Springer, Dordrecht.

Słodczyk J. 2001. *Przestrzeń miasta i jej przeobrażenia*. Uniwersytet Opolski, Opole.

Stryjakiewicz T. (red.), 2014. *Kurczenie się miast w Europie Środkowo-Wschodniej Bogucki Wyd*. Naukowe, Poznań, p.155.

Sobczyński M. 2000. Historia powstania i przemiany administracyjne województwa łódzkiego. [w:]: Województwo łódzkie na tle przemian administracyjnych Polski. W osiemdziesiątą rocznicę utworzenia województwa, PTG, Rządowe Centrum Studiów Strategicznych, Biuro Rozwoju Regionalnego w Łodzi, Łódź, pp. 7–21.

Suliborski A., Wójcik M., Kulawiak A., 2011. Przemiany funkcjonalne Łodzi na przełomie wieków – uwarunkowania systemowe [w:] S. Kaczmarek (red.), *Miasto. Księga jubileuszowa w 70.rocznicę urodzin prof. Stanisława Liszewskiego*, Wydawnictwo UŁ, Łódź, pp. 227–243.

Suliborski A. 2001. *Funkcje i struktura funkcjonalna miast. Studia empiryczno – teoretyczne*. Wydawnictwo UŁ, Łódź.

Sýkora L. 2009. Post socialist cities. In: Kitchin, R., Thrift, N. eds. *International Encyclopedia of Human Geography*, pp. 387–395, Elsevier, Oxford.

Sýkora L., Bouzarovski S. 2012. Multiple transformations: Conceptualising the post-communist urban transition. *Urban Studies*, 49(1), 43–60.

Young C., Kaczmarek S. 2008. The socialist past and postsocialist urban identity in Central and Eastern Europe. The Case of Łódź, Poland. *European Urban and Regional Studies*, 15, 53–70.

11 The socio-economic transformation of the Katowice conurbation in Poland

Robert Krzysztofik

Introduction

The socio-economic transformation begun in Poland in 1989 had a significant impact on development and changes in towns and agglomerations. These changes were highly dynamic due to the speed at which economic reforms were introduced. They were additionally compounded by the consequences of the so-called second demographic transition, already visible in the 1980s.

The transformation of the political and economic system consisting in the transition from socialism to capitalism has become evident in Poland among other phenomena and attributes such as economic and political freedom, privatisation, self-government, personal freedom, the inflow of foreign capital and monetary reforms. In addition to the undoubted social and economic benefits, these long-awaited changes have also had negative consequences.

At least formally unknown phenomena in socialist Poland have appeared, such as unemployment, social exclusion, deurbanisation and urban decay, spatial mismatch or the strongly increasing stratification of society (elitism). In the Katowice conurbation, these phenomena additionally contributed to the problem of reforms in mining (employment overgrowth high rate of employment in mining sector, low hard coal sales prices, environmental devastation) and metallurgy. Due to the political strength of the mining sector, the reforms implemented were socially responsible (although at the cost of underinvesting other industries in need of social protection—the machinery industry and, above all, the cotton and clothing industry).

The reaction to these changes was undoubtedly "transformation shock", which later resulted in varying degrees of adaptation to the completely new challenges that appeared with respect to urban policy, economic development and social transformation (i.e. Birch et al. 2010; Neffke et al. 2011; Hassink 2010; The City After Abandonment… 2013). The urban region in which these differences became particularly apparent is the Katowice conurbation in southern Poland.

The aim of the chapter is to explain the essence of the development of one of the largest urban agglomerations in Europe in several key aspects. First, the background for economic and functional changes in the last 30 years has been pointed out. In turn, these issues constituted one of the most important drivers and effects of the observed social and population changes in the region. An important part of the chapter is devoted to issues surrounding the settlement structure of the

DOI: 10.4324/9781003039792-11

conurbation, the specificity of the regional governance and, particularly important from the point of view of sustainable development, the human–environment relationship. All the mentioned aspects of changes are presented in a narrative that exposes the most important cause-and-effect relationships.

The administratively and partly morphologically polycentric Katowice conurbation has a surface of 3,329 km² (Figure 11.1). This area of around 2.4 million

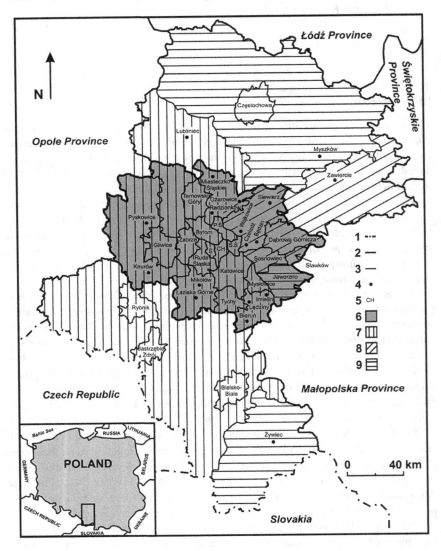

Figure 11.1 Katowice conurbation on the background of region.

Explanations: 1—borders of countries; 2—boundary of province; 3—boundaries of counties, including urban counties; 4—other towns; 5—city names abbreviations; 6—Katowice conurbation (CH—Chorzów, P.Ś.—Piekary Śląskie, S.Ś.—Siemianowice Śląskie; Ś—Świętochłowice); 7—historical region of The Upper Silesia; 8—historical region of The Zagłębie Dąbrowskie (The Dąbrowa Basin); 9—the eastern part of historical region of The Small Poland. Source: Own elaboration.

inhabitants (2018) and 54 communes developed mainly around coal mining and other traditional industry branches (Spórna et al. 2016). The conurbation core comprises 17 large and medium-sized towns populated by about 1.6 million inhabitants. In demographic terms, the Katowice conurbation is the second-largest urban region in Central Europe (after Berlin). The region's peculiar spatial distribution, in which several towns play a dominant function, means that the largest and most important of them, Katowice, has only 280,000 inhabitants. Meanwhile, Sosnowiec, Gliwice, Zabrze and Bytom have between 150,000 and 200,000 inhabitants (Spórna et al. 2016).

The Katowice conurbation is divided into two zones—the core is inhabited by 1.8 million, while the peripheral zone is inhabited by 0.6 million people. The region's core is metropolitan and urban in character, while the peripheral zone comprises mainly small towns, suburban settlements and even still partly traditional rural villages (Runge and Kłosowski 2011; Spórna 201-8).

The peculiar spatial distribution of the conurbation in the form of highly urbanised area results from the fact that it is an agglomeration of many large and medium-sized towns that were formed by the earlier integration of smaller settlements and towns, which mostly grew around large industrial plants and mines. This characteristic polycentric distribution of the Katowice conurbation[1] and its mining and industrial origins actually had a fundamental significance in the region's recent transformation process. It was these which determined the features of certain socio-economic and spatial phenomena that in Central Europe were limited only to this region. The most characteristic of these include: urban policy for one urban region that is drafted in dozens of municipal offices; existence of towns which have lost their exogenous functions; dramatic depopulation, small role of urban sprawl and spectacularly feature—internal suburbanisation (Krzysztofik et al. 2012, 2016, 2017; Spórna 2018; Runge et al. 2018).

Katowice conurbation—formation of strong monofunctionality

The shaping of the Katowice conurbation as an urban region started at the turn of the 18th and 19th centuries. It was politically divided at the time. Its western and central part (Górny Śląsk—Upper Silesia) located in historical Silesia was located in Prussia, while its eastern part (Zagłębie Dąbrowskie—the Dąbrowa Basin and surroundings of Jaworzno) was divided between Russia and Austria-Hungary. While Silesia belonged to Prussia from the mid-18th century, Russia and Austro-Hungary occupied this part of Poland from the first half of the 19th century (at the turn of the 18th and 19th centuries—Prussia and Austro-Hungary). Within today's conurbation (between Sosnowiec and Mysłowice) was located the famous in Europe tri-border of three great empires—Prussia, Russia and Austria-Hungary (so-called Drei-Kaiserreich-Ecke).

Irrespective of state affiliation, in the 19th century the area became one of Central Europe's largest mining and industrial centres. It is, among others, towards this region that the first railway lines from Berlin, Vienna and Saint Petersburg were built. This was when dozens of new settlements located around mines, smelters and other large industrial plants then grew in the vicinity of several small mediaeval towns (Pound 1958; Gwosdz 2004; Mihaylov et al. 2019). At the start of the 20th century the region was already a relatively compact urban space, and individual settlements started merging to form administratively integrated towns. In over 90% of settlements, the economic base of the settlement or town was mining, industry and railway transport. The few service towns, with the exception of Katowice, Gliwice, Sosnowiec and partly Bytom, played a secondary role. In view of the mining and industrial landscape of the dynamically growing conurbation, Katowice and Sosnowiec had unusual origins. These two quite new twin towns emerged as important border centres (the region was divided between Germany, Russia and Austria-Hungary at the time), whose development was based on rail transport, shipping, wholesale trade, administrative functions in industry and services directly related to the existence of political borders (Gwosdz 2004; Murzyn-Kupisz and Gwosdz 2014). The specific nature of these towns quite strongly calls to mind the gateway city model by Burghardt (1971). Apart from these two cities, such a role in the region was also played by Mysłowice, Szczakowa and Granica (Dragan 2016).

In the inter-war period (1918–1939), the region in question was divided between Poland (central and eastern part) and Germany (western part). The most tangible effect of this political division of the region before the Second World War was the development of Katowice as a regional administrative centre—the seat of the Silesian province. Although the rivalry between Katowice and Chorzów in Poland and Gliwice and Bytom in Germany reinforced the entrenched polycentricity, it also had one positive aspect. Namely, it fostered a kind of international rivalry, which was reflected in monumental architectural and urban projects and the development of new non-industrial functions. Urban heritage and administrative, cultural and service functions served in this period by some towns, particularly Katowice and Gliwice, became after The Second World War a significant supplementary attribute of growth. After the socio-economic transformation of 1989, meanwhile, they were an important factor of economic success and contributed to these towns' status as the two greatest winners of the transformation period. Another perspective indicates that for Katowice and Gliwice the first half of the 20th century was an important stage in establishing a positive, self-confirming path dependency (Gwosdz 2004).

The region's strong specialisation in mining and industry developed after the Second World War. The whole area of what is now the Katowice conurbation became part of Communist Poland (The Polish People's Republic).[2] After 1945, Poland's Communist authorities deemed the region to be a key centre of Poland's economic development, which provided the two most important export goods: coal and steel. The two-stage growth of the region's economic potential: 1950s to the start of the 1960s and the 1970s to the start of the 1980s, caused rapid development of industrial functions, population and new high-rise building.

In 1990, the conurbation was inhabited by 2.8 million, and this was its highest-ever population (Table 11.1). However, the development of services, communal facilities, transport and higher-order services did not meet the pace of these priorities. As it turned out at the end of the 20th century, these contrasts were significant for the model of the region's socio-economic transformation. In the 1970s and 1980s, along with the expansion of the mining and metallurgy sectors and the

Table 11.1 Population in the area of the Katowice
Conurbation, 1850–2018

Year	Number of population (in mln.)
1950	1.7
1960	2.0
1970	2.4
1980	2.7
1990	2.8
2000	2.6
2010	2.5
2018	2.4
2025	2.3
2030	2.2
2040	2.0

Source: Author elaboration on the basis of Spórna (2012).

cotton industry already functioning there in the 19th century, several other large industrial plants of other sectors: electrical engineering, automotive and food, were also built. The development of industry, the region's strong monofunctionality and lack of a clearly distinct large city caused the area in question to be categorised as an industrial and urban region rather than a "typical" urban region. In this context, its name at the time, the Upper Silesian Industrial Region (Górnośląski Okręg Przemysłowy) was symbolical. Emphasis on the economical factor, which over-shadowed the urban element in the region's name and was de facto visible in central policy towards the conurbation in question at the time became, 20–30 years later, one of the barriers to leaving behind transformation trauma for many towns located there (Riley and Tkocz 1999; Tkocz 2001; Gwosdz 2004).

1989's political and economic crisis and its consequences

Political and economic crisis in the face of traditional economy

The consequences of the transformation of Poland's political system that came after 1989 were the most prominent in monofunctional and industrial towns and regions, and those where higher-order services were relatively weak. In Poland, the Katowice conurbation contained the highest number of such towns and was itself Poland's largest industrial region. Naturally, this was exactly the region in which the momentum of the changes would cause transformation shock, especially as the political and economic changes were compounded by two other factors. First, from the mid-1980s in some towns (i.e. Chorzów) there appeared symptoms of a group of phenomena defined within the concept of the second demographic transition. Among others, there was a fall in the number of births, in inhabitant numbers, and in the number of persons of pre-working age. The emergence of unemployment additionally increased trauma related to the transformation, and this was exacerbated during each succeeding year of the 1990s. This factor also significantly contributed to the deterioration of demographic indicators. Until the mid-2000s, each year around 10 large or average industrial plants employing 300–5,000 each in the region were closed or underwent deep restructuring. For example, in Sosnowiec three large textile and cotton mills were closed. Over several years, around 6000 people, mostly low-educated women, lost their jobs (Tkocz 2001; Runge et al. 2003; Krzysztofik et al. 2012). In the Dąbrowa Basin, the historical eastern part of the Katowice conurbation, all nine coal mines closed. In Bytom, of the six mines in operation in 1988 only one remains (2018). The unemployment rate rose to over 20%, and it exceeded 30% in some districts. The reduction in jobs was also connected to mine closures due to the depletion of the fields. Global fluctuations in coal and steel prices also played a part.

The existing hard coal mining restructuring programmes (implemented since the 1990s) have set themselves the basic objectives of adapting mining to market economy conditions and international competition. This policy consisted in closing down some mines and limiting employment in others. At the same time, efforts were made to increase the profitability of mines by increasing production and limiting the other activities of mining plants (Tkocz 2001; Fornalczyk et al. 2008). The effect of this was the liquidation of mines in many cities and traditional mining regions (Wałbrzych Basin, Dąbrowa Basin or city of Siemianowice Śląskie, city of Świętochłowice and others). The severe coal mining restructuring programme in Poland was similar to that which took place in Germany (Ruhrregion) and was based on the assumptions of the transition in social justice. This model is continued today. Plans for the neoliberal treatment of the mining sector (as in Great Britain) formulated by some authorities have never gained an advantage in Poland. To a large extent, this is due to the degree of coal dependence on the Polish energy sector.

It should be emphasised, however, that mining restructuring programmes, mainly due to the lack of consistency in their implementation, did not lead to the achievement of the basic goal of achieving profitability across the entire industry. The factor influencing this fact is also increasing imports of hard coal from other countries.

Around 2010, in the traditional, heretofore large mining centre of Chorzów, more people were employed in agriculture than in mining. In another large town, Dąbrowa Górnicza, whose very name contains an adjective indicating its origins and function (górnicza—mining), the last coal mining facility closed in 1996 (Domański 2002; Kłosowski et al. 2013; Krzysztofik et al. 2016). The presented examples of closures of the previous economic base and functional changes caused the region to enter deep regression. Closure and restructuring of industry caused a rapid growth of brownfields, which were often located between residential estates. Brownfield spaces, bad for the towns' images, were additionally a visible "monument" to the consequences of transformation. The closure of industrial facilities also caused taxes paid to the local governments to fall drastically. These restrictions caused the decapitalisation of communal, transport and residential infrastructure to deepen. The negatively perceived brownfield landscape came to surround decapitalised residential estates and urban infrastructure. Towns' budgetary problems also became, in some cases, the cause of territorial disintegration. Residents of certain districts of towns such as Tychy, Bytom, Będzin and Mysłowice decided in referenda to secede in the hope that it would be easier to survive the functional crisis as part of smaller towns (Krzysztofik et al. 2012; Rink et al. 2014). From the 1990s, both the conurbation as a whole and individual towns started to experience urban shrinkage (Kantor-Pietraga, 2014; Krzysztofik et al. 2017). For some (e.g. Gliwice) it was only an episode; in others (e.g. Bytom, Świętochłowice) it is a long-term state of affairs. Apart from closure of industrial plants, demographic decline was the most characteristic. Its speed and magnitude were and are different in individual towns; however, most of them have already lost 20% of residents, and some (including Bytom, Katowice and Sosnowiec) even 30%. As demographic forecasts indicate (*Prognoza ludności...* 2014), this process will not slow down in any of them, and in some it will increase further. It is estimated that by 2050 most of the region's towns will lose even over 60% of their inhabitants as compared to 1989.

Regional crisis into the agenda setting

The mounting problems of the conurbation, but also of the province as a whole, were rapidly noticed both on the country and the regional forum. However, two key barriers to overcoming these problems appeared. First, the means for preventing all the consequences of the transformation were dramatically short. Second, legal and administrative conditions did not keep up with the changes. The rapidity of some phenomena often caused administrative action to be paralysed. Against this background, the first post-1989 regional agreement in Poland that drew attention to the scale and variety of problems in the conurbation and the province, the Regional Contract for the Katowice province of 1998 (Kontrakt... 1999) was a great success. The participants of this agreement included province authorities, urban authorities, trade unions, experts, representatives of civil associations, business and large enterprises. It was undoubtedly one of the most important and earliest acts of regional governance in that period not only in Poland, but also in Central Europe. The most important achievement of the Contract was indicating a pathway for the region's re-industrialisation. Re-industrialisation was very important for at least reasons in the analyzed region. First of all, it was difficult to change the profession (especially towards new services) of hundreds of thousands of people working in the mining or industry to date. Secondly, in creating new jobs in services, this region was relatively less competitive than other large Polish metropolises. It only began to change in 2010.

Re-industrialisation and the Special Economic Zone as a way to solve different problems

A tangible effect of setting a direction for re-industrialisation, meanwhile, was the creation of the Katowice Special Economic Zone (KSEZ). The Katowice Special Economic Zone became (since 1996) a fundamental initiator of development of new industry branches and an active actor on the road to obtaining new investments. KSEZ is one of 15 special economic zones in Poland. The idea of this enterprise is based on a specific public–private partnership, in which new investments in selected areas can count on support in the field of promotion, management, infrastructure facilities and tax exemptions. Although since 2016 similar legal and tax solutions apply to the entire country, special economic zones are still one of the key instruments of the investment policy of the state and regions.

As early as the start of the 2000s, 4 subzones of KSEZ were formed in the conurbation, and over 50 companies were located in over a dozen investment areas. The dynamics of new investments grew particularly after General Motors (Opel) started operations in Gliwice in 1998. In addition to having the tangible effect of increasing employment in the region, KSEZ was also an inspiration for many local governments in obtaining investors who wanted to operate outside KSEZ structures. For companies, meanwhile, it was a guarantor of suitability of the location selected for business activity and good cooperation with local government. From the early 2000s, some economic recovery became visible in the conurbation's towns. New companies started appearing also in small and medium-sized towns (including Piekary Śląskie, Siemianowice Śląskie, Czeladź, Siewierz and Bieruń). The number of new jobs, however, still did not eliminate the 10–20% unemployment rate in the region's towns. In this period, a policy for recultivating brownfields as the most anticipated investment areas in the conurbation was also worked out. In fact, most new companies started selecting brownfields for their location. Brownfields also rapidly became key areas for other investments: service, communal and leisure (Krzysztofik et al. 2016). Unfortunately, successes in attracting new investors and even minimising unemployment in some cities (Katowice, Gliwice) did not cause depopulation to slow. The process continued. Facts such as the decrease in the number of Bytom's inhabitants by 11,000 from 170,000 to 159,000 over 4 years (2009–2013) or the annual loss of 3–4000 inhabitants in Sosnowiec (200,000 inhabitants) indicate that the challenges for urban policy are still valid, and the processes disintegrating towns' social and demographic structures are still very strong. Paradoxically, a significant demographic loss was recorded even in those towns in the region (such as Katowice, Gliwice) which had a very well-developed labour market with a very low unemployment rate, significant surplus of incoming commuters over outgoing commuters, a very diverse manufacturing and service sector and also a broad offer of leisure options. Another paradox was that depopulation, pervasive in the conurbation's towns, was accompanied by the lowest average prices of residential real estate in Poland. They were four or even five times lower than those in Warsaw or Cracow.

Social dimension of changes

The evident social polarisation also became a great problem for the conurbation. Due to the conurbation's complicated polycentric system of settlements, social diversification had multiple nuclei, almost perfectly described by Harris Ullman's now-classic model. Zones of social exclusion in the region are clearly linked to older workers' housing workers and inner-city tenement houses (Lokalny wymiar rozwoju społecznego Polski, 2018). In many towns, especially mid-sized ones, a centripetal model of change in ground rent level practically does not function. Buildings, and especially flats, in the centres of these towns are the cheapest, and their prices grow further away from the centre. This phenomenon had at least two causes. Namely, industrialisation caused a strong expansion of workers' housing in traditional town centres and along main streets. In the socialist period, because of the enlargement of large residential estates for workers, old buildings in town centres were settled largely by the socially excluded. Many buildings were also under municipal administration as property without established ownership. In the Silesian part of the conurbation this was usually related to migration of the German or Silesian minority to Germany; in the eastern part of the region, the Dąbrowa Basin, the properties in question had been owned by Jews before the War.[3] Zones of old, decapitalised, often-substandard buildings in the centres of the conurbation's towns are an area traditionally concentrating social problems and exclusion. Areas of social exclusion in the Katowice conurbation were also located in former workers' estates. While town centres became such zones already in the socialist period, former workers' estates started turning into them mainly from the 1990s. The most important reason for this state of affairs was severance of traditional ties between the local community and the workplace, which was closed or restructured. Lethargy, unemployment, alcoholism, and lack of perspectives in the communities of these neighbourhoods caused most of them to shortly become identified as the most dangerous and repulsive parts of towns. Revitalisation projects, put in place particularly after 2010, could not completely change the image of these settlements. Nikiszowiec in Katowice and the Ficinus estate in Ruda Śląska are some of the few cases in which it was possible to transform the region's traditional heritage into a tourist attraction, and where gentrification is even being observed in the estates themselves (Lamparska 2013). After 2010, the process of social exclusion also started to affect some block-of-flat estates from the 1950s and 1960s. The reason for this is the permanent depopulation of towns, which causes the competitiveness of this type of buildings to fall compared to block housing estates from the 1970s and 1980s. These buildings are more attractive to young families with children than buildings from the 1950s.

Metropolisation of conurbations as an attribute of the expected direction of changes

The development of the Katowice conurbation, and also counteracting the negative phenomena affecting it, was also the objective of Upper Silesian Metropolitan Union (Górnośląski Związek Metropolitalny, GZM) formed in 2007. The union encompassed only 15 towns that held the status of urban county. Three towns later resigned from participation in the endeavour for various reasons. Towns of GZM were located in the strict conurbation core, but several other towns in the core could not due to their formal status (they were not urban counties) participate in the Union. This was one of the disadvantages of this metropolitan union, causing numerous obstacles and barriers to sustainable and dynamic development. The limited territorial scope and growth of social and institutional expectations as to the role of the institution organising the region's metropolitan functions led to the foundation of the Upper Silesian-Basin Metropolis (Górnośląsko-Zagłębiowska Metropolia) in 2017 (Zuzańska-Żyśko 2016). This institution, thanks to a special government regulation and partial renouncement of competencies by the communal self-government, gained a key role in the shaping of local policy in the conurbation. As its priority the Upper Silesian-Basin Metropolis set issues related to the organisation of mass transport, spatial planning, environmental protection, and promotion and marketing of the region. From the perspective of the region's development, the principal objective of GZM is, however, increasing the conurbation's competitiveness not only compared to large Polish metropolises, but also European ones. GZM's functioning, and particularly its effectiveness up to now, is variously assessed, with criticism being directed mainly towards excessive focus on performance of endogenous and local tasks instead of exogenous, superregional tasks and those increasing its competitiveness. Emphasis is also placed on the "genetic" disadvantage of the metropolis, i.e. its polycentricity. Administrative polycentricity contributes to discrepancies with respect to GZM's strategic purposes, former historical and cultural divisions that are becoming apparent, doubts as to the justifiability of also including rural communes, inter-urban competition for investments or metropolitan-level institutions. Irrespective of the current assessment of the "new metropolis'" activity, attention should be drawn to the undeniable legitimacy of its creation and its coordination of much pro-developmental activity needed by the metropolis (Zuzańska-Żyśko 2016).

The Katowice conurbation after 30 years of transformation— what next?

Modelling the economic transformation process

After 30 years of transformation the Katowice conurbation is at a fairly significant point in its development. On the one hand, processes started after 1989 have not yet come to a close; on the other, quite new challenges caused by scientific and technological advances, globalisation, European Union policy and changes in the residents' social and individual needs are appearing. Mining and heavy industry are still important elements shaping economic and functional identity. These sectors dominate as the most important in the exogenous group in towns including: Ruda Śląska, Bytom, Jaworzno, Piekary Śląskie, Mysłowice, Knurów, Łaziska Górne, Bieruń, Lędziny, Miasteczko Śląskie and others. In many others, e.g., Dąbrowa Górnicza, Zabrze and Sosnowiec, the potential of old and new industry is equal. In many towns traditional industry has been completely dismantled (Figure 11.2) or is in decline (Klasik and Heffner 2001). These include towns of varying size that have managed to replace traditional industry with new or modernized sectors (including Gliwice, Tychy, Czeladź and Radzionków) and those that are faced with adaptation problems of the new sectors (such as Świętochłowice and Wojkowice). Completely new economic centres are also appearing, having been pulled out of their previous economic lethargy by the transformation (e.g. Siewierz, Ożarowice). Towns which concentrate dynamic new investments in modern BPO (Business Process Outsourcing), R&D (Research & Development) and SSC services (Shared Services Centre) should also be mentioned. Katowice is definitely the leader in this respect, but many companies are also located in

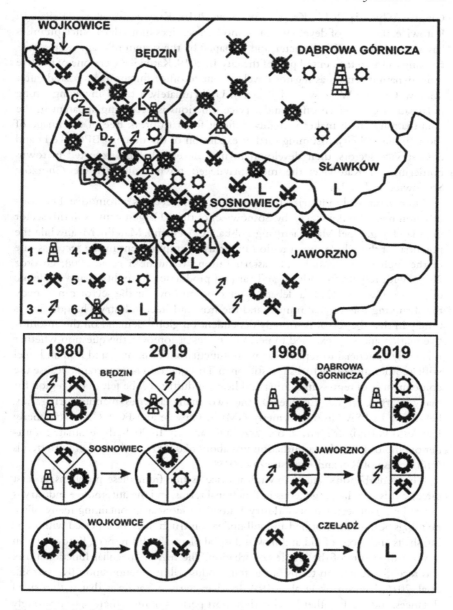

Figure 11.2 Models of economic transformation in towns of the eastern part of Katowice conurbation, 1980–2019. (a) localisation of the biggest industrial plants and coal mines, (b) changes of structure, 1980–2019.

Explanations: 1—existing steel works; 2—existing coal mines; 3—power stations; 4—existing machinery, electro-technical or chemical works; 5—closed coal mines; 6—closed or strongly restructured steel works; 7—closed or strongly restructured machinery, electro-technical or chemical works; 8—modern industry after 1990 (automotive, electro-technical or chemical plants); 9—huge logistics complexes. Source: Own elaboration.

Gliwice (Gwosdz 2014; Krzysztofik et al. 2016; Krzysztofik et al. 2019a). In Katowice, the zone of development of modern services in modern office premises has a concentric, radial character, and is shaped by the system of main expressways running through the central part of the city. In 2019, Katowice were one of the five most dynamic centres as regards development of office functions in Poland (after Warsaw, Cracow, Wrocław and Poznań). Unfortunately, a trend involving strong polarisation of modern office and service functions is becoming apparent in the conurbation, which makes it highly difficult for it to break free of the image of (post-)industrial city. Assuming further expansion of the modern BPO, R&D and SSC service sectors, their development may nevertheless be expected in towns bordering on Katowice; the most privileged towns will then be Chorzów, Sosnowiec and Tychy.

After traditional industries, the most important sector is automotive. The conurbation hosts 3 (Gliwice, Tychy, Sosnowiec) of the 10 largest centres in this sector in Poland (e.g. Opel Manufacturing Polska, FCA Poland, Marelli). Meanwhile the potential of the whole urban region in terms of employment in the sector is one of the highest in Central and Eastern Europe (Domański et al. 2008; Gwosdz 2014), with respect to both car and car part production. In political and academic discourse (Gwosdz 2014) a debate is ongoing on how far the sector can remedy the shrinking role of coal mining and the iron and steel industry. The question is timely in that the sector is strongly dependent on global demand on the automotive market and thus exposed to periodic regress. Moreover, the question whether it is really beneficial to replace the monofunctualism of mining and iron and steel with another monofunctualism is still open. This question is important because the consequences of regress in mining and heavy industry will be felt in the region for many years to come. Changes in the two main factories manufacturing cars, belonging to PSA Group (formerly GM) in Gliwice and FCA Group (formerly Fiat Auto Poland) in Tychy, also always cause anxiety. In 2019, the estimated number of employees in the conurbation was about 650,000 and showed growth trends (on the base of Rocznik Statystyczny 2019).

The conurbation's largest town, Katowice, is free from these problems, at least directly. It is very heterogeneous in functional terms, and the automotive industry is not a dominant sector of its industry. Katowice's cityscape, containing many office buildings both newly finished and still under construction, is interspersed with mining shafts and shops of old and new industrial plants. In this respect, Katowice is in a way a miniature of the whole conurbation. The region's peculiar functional mix was recently defined in terms of the trans-industrialism phenomenon (Krzysztofik et al. 2016; Krzysztofik et al. 2019a). Its characteristic feature is that within small distances, and on the other hand within short periods of time, there appear strongly pronounced changes in economic structure. In this region "old industry", new industry and re-industrialisation, modern services as well as post-industrialism shift fluidly over time and space. Despite numerous successful investments new industrial plants are not being built everywhere, modern services take root in few places, and in many places traditional industry has (theoretically) very good perspectives of development. In 2019 it cannot be said of the region that it is at any particular point of the transformations defined for example in Phelps and Ozawa's (2003) typology.

Another well-developing group of towns in the urban region in question are those whose development is based on the power sector or economically profitable coal resources. From the perspective of sustainable development this fact is dissonant (Runge et al. 2018). However, it should be noted that these settlements are characterised by a good economic situation, relatively less visible demographic problems and a very legible investment policy with respect to communal, transport and leisure projects. The leader in taking advantage of financial assets derived from the presence of traditional industry within town limits and a creative policy for spending them is undoubtedly Jaworzno.

A functional changes in the conurbation's economy may, however, soon occur as a result of mounting pressure by the European Union to restrict fossil fuel usage, and also the imposed limits of CO_2 emission. The consequences of this policy may already be seen in the iron and steel industry, which in 2019 was on the brink with respect to further operation. Because of the EU and UN environmental and health protection policy, it is also quite likely that coal mining will at least partly and prematurely (taking into account the size of economically viable deposits) be reduced in the coming years. This scenario, particularly of a rapid and premature closure of coal mining in a situation where the situation of the iron and steel industry is uncertain, would most probably result in the deterioration of the region's economic situation. This uncertainty is additionally aggravated by the slowdown in large and medium-sized productive investments in the region. This fall is not counterbalanced by the dynamic influx of logistics investments (Krzysztofik et al. 2019b). As mentioned, the situation of the local automotive industry may, in a longer perspective, also be quite unstable. The remedy to problems in the industrial sector is the development of services. While a distinct inflow of investments of this type is being observed, it is very uneven. Employment reductions in basic services and commerce, also large-surface, are another problem. In this case the problem is ongoing depopulation, and partly also the increase in the share of persons of post-working age, who have comparatively smaller financial resources. In general, the increase in town residents' purchasing power in recent years is only a partial compensation in view of this sector's problems.

Socio-demographic transition

Changes in the age structure of the population in the region in question and its ongoing depopulation provide still further challenges. Namely, around 2014–2018 in the Katowice region the unemployment problem was replaced by a worker shortage. The lack of workers particularly concerns the manufacturing sector and some basic services and commerce. Opening the labour market to workers from outside the European Union, mainly from Ukraine, Belarus and Far Eastern countries, was to be a solution to the problem. Immigration did, in fact, partly assuage these shortages, but they still seriously restrict business development in the region. Immigration of workers from other countries to the Katowice conurbation's towns revealed a rather strange phenomenon, which may somewhat indicate that the traditional industries are still highly important. That is, immigration from other countries to individual towns in the region is inversely proportional to the share of employed in traditional industries in these towns (Krzysztofik et al. 2019c). The explanation for this phenomenon is high salaries in the coal, fuel and iron and steel sectors, often 50% higher than those offered by the manufacturing sector. Thus, demand for jobs in these companies is high. Furthermore, trade unions hold a strong position in large plants in traditional industries, safeguarding the interests of their members—current employees. Trade unions also support employing workers directly by the large industrial plant, and not via outsourcing by other intermediaries, which offer lower wages—satisfactory for economic migrants but not for local Polish workers.

Poland's unsatisfactory immigration policy will be most heavily felt primarily in depopulating regions such as the Katowice conurbation. Particularly if hard data are taken into account: according to Statistics Poland's forecasts and local forecasts for local governments, it may namely be expected that the fall in population numbers will lead to the region in question having around 1.0–1.2 million inhabitants. In the years 2015–2018 the Katowice conurbation lost around 30,000 inhabitants annually, and due to a permanently low birth rate (around 1.4) and its specific age structure (low share of pre-working age population and rising share of post-working-age population) the magnitude of the loss will stay stable or even grow in some sub-periods.

The region's growing depopulation impacts and will impact its spatial structure, and most of all its residential and communal infrastructure. Already now in some block estates the average occupancy of flats is 1.2–1.5 and this indicator is deteriorating. Furthermore, around 30% of flats in block estates from the 1970s and 1980s are inhabited exclusively by persons aged 70+ (own research partly published in: Krzysztofik et al. 2019c). These conversions will aim to adjust building numbers and size to real population and household structure, which will involve partial demolitions and reduction of upper storeys. There are also great challenges facing the local government with respect to improving the flexibility of the school and kindergarten network. For example, 5 of Sosnowiec's 40 schools have already closed due to depopulation; a further 7 are endangered. Currently, an attempt is being made to avert closure of some schools with a policy to reduce pupil and kindergartener numbers in groups. The region's depopulation is largely caused by a low birth rate, but also by outmigration exceeding immigration. Outmigration, particularly of young inhabitants, has two causes. First, it is the result of difficulties in finding satisfactory jobs in the conurbation. Additionally, the negative image of the region as post-industrial and the so-called "scattered metropolis" phenomenon are not negligible. Metropolitan functions, facilities and spaces are located in many towns of the conurbation, often divided by tens of kilometres. This also pertains to tourist attractions and leisure space for inhabitants. There is no synergy effect between them, which in large monocentric metropolises creates a certain genius loci and city identity. Thus, migration to other Polish towns is conspicuous, as is the greatest share of international migration in Poland (Krzysztofik et al. 2019c). The large share of foreign migration is contrasted with limited suburbanisation, meanwhile. It is so small that in the zone of its influence a demographic decline has been recorded for over a dozen years. This is the only such case in Central European metropolises (Krzysztofik et al. 2017). The uniqueness of the Katowice Central

Europe conurbation in relation to the phenomenon of depopulation and partly urban shrinkage lies in the fact that it is the most populated urban region in this part of the continent. The quantitative scale of observed phenomena and the resulting socio-economic consequences are particularly visible.

Spatial patterns

Another characteristic feature of the Katowice conurbation related to its inhabitants' place of residence is so-called inner suburbanisation (Spórna and Krzysztofik 2020). This phenomenon has interesting origins dating back to the industrial and socialist period. Namely, the polycentric nature of the region prevented some districts, larger estates and towns from becoming a fully unified urbanised area. These estates grew over the years but at some point in their history they stopped developing, which resulted in non-built-up areas in their close proximity. Paradoxically, most of them were not connected to mining, but to mostly individual, and also state agriculture (in the socialist period). At the turn of the 20th and 21st centuries these areas, lying in the geometric centre of a given town or the conurbation as a whole, became an extraordinarily attractive area for residential and service and residential investment. In the environs of high-rise buildings from the socialist and earlier period, characteristic enclaves of low-rise buildings formed, typical of suburbs located around monocentric metropolises. Their surface is sometimes up to 2 km², and as studies of their populations show, their communities settled there for all the reasons that residents move to "classic" suburbs (Sosnowiec-Kukułki, Siemianowice Śląskie—Bytków, Zabrze—Dwór, Katowice—Podlesie). Because in the Katowice conurbation chimney stacks of large industrial plants and mine towers are visible also from areas outside the boundaries of large cities, internal suburbanisation wins in the rivalry by offering a significantly shorter commute to work and school instead.

The described phenomenon is also partially a remedy for depopulation of the town's regions. On a regional scale its role in slowing the demographic decline is small, however. This is because of the limited number of this type of zones and the concentration of a comparatively small population. Independently of low-rise buildings, high-rise construction investments are in progress. In many cases reservations appear concerning simultaneous building of new block estates in the close neighbourhood of dramatically depopulating and ageing older estates from the socialist period. Despite unfavourable demographic phenomena, the latter are being modernised and insulated. Within 10–20 years, some of these buildings will become superfluous. Additionally it should be mentioned that, as may be seen from studies of younger generations, families with children prefer smaller-scale housing. Questions about the incompletely thought-through strategic spatial planning are therefore asked in particular with respect to the needs of the Katowice conurbation's sustainable development (Runge et al. 2018). Smog, which is present for around six months of the year, also has an impact on the region's depopulation, or at least lowers the comfort of life in the region. The phenomenon is indirectly related to coal mining itself. It is present in all of Poland, and particularly in hilly areas. This includes the Katowice conurbation, and the parties responsible for the phenomenon are individual households and tenement buildings erected in the industrial and socialist periods. Interestingly, in the (post-)industrial Katowice conurbation the industrial sector is the least responsible for smog. This does not mean that there do not exist large plants that negatively impact the environment in certain towns (i.e. Dąbrowa Górnicza, Jaworzno, Będzin, Katowice, Zabrze, Bytom, Łaziska Górne).

Environmental challenges

With respect to challenges related to environmental protection and combating smog, many initiatives to reduce the problem exist. These are financial, educational, monitoring and organisational initiatives. In practice, the most effective course of action is changing outdated coal-based energy systems to new installations that use gas and photovoltaic panels. This is being carried out as part of state, province and metropolis policy and by the communes themselves. Effectiveness is still insufficient, which is caused by the impecuniousness of part of the local communities, manifesting itself in energy poverty.

Environmental and transport problems and demographic challenges are key elements of urban policy in the region. They are visible both in the policy of individual towns, and in regional policy by the provincial authorities and the authorities of Górnośląsko-Zagłębiowska Metropolia (Upper Silesian-Basin Metropolis). An analysis of strategic and planning documents defining future development almost unanimously indicates a high consensus as to the causes of crisis phenomena and the key to solving them (Zuzańska-Żyśko 2016). Some differences may be seen, however, as regards problem-solving methods, sources for obtaining funds and participation and roles of individual stakeholders. Future metropolitan policy is thus to be an attempt to avoid these obstacles, harmonising and above all unifying, step by step, the methods of correcting transport and environmental problems, as well as those arising from demographic loss. The successes of GZM authorities undoubtedly include unification of the public transport offer in the metropolis (buses, trams, trolleys) and strong efforts towards further consolidation that would also include rail transport. However, environmental issues, and particularly combating smog, are a great challenge. In this respect, due to limited funds held, GZM will offer support to local governments in individual towns rather than undertaking large investment projects. The exception is the attempt to build a large metropolitan waste incinerator. This issue is interesting in that the discrepancy in positions as to its location is not only the result of the relationship between target town and metropolis, but also between target town and neighbouring towns. The latter are often several in number and their opinion as to the location of this type of facility is no less important. The situation is certainly not improved by numerous local conflicts related to existing facilities such as waste dumps or sewage treatment plants. As a rule, these conflicts concern two or more towns. Solved conflicts (e.g., Katowice-Chorzów) are replaced by new ones (Zabrze-Ruda Śląska, Bytom-Zabrze, Katowice-Sosnowiec-Siemianowice Śląskie-Czeladź).

Yet another dimension is presented by coal mining conflicts or, more precisely, the negative effects of this industry on the built and natural environments. Here the role of metropolis authorities as a potential mediator is both indicated and extremely difficult. Particularly as concerns cases of coal mining by a mine located in a different town than the one in which the negative consequences of extraction would be felt. The two most significant conflicts, which will come to a fore after 2020, will concern the towns Lędziny (mine)—Imielin (endangered buildings) and Jaworzno (mine)—Mysłowice (endangered buildings). The difficulty of mediation results from the fact that in each case and by adopting one of the two possible solutions one of the towns will always be the beneficiary, and the neighbouring one the victim (operation of the mine—mine closure, built-up area—devastated built-up area). While in the case of Mysłowice and Jaworzno the town supported by GZM could be Mysłowice (Jaworzno is not a member of the metropolis), the case of Imielin and Lędziny is very difficult as both towns are part of the metropolitan union formed (i.e. Krzysztofik et al. 2019a).

The derelict areas and brownfields located inside of the core of the conurbation are also a significant challenge in regional policy focused on the question: how to manage such areas? In general, two alternative ideas are considered—further re-industrialisation or transforming them into recreational areas and so-called urban forests. The implementation of adopted policies is difficult because the opinion of the public and local decision-makers on this issue is divided and changes depending on the situation on the labor market or other conditions (Spórna and Krzysztofik 2020).

Dilemmas related to slowing depopulation are no less important to metropolis authorities. Paradoxically, financing infrastructural projects that improve quality of life and transport accessibility in suburban communes of the GZM is conducive to migration to them from large cities of the region's core. While in itself the phenomenon of raising the competitiveness of space and living conditions in the suburbs in a "fight for residents" is not original, a situation in which such support is systemic and intentional in the name of sustainable development of all the communes (also suburban ones) in the metropolis causes some dissonance. This is all the greater when it is in fact the large, dramatically depopulating towns of the core that finance investments in the suburbs. The explanation for this paradox is that in order to formally exist, the metropolis has to fulfil the criterion of minimum population (2.0 million inhabitants)—and without integrating neighbouring suburban communes this condition would not be met. Due to the fact that in 2019 the number of population of GZM was 2.2 million and kept falling, it seems that in future this fairly curious regional development policy will be continued.

Conclusion

The transformation of the Katowice conurbation as one of the largest urban regions in Poland had at least three basic dimensions: functional, state and political, and geographical. The determining factor for the first of them was the specific monofunctionalism based on coal mining and other sectors of heavy industry. In practice, this factor makes the region similar to other areas of this type in Central Europe, such as the Ostrava Agglomeration; nevertheless, due to the size of the Katowice region the scale of the changes and related challenges was immeasurably larger (Martinát et al. 2014; Rumpel and Slach 2012). Compared to the post-industrial regions of Great Britain and Germany, despite many similarities, it should be emphasised that the transformation often began in them half a century ago, it was not always so dynamic (Ruhrregion) and did not overlap with the phenomena defined as post-socialist development (Neffke et al. 2011; Schwarze-Rodrian 2016).

On the other hand, the Katowice conurbation was part of changes characteristic only of Poland. While they were similar to those taking place in other countries in the region, they certainly differed as to the diverse pace of introducing reforms and the placement of emphasis on selected priorities (Grabher 1993; Marot and Černič Mali 2012; Schwarze-Rodrian 2016). An important distinguishing feature of the region's transformation is the third aspect, related to the clear morphological and administrative polycentricity of the Katowice conurbation. This seemingly uniform, territorially integrated area is really a collection of settlements characterised by often-differing policies or visions for development, and additionally competing for investments that could form a new basis for their development. These determinants in general, both those unifying the Katowice conurbation's development and those magnifying its differences, shaped the region's fairly characteristic transformation at the start of the 21st century (Krzysztofik et al. 2016). It was undoubtedly strongly polarised with respect to the directions of changes, as well as the qualitative assessment of these processes.

The economic transformation of the region is its undoubted success, if the process is taken as a whole, of course. Due to depletion of resources and competition in a globalising world, many mining and industrial plants were sentenced to closure or complete restructuring. Both these processes took place, but at the same time it was possible to ease the negative consequences of economic transformation via an inflow of new investments, including direct foreign investments. As mentioned, this was not unaccompanied by problems and was not an unequivocally positive process. Nevertheless, from a perspective of 30 years of transformation the process should be seen as generally positive. Importantly, however, it is not finished yet and will last at least two or three more decades. It seems, nevertheless, that it will have a very balanced nature and will not be generally very noticeable from the perspective of the region's labour market. It is rather the fall of competitiveness resulting from the region's depopulation that should be seen as a threat, and not the only recently pervasive unemployment.

The second positive element is limiting of industrial pollution and the expansion of green spaces. While the Katowice conurbation, like all urban regions, is still struggling with smog, it should be emphasised that its origin is municipal management and not industry. This is indicated, for example, by the fact that it occurs almost exclusively in cold months. Over 30 years it was possible to mitigate the most negative effects of industry—emission of dust and gasses and to the production of waste and sewage. Since the start of the 21st century, the number of closed and reorganised waste dumps has significantly exceeded the number of those newly built or expanded. In some of the region's rivers fish have appeared after over 150 years. Brownfields, a great problem of the 1990s, are now valuable investment areas for newly created economic operators and where this is impossible, they are valuable natural areas with visible ecological succession.

Nevertheless, the transformation of the Katowice conurbation at the turn of the 20th and 21st centuries also caused a number of changes whose negative effects are still a great challenge for local authorities and residents. Undoubtedly the greatest challenge is slowing depopulation and reducing its consequences. As mentioned, in demographic terms the region is also characterised by growing ageing. Both factors are currently a fundamental challenge for local and regional policy, though still something of a taboo subject. In any case, there is certainly no vision for a Katowice conurbation that will in 30 years' time number around 1 million inhabitants fewer than at present. This is important because many investments currently being carried out do not take this fact into account, often being completely removed from the "built to suit" concept. Ultimately, this also leads to actions that turn out to be unnecessary, and certainly discrepant with sustainability policy (Runge et al. 2018). One of the causes of depopulation is also the state of the environment. While a great improvement has been noted with respect to lowering the level of pollution and negative impact on water and soil, compared with most large urban agglomerations in Poland there are still significant challenges in this respect.

Over the last 30 years, the Katowice conurbation, currently one of the largest urban regions in Central Europe, has undergone probably the deepest transformation in the region. Due to its specific nature this process is not only unfinished, but will most likely last the longest in this part of Europe. Also from this perspective it is apparent that the "built to suit" concept implemented through careful municipal planning and sustainability should the basis for further transformation in such a region.

Notes

1 The term conurbation should be understood as a region comprising several large and medium-sized cities, which through physical expansion merged to form one morphologically continuous urban area. The most common factor in the formation of the conurbation is the multicentre development of cities with a strong industrial, mining, port and tourist function. Conurbations are characterised by morphological continuity, while from a formal (administrative) point of view they consist of separate cities and urban areas with their own municipal authorities (it is administrative, historical and identity polycentrism). In Poland, it is also referred to as the urban region of many cities, none of which significantly dominates the second largest or subsequent ones.
2 Republic of Poland in 1944–1952.
3 The extermination of over 90% of Jews in the Dąbrowa Basin caused a very complicated legal situation regarding real estate. In addition, these properties were communalised and/or nationalised during the socialist period. In Upper Silesia, on the other hand, there were many buildings belonging to the German population before the Second World War. They have also been largely communalised and/or nationalised. After 1989, heir recovery processes, partial commercialisation and privatisation of buildings began (where the legal status was established).

References

Birch, K., MacKinnon, D., & Cumbers, A. (2010). Old industrial regions in Europe: A comparative assessment of economic performance. *Regional Studies*, 44 (1), 35–53.

Burghardt, A. (1971). A hypothesis about gateway cities. *Annals of the Association of American Geographers*, 61 (3), 269–285.

Domański, B. (2002). The economic restructuring of Upper Silesian Region. In G. Stöber (ed.), *Polen. Deutschland und die Osterweiterung der EU aus geographischen Perspektiven* (pp. 91–103). Isernhagen: Verlag Hahnsche Buchhandlung.

Domański, B., Guzik, R., & Gwosdz, K. (2008). The new international division of Latour and the changing role of the periphery; the case of the Polish Automotive industry. In C. Tamasy and M. Taylor (eds.), *Globalising worlds and new economic configurations* (pp. 85–100). Farnham: Ashgate.

Dragan, W. (2016). *Przemiany funkcjonalno-przestrzenne Starego Miasta w Mysłowicach w latach 1913–2013*. Katowice: Wydawnictwo Uniwersytetu Śląskiego.

Fornalczyk, A., Choroszczak, J., & Mikulec, M. (2008). *Restrukturyzacja górnictwa węgla kamiennego – programy, bariery, efektywność, pomoc publiczna*. Warszawa: Poltext.

Grabher, G. (1993). The weakness of strong ties: The lock-in of regional development in the Ruhr area. In G. Grabher (ed.), *The embedded firm: On the socioeconomics of industrial networks* (pp. 255–277). London: Routledge.

Gwosdz, K. (2004). *Ewolucja rangi miejscowości w konurbacji przemysłowej. Przypadek Górnego Śląska (1830–2000)*. Kraków: IGiGP Uniwersytet Jagielloński.

Gwosdz, K. (2014). *Pomiędzy stara a nową ścieżką rozwojową. Mechanizmy ewolucji struktury gospodarczej i przestrzennej regionu tradycyjnego przemysłu na przykładzie konurbacji katowickiej po 1989 roku*. Kraków: IGiGP Uniwersytet Jagielloński.

Hassink, R. (2010). Locked in decline? On the role of regional lock-in in old industrial areas. In R. Boschma and R. Martin (eds.), *The handbook of evolutionary economic geography* (pp. 450–468). Cheltenham: Edward Elgar.

Kantor-Pietraga, I. (2014). *Systematyka procesu depopulacji miast na obszarze Polski od XIX do XXI wieku*. Katowice: Wydawnictwo Uniwersytetu Śląskiego.

Klasik, A. & Heffner, K. (2001). Polish regional policy and the problems of Upper Silesia ten years into transformation In A. Klasik and K. Heffner (eds.), *Restructuring heavy industrial regions. Some experience from Scotland and Upper Silesia* (pp. 11–34). Katowice: Wydawnictwo Akademii Ekonomicznej im. Karola Adamieckiego.

Kłosowski, F., Pytel, S., Runge, A., Sitek, S., & Zuzańska-Żyśko, E. (2013). *Rynek pracy w subregionie centralnym województwa śląskiego*. Sosnowiec: WNoZ Uniwersytet Śląski.

Kontrakt regionalny dla województwa katowickiego: dorobek i opinie. (1999). Barański, M. (ed.) Katowice: Uniwersytet Śląski.

Krzysztofik, R., Dulias, R., Kantor-Pietraga, I., Spórna, T., & Dragan, W. (2020). Paths of urban planning in a post-mining area. A case study of a former sandpit in southern Poland. *Land Use Planning*, 99 (3), 104801.

Krzysztofik, R., Kantor-Pietraga, I., Runge, A., & Spórna, T. (2017). Is the suburbanisation stage always important in the transformation of large urban agglomerations? The case of the Katowice conurbation. *Geographia Polonica*, 90 (2), 71–85.

Krzysztofik, R., Runge, J., & Kantor-Pietraga, I. (2012). *An introduction to Governance of urban shrinkage. A case of two Polish cities: Bytom and Sosnowiec*. Sosnowiec: WNoZ Uniwersytet Śląski.

Krzysztofik, R., Kantor-Pietraga, I., Spórna, T., Dragan, W., & Mihaylov, V. (2019a). Beyond 'logistics sprawl' and 'logistics anti-sprawl'. Case of the Katowice region, Poland. *European Planning Studies*, 27 (8), 1646–1660.

Krzysztofik, R., Runge, J., Runge, A., & Kantor-Pietraga, I. (2019b). Pomiędzy demograficzną przeszłością a wyzwaniami przyszłości. Czy w ogóle chcę mieszkać w Polsce, w mieście? In R. Krzysztofik (ed.), *Przemiany demograficzne miast Polski. Wymiar krajowy, regionalny i lokalny* (pp. 81–100). Warszawa-Kraków: Instytut Rozwoju Miast i Regionów.

Krzysztofik, R., Kantor-Pietraga, I., & Kłosowski F. (2019c). Between Industrialism and postindustrialism—The case of small towns in a large urban region: The Katowice Conurbation, Poland. *Urban Science*, 3 (3), 68.

Krzysztofik, R.,Tkocz, M., Spórna,T., & Kantor-Pietraga, I. (2016). Some dilemmas of post-industrialism in a region of traditional industry: The case of the Katowice conurbation, Poland. *Moravian Geographical Reports*, 24 (1), 42–54.

Lamparska, M. (2013). Post-industrial cultural heritage sites in the Katowice conurbation, Poland. *Environmental and Socio-economic Studies*, 1 (2), 36–42.

Lokalny wymiar rozwoju społecznego Polski. (2018). Krzysztofik, R. (ed.) Kraków: Krajowy Instytut Polityki Przestrzennej i Mieszkalnictwa.

Marot, N., & Černič Mali, B. (2012). Using the potentials of post-mining regions—A good practice overview of Central Europe. In P. Wirth, B.C. Mali and W. Fischer (eds.), *Post-mining regions in central Europe problems, potentials, possibilities* (pp. 130–147). München: Oekom.

Martinát, S., Navrátil, J., Dvořák, P., Klusáček, P., Kulla, M., Kunc, J., & Havlíček, M. (2014). The expansion of coal mining in the depression areas – A way to development. *Human Geographies – Journal of Studies and Research in Human Geography*, 8 (1), 5–15.

Mihaylov,V., Runge, J., Krzysztofik, R., & Spórna,T. (2019). Paths of evolution of territorial identity. The case of former towns in the Katowice Conurbation. *Geographica Pannonica*, 23 (3), 173–184.

Murzyn-Kupisz, M., Gwosdz, K. (2014). The changing identity of the Central European city: The case of Katowice. *Journal of Historical Geography*, 37 (1), 113–126.

Neffke, F., Henning, M., & Boschma, R. (2011). How do region diversify over time? Industry relatedness and the development of new growth paths in regions. *Economic Geography*, 87 (3), 237–265.

Phelps, N.A., & Ozawa, T. (2003). Contrasts in agglomeration: Proto-industrial, industrial and post-industrial form compared. *Progress in Human Geography*, 27 (5), 583–604.

Pound, N.J.G. (1958). *The Upper Silesian industrial region.* Bloomington: Indiana University Press.

Prognoza ludności na lata 2014–2050. Population projection 2014–2050. (2014). D. Szałtys (ed). Warsaw: GUS.

Riley, R., & Tkocz, M. (1999). Local responses to changes circumstances: Coalmining in the market economy in Upper Silesia, Poland. *Geojournal*, 48, 279–290.

Rink, D., Couch, C., Haase, A., Krzysztofik, R., Nadolu, B., & Rumpel, P. (2014). The governance of urban shrinkage in cities of post-socialist Europe: Policies, strategies and actors. *Urban Research & Practice*, 7 (3), 258–277.

Rocznik Statystyczny Województwa Śląskiego 2019. (2019). Katowice: Urząd Statystyczny w Katowicach.

Rumpel, P., & Slach, O. (2012). *Governance of shrinkage of the city of Ostrava.* Praha: European Science and Art Publishing.

Runge, A., Kantor-Pietraga, I., Runge, J., Krzysztofik R., & Dragan, W. (2018). Can depopulation create urban sustainability in postindustrial regions? A case from Poland. *Sustainability*, 10 (12), 4633.

Runge, J., & Kłosowski, F. (2011). Changes in population and economy in Śląskie voivodship in the context of the suburbanization processes. *Bulletin of Geography. Socio-economic Series*, 16, 89–106.

Runge, J., Kłosowski, F., & Runge, A. (2003). Conditions and trends of social-economic changes of Katowice region, *Bulletin of Geography. Socio-economic Series*, 2, 85–102.

Schwarze-Rodrian, M. (2016). Ruhr region case study. In D.K. Carter (ed.), *Remaking postindustrial cities lessons from North America and Europe* (pp. 187–209). NewYork, Abingdon: Routledge.

Spórna,T. (2012). *Model przemian urbanizacyjnych w województwie śląskim.* Sosnowiec: WNoZ Uniwersytet Śląski.

Spórna, T. (2018). The suburbanisation process in a depopulation context in the Katowice conurbation, Poland. *Environmental & Socio-economic Studies*, 6 (1), 57–72.

Spórna, T., Kantor-Pietraga, I., & Krzysztofik, R. (2016). Trajectories of depopulation and urban shrinkage in the Katowice Conurbation, Poland. *Espace Populations Sociétés*, 2015 (3)-2016 (1): Espaces en dépeuplement.

Spórna, T., & Krzysztofik, R. (2020). 'Inner' suburbanisation – Background of the phenomenon in a polycentric, post-socialist and post-industrial region. Example from the Katowice conurbation, Poland. *Cities*, 104, 102789.

The City After Abandonment. (2013). M. Dewar and J.M. Thomas (eds.). Philadelphia: University of Pennsylvania Press.

Tkocz, M. (2001). *Restrukturyzacja Przemysłu Regionu Tradycyjnego*. Katowice: Wydawnictwo Uniwersytetu Śląskiego.

Zuzańska-Żyśko, E. (2016). *Procesy metropolizacji. Teoria i praktyka*. Warszawa: PWN.

12 Population ageing processes in towns and cities situated in peripheral areas

An example of urban centres in Eastern Poland

Wioletta Kamińska and Mirosław Mularczyk

Introduction

The turn of the 21st century was the time of negative demographic changes, which have affected many European countries and regions. The systematically decreasing fertility and birth rates, along with the growing life expectancy and decreasing death rates, have resulted in a growing number of senior citizens in European societies. During the whole post-war period, the pace of population ageing was not as fast as it is nowadays. The percentage of people aged 65+ in the 28 countries forming the European Union in 2019 had nearly doubled in 1950–2019, going up from 8.9% to 19.2%. Based on demographic forecasts, it may be assumed that population ageing will become a global phenomenon in the coming decades. The percentage of people aged 65 and over will increase from 8.3%, as it stands now, to 16% by 2059 and to 23% by 2100 (Population trends 1950–2100…, 2020).

The transformations of demographic structures towards an increase in the old-age population have their consequences, mostly negative, in nearly all domains of life. Population ageing usually means smaller state and regional budget revenues, shrinking technical infrastructure as well as economic and social stagnation. Therefore, demographic changes require continuous monitoring and consideration in socio-economic development strategies.

Dynamic population ageing processes were noticed in the early 1990s. Since that time, an average inhabitant of Poland has aged by nearly 9 years. In 2018, the median age for the whole population of Poland was nearly 41 years. The percentage of elderly people (65+) increased considerably, from 10.1% in 1990 to 17.5% in 2018. Negative changes in the Polish population structure are becoming increasingly profound due to the growing emigration of young people. By 2060, Poland will have had one of the oldest populations in the European Union—the process seems inevitable and irreversible in the most recent decades (Okólski, 2010).

Population ageing takes place with varying intensity in time and space. It can be generally assumed that it occurs at a higher rate in developed countries, in cities rather than in the countryside and in peripheral rather than central areas (Rosenbers and Wilson, 2018).

DOI: 10.4324/9781003039792-12

Changes in the age structure of the Polish society vary from region to region. Negative demographic processes are particularly visible in peripheral areas, such as Eastern Poland. It is a problem area, demographically and economically (Bański, 2002), perceived as a peripheral region, both of Poland and of the European Union. Historical, environmental, economic and social conditions were hampering sustainable development of the borderland areas for decades: before 1991—those bordering with the Soviet Union and presently—with Lithuania, Ukraine, Belarus and Russia. As a consequence of slow development, the region has been abandoned by large numbers of inhabitants, which contributed to the shrinking of the economy. It was the cause of and reason for a spiral of unfavourable demographic and economic processes. Hoekveld (2012) refers to this process as a "vicious circle", leading to a point of no return.

From the cognitive point of view, the population ageing processes in the towns and cities of Eastern Poland are very interesting. All services and high-quality human capital concentrate in urban centres, and their development radiates over the whole region. The economic growth of every region largely depends on how advanced its urban settlement system is. Sustainable urban development results in the economic prosperity of the whole region and improves its competitiveness. Also, local demographic processes have a key importance as regards the socio-economic development of the whole region.

The population potential of towns and cities depends on natural population growth and migration balance, which determine the demographic situation of these settlement units to different degrees. In other words, they have a varying influence on demographic structures, including population ageing.

The purpose of the chapter is to define the pace of population ageing in towns and cities situated in the peripheral areas of Poland and the European Union at the same time, in the early 21st century. The authors' purpose is also to define the relationship between natural growth and migration balance, as well as to evaluate the impact of these elements on contemporary population ageing processes. They have also focused their attention on the correlations between the size, the location of individual urban settlement units and the ageing processes and elements of population dynamics. The study was conducted with reference to the historical and modern context of the development of Eastern Poland.

The added value of this chapter is the diagnosis of urban population ageing processes in peripheral areas of Poland and European Union at the beginning of the 21st century, as well as the identification of the main factors shaping these processes. The research results may refer to other peripheral areas, especially in the countries of Central Europe which used to belong to the Eastern Bloc and represented similar historical conditions of socio-economic development.

In order to define the relationship between natural growth and migration balance in the towns and cities considered in the study, the researchers used Webb's method. Eight demographic types were distinguished and marked with letters from A to H. The first four letters (A, B, C, D) are progressive types, where the number of inhabitants increases, and the remaining four types (E, F, G, H) are regressive, where this number decreases. Individual types are characterised by different relationships between natural growth and migration balance:

A – positive natural growth outweighs the negative migration balance,
B – positive natural growth outweighs the positive migration balance,
C – positive natural growth is lower than the positive migration balance,
D – positive migration balance outweighs the negative natural growth,
E – negative natural growth is not compensated by the positive migration balance,
F – negative natural growth outweighs the negative migration balance,
G – negative natural growth does not outweigh the negative migration balance,
H – negative migration balance is not compensated by the positive natural growth.

In order to exclude accidental, one-year changes, partial indexes referring to natural growth and migration balance were calculated, based on mean values for three subsequent years: 2017, 2018 and 2019.

Demographic analyses related to population ageing in the towns and cities of Eastern Poland were conducted in three stages, in accordance with the procedure proposed by E. Rosset (1978). At the first stage, the demographic old age was defined. Secondly, appropriate indexes were selected and, finally, the ageing scales were established (Rakowska, 2016). Following the UN and Eurostat recommendations, it was assumed that the demographic old age is 65 years and over.

To measure ageing processes, the researchers used the population ageing ratio (AR) in 2008–2019 (for the set of 203 settlement units with municipal rights in 2008). The ageing ratio is based on the differences between the percentages of young and old populations (Długosz, 1997) and was calculated according to the following formula:

$$AR = \left[U(0-14)_t - U(0-14)_{t+n} \right] + \left[U(\geq 65)_{t+n} - U(\geq 65)_t \right]$$

where:

$U(0-14)_t$ – percentage of population aged 0–14 at the beginning of the studied period;
$U(0-14)_{t+n}$ – percentage of population aged 0–14 at the end of the studied period;
$U(\geq 65)_t$ – percentage of population aged 65+ at the beginning of the studied period;
$U(\geq 65)_{t+n}$ – percentage of population aged 65+ at the end of the studied period.

The higher the ratio value, the faster the pace of population ageing.
The towns/cities were divided into four groups:

- $AR \leq 0$ – rejuvenating cities – group I;
- $0 < AR < 4,12$ (*value lower than average, minus one standard deviation*) – towns with a low ageing rate, below average - group II;
- $4,13 < AR < 9,65$ (*mean value plus/minus one standard deviation*) – towns with an average ageing rate - group III,
- $AR > 9,65$ (*value higher than average plus/minus one standard deviation*) – towns with a high ageing rate, above average – group IV.

In order to establish the intervals, the authors used the mean value (6.89) and the standard deviation value (2.76) of the calculated population ageing ratio for all the settlement units in Poland which had municipal rights in 2008.

The choice of this measure (AR) enabled the authors not only to define the percentage of old people living in the towns and cities in question, but also to demonstrate the dynamic relationship between the oldest and the youngest part of the population. It also allowed them to investigate the pace of population ageing in the urban centres of Eastern Poland, in comparison to other areas of the country.

The primary source of information was the data obtained from the Central Statistical Office (GUS), regarding the population, the migration balance and natural growth (https://stat.gov.pl/).

The research encompassed towns and cities in five voivodeships (provinces) of Eastern Poland: Lubelskie, Podkarpackie, Podlaskie, Świętokrzyskie and Warmińsko-Mazurskie (Figure 12.1). In 2019, they constituted ca. 25% of all urban settlement units in Poland. They were inhabited by nearly four million people, which is about 17% of the urban population in the country (Table 12.1).

Figure 12.1 Location of the voivodeships (provinces) of Eastern Poland in Poland and in the EU.

Source: Authors' elaboration.

Table 12.1 The size structure of urban centres in Poland, 2019

Size of the city Number of inhabitants	Inhabitants				Cities in the remaining voivodeships				Total			
	No of towns	Towns (%)	No of inhabitants (in thousands)	Inhabitants (%)	No of towns	Towns (%)	No of inhabitants (in thousands)	Inhabitants (%)	No of towns	Towns (%)	No of inhabitants (in thousands)	Inhabitants (%)
>100,000	6	2.6	1319.6	33.2	32	4.5	9387.2	49.3	38	4.0	10706.9	46.5
50,000–100,000	10	4.3	627.2	15.8	35	4.9	2425.7	12.7	45	4.8	3052.8	13.3
20,000–49,999	28	12.1	895.1	22.5	107	15.1	3387.7	17.8	135	14.4	4282.8	18.6
10,000–19,999	41	17.7	591.5	14.9	139	19.6	2028.9	10.6	180	19.1	2620.4	11.4
<10,000	147	63.4	545.7	13.7	395	55.8	1824.6	9.6	542	57.7	2370.2	10.3
Total	232	100.0	3979.1	100.0	708	100.0	19053.9	100.0	940	100.0	23033.1	100.0

Source: stat.gov.pl

The size structure of the urban centres varied, with the smallest element of the set populated by 338 inhabitants (Opatowiec, Šiętokrzyskie Voivodeship), and the largest—by 339,784 (Lublin, Lubelskie Voivodeship).

The predominant group (147) consisted of the smallest urban centres, with a population of up to 10,000 people. They constituted ca. 63% of the overall number of the urban settlement units in Eastern Poland, which was more than the percentage of similar towns situated in the remaining areas of the country (ca. 56%) (Table 12.1). They were inhabited by about 546,000 people, which was over 13% of the urban population of Eastern Poland. This percentage was higher than in the remaining area of Poland, where it oscillated around 10%.

The second most numerous group included 41 small towns with 10–20,000 inhabitants (ca. 18% of all urban centres in Eastern Poland). Jointly, they were populated by over 591,000 registered residents, which was nearly 15% of the whole urban population. The percentage of settlement units of that size in the remaining area of Poland was higher (nearly 20%), and the percentage of people inhabiting them was smaller (ca. 11%). In Eastern Poland, there were 28 towns with a population of 20–50,000 people (ca. 12%). Jointly, they were inhabited by ca. 895,000 people, which was over 22% of the urban population in this region. In the remaining area of the country, the percentage of similar towns in the overall number of settlement units was higher (ca. 15%) and the percentage of urban population was smaller (ca. 18%).

In Eastern Poland, there were 10 towns with a population of 50–100,000 inhabitants (ca. 4%). Jointly, they were inhabited by about 627,000 registered residents (ca. 16% of the urban population of Eastern Poland voivodeships). In the remaining area of Poland, the percentage of similar towns was slightly higher (4.9%) and of their residents – lower (ca. 13%).

The least numerous group included the six largest cities, inhabited by over 100,000 people: Lublin—340,000, Białystok—298,000 Rzeszów—196,000, Kielce—195,000, Olsztyn—172,000 and Elbląg—119,000. They constituted about 2.5% of all urban settlement units in Eastern Poland and were inhabited by nearly 1,320,000 people, which was ca. 33% of the region's urban population. In the remaining area of Poland, the percentages of the largest cities, (over 100,000 inhabitants) and their registered residents were higher (4.5% and over 49%, respectively) (Table 12.1).

Peripheral areas

The term "peripheral areas" has not been clearly defined. Peripheries are usually associated with being distant, different and dependent (Olechnicka, 2004). As a rule, it is assumed that peripherality means poor transport infrastructure, lack of access to modern means of transport, as well as extraordinarily high costs of connections with other regions (Zarębski, 2012; Hall and Harrison, 2015).

Many authors believe that peripherality may be analysed with reference to the key research approaches: anthropological, geographical, economic, socio-demographic, socio-cultural and political-administrative (Gren, 2003; Miszczuk, 2013).

In the anthropological approach, remote areas are analysed from the point of view of their connection with central areas (Ardener, 2012). It is a relational approach, as the analysis depends on what we understand by a central area and a remote area (Ardener, 2012). Here, remoteness (peripherality) does not result from the co-occurrence of certain characteristics but is a form of a relationship with the central area, which is expressed solely by a shorter or larger distance (Pezzi and Urso, 2016).

Geography-wise, peripherality can be defined in terms of poor accessibility and economic and political insignificance of a given territory, compared to the central area (Copus, 1999). In this approach, the relative position of every place is the factor which explains the key features of a peripheral area, i.e. the higher cost of transport and remoteness (Ferrao and Lopes, 2004). In this sense, a peripheral character can be attributed to an area which is situated far from any centres of the political-administrative and socio-economic life, as well as being hard to access by transport (Miszczuk, 2013). Unfavourable location causes other negative phenomena, such as high production costs, limited access to markets, lack of innovativeness and entre-preneurship, exogenous control over the means of production, concentration of power in the central area, lack of competition, etc. (Hall and Harrison, 2015).

Looking at peripherality from the economic point of view, the emphasis is put on the economic aspects of peripheral areas. Peripheries are treated as economi-cally backward areas, compared to central territories. They are characterised by an underdeveloped production structure, insignificant exports and dominance of imports (Davies and Michie, 2011). Peripheries are delimited based on economic development measures, usually GNP per capita (Miszczuk, 2013).

In the socio-demographic approach, the elements which are considered include small population density in peripheral areas, low level of urbanisation and clearly noticeable depopulation processes, which cause deformations of demographic structures. Long-term population outflow, especially as regards young and well-educated people, results in rapid population ageing and an erosion of human and social capital (Bański, 2005; Proniewski, 2014). Faulty demographic structures are at the same time the cause and the outcome of further population and economic problems, contributing to further peripheralisation (Hoekveld, 2012).

From the sociological-cultural perspective, "peripherality" is treated as a state of social identity, understood as people's emotional attitude to a given territory, its landscape, and tangible and intangible culture products (Szul, 1991). Here, peripheries cannot be treated in terms of economic backwardness, as a problem area or an area of cultural degradation. Remoteness, first of all, signifies cultural potential, social and territorial roots, and a set of traditional values (Zarycki, 2007; Wójcik, 2009).

Taking the political-administrative approach, peripherality is related to areas of small strategic significance in international and national terms. It is visible in their weak representation in central authority bodies, lack of elites, limited competences of regional authorities and a small financial potential of public authorities (Olechnicka, 2004; Miszczuk, 2013). From the political point of view, peripheral regions are perceived as "backward" areas due to the poor access to large markets as well as low or decreasing population density (Proniewski, 2014).

The above-mentioned aspects of peripherality are usually interrelated and ana-lysed as common features of areas situated far from central areas. Peripherality is a relative notion, depending on the chosen criterion (Wilkin, 2003) and strictly connected with unfavourable environmental, cultural, social, economic and political factors (Hall and Harrison, 2015).

Eastern Poland as a peripheral area

In the European Union, the basic indicator of peripherality is the economic criterion (Figure 12.1). The makers of the *Program Polska Wschodnia 2014–2020* (2014) *(Eastern Poland Program 2014–2020, 2014)* distinguished the following areas in need of support:

- the "less developed", peripheral regions—GDP per capita (PPP) does not exceed 75% of the European Union's average GDP;
- the regions in the "transition phase"—GDP per capita ranges from 75% to 90% of the European Union's average GDP;
- the "more developed" regions—GDP per capita exceeds 90% of the European Union's average GDP.

Eastern Poland belongs to the less developed, peripheral areas, both in the European Union and in Poland. Before 2019, GDP per capita there had not exceeded 50% of the mean value for the EU.

Many researchers have underlined the low level of development of Eastern Poland, compared to other regions of the country. They have emphasised that already in the central management period, all indicators of economic development at the level of *powiats* and *gminas* were below the average values for Poland (cf. Swianiewicz, 2007). Gorzelak (2007) claimed that the difference between the level of development in Eastern Poland and the rest of Poland, measured with GDP per capita, increased in the first years of the 21st century. The worst economic situation of Eastern Poland was confirmed by the research conducted by Kamińska and Mularczyk (2014a). The disproportions were smaller only in the domains whose development depended on EU funds.

The unfavourable socio-economic situation was also confirmed in *Strategia rozwoju Polski Wschodniej 2020* (2013) (*The strategy for the development of Eastern Poland 2020, 2013*). In the diagnostic part of this government document, a number of retardations were indicated, typically observed in the voivodeships of Eastern Poland: the considerably smaller GDP income per capita than in the European Union, poorer regional assets (infrastructure, technological and innovative potential, work standard), high poverty rates, a lower level of entrepreneurship, urbanisation, and the science sector than the national average, larger involvement of labour resources in agriculture and considerably smaller in the sector of advanced technologies than in other parts of the country. Other factors mentioned in the documents included: a lower percentage of people with university education than the average value for Poland, poor transport accessibility, and backwardness in telecommunication and electric power infrastructure (*Strategia rozwoju ...*, 2013). The unfavourable social and economic situation in the voivodeships of Eastern Poland, worse than in other parts of Poland, was visible in rural and urban areas alike.

Retardations in the analysed settlement units resulted primarily from the economic policy of subsequent governments, the agricultural character of the area, its historical conditions and borderland location. In the period of the socialist economy, the predominant branch of economy in the voivodeships of Eastern Poland was agriculture. Warminsko-Mazurskie Voivodeship was based on large-area state collective farms (PGR), while Podlaskie, Lubelskie, Podkarpackie and Świętokrzyskie Voivodeships—on small individual farms. Towns and cities mostly performed service functions for the surrounding farming areas. Individual industries developed mostly in the largest urban centres of the region. The branches which were more significant than in the remaining part of the country included the food, wood, furniture, rubber and plastics industries, while the unique branches were the tobacco and aviation industries. The number of industrial plants was smaller than in the remaining parts of the country (*Strategia rozwoju...*, 2013). During the system transformation, the towns and cities of Eastern Poland were strongly affected by the economic crisis. The majority of the few surviving industrial plants were unable to adjust to the new market rules, which led to their liquidation and increasing structural unemployment. Therefore, after 1989, self-government authorities attempted to redefine the role and significance of cities as the centres of regional growth, and looked for new development impulses and chances to improve the living conditions of local communities. The outcomes of these activities varied—some towns achieved economic success and some plunged into recession.

Historical factors referred most of all to the times of the partition of Poland (1772–1918). Before Poland regained independence in 1918, the area of the voivodeships where the analysed towns were located had been under the administration of Prussia (Recovered Territories: Warmińsko-Mazurskie Voivodeship), Russia (Lubelskie and Podlaskie Voivodeships) and Austria (Podkarpackie Voivodeship). The social and economic consequences of belonging to different states are still visible in today's Eastern Poland. They refer to the functioning of three different legal, administrative and educational systems, which shaped diverse models of economic and social development. As a result, the economic development of the areas remaining under the administration of individual partitioning powers varied considerably. The best-developed areas were those belonging to the Prussian territory, and the least advanced were those under Austrian rule, which was clear from the considerable differences in the amount of the gross national income per capita in Swiss francs. In 1911, it was 242 FF in Galicia (Austrian section), 400 FF in the Kingdom of Poland (Russian section) and 718 FF in the Prussian section and Upper Silesia (Talarowski, 2017). The differences in economic development were also reflected in the employment structure. At the turn of the 20th century, the largest percentage of people working in agriculture was recorded in the territory under Austrian administration (ca. 77% of the working population), a smaller percentage—in the Prussian (ca. 61%) and Russian section (ca. 57% in the Kingdom of Poland and a considerably higher percentage in the territories of Lithuania and Belarus). The highest percentage of people working in crafts and industry (ca. 22%) was observed in the Prussian section, a lower one (15%) in the Russian part and the lowest (8%)—in the Austrian part (*Historia Polski* ..., 1994).

The urban network on the Polish lands annexed by the partitioning powers was diversified, mostly due to the varying rate of industrialisation and interior policy of the occupant authorities. They were not interested in developing towns and cities in the subordinate territories of former Poland. Actually, urban settlement units were often deprived of municipal rights. The most restrictive in this respect were the Russian authorities

The varying rate of industrialisation and different policies of the occupant authorities had an influence on the development of transport infrastructure, especially the railway network. In the area of the Kingdom of Poland, on the west bank of the Vistula river, the tsarist authorities hampered the development of rail. In this way, they were preparing an arena for the anticipated war with Prussia. The policy of the Prussian authorities was different. In order to intensify the industrialisation processes, they supported the development of rail to connect industrial centres and the largest cities. A similar policy was pursued by the Austrian authorities. However, in this case, limited possibilities of financing investments were a serious impediment (Wilczek-Karczewska, 2015).

The partitions resulted in a number of socio-economic retardations, listed in the Strategy for the Development of Eastern Poland 2020. For instance, the low level of entrepreneurship may be traced back to the weaker self-discipline among the population of the Russian and Austrian sector than was to be found in the Prussian one, due to their less responsible attitude to work. A lower level of education among the inhabitants of Eastern Poland voivodeships, compared to the remaining areas, may have resulted from the lack of compulsory school in the Russian partition sector (Schmidt, 2007).

An important factor of urban development is the borderland location. The territory of the Warminsko-Mazurskie Voivodeship borders with the Kaliningrad *Oblast* (Russian territory), Podlaskie Voivodeship—with Lithuania and Belarus, Lubelskie Voivodeship—with Belarus and Ukraine, and Podkarpackie Voivodeship—with Ukraine and Slovakia. The borders with Russia, Belarus and Ukraine are at the same time the external border of the European Union.

The external borders of Eastern Poland territories vary as regards permeability. The borders with Lithuania and Slovakia (EU member states and countries of the Schengen Area) are open and not a significant spatial barrier. They are freely crossed by people and goods. Borders with Russia, Belarus and Ukraine should be regarded as partly permeable, hampering and controlling cross-border flows.

Both types of borders may be seen as beneficial in terms of location, as they stimulate local development, particularly near border crossing points. The growth of frontier trade, customs duties, inexpensive services, hospitality industry and gastronomy, as well as transit traffic services may contribute to the development of borderland areas (Bański, 2010).

Relations between natural growth and migration balance in the towns of Eastern Poland (Webb's demographic types)

In 2019, the towns and cities of Eastern Poland represented all demographic types. The majority of them represented regressive types (E, F, G, H), characterised by the declining number of inhabitants. Regressive types were identified in 87% of the settlement units included in the study, while in the remaining area of the country, this percentage was smaller (nearly 82%). Progressive types (A, B, C, D), character-ised by an increasing number of population, included about 13% of all the towns and cities in the studied peripheral area. In the remaining area of the country, the percentage of progressive urban centres was higher (over 18%) (Figure 12.2 and Table 12.2).

In Eastern Poland the majority (48.3%) of urban centres represented regressive type G. The same type dominated in the remaining area of the country as well, though to a smaller extent (35.5%) (Table 12.2). Type G was characterised by natural population decline and negative migration balance, with the latter having a stronger impact on the decreasing number of population. Natural population decline in the analysed settlement units in Eastern Poland ranged from $-10.9‰$ to $-0.003‰$, and the migration balance—from $-18.1‰$ to $-0.03‰$. Type G was typical of small and medium-sized towns, which fell within the range of 40% among urban centres inhabited by 50–100,000 people to 64% among towns with

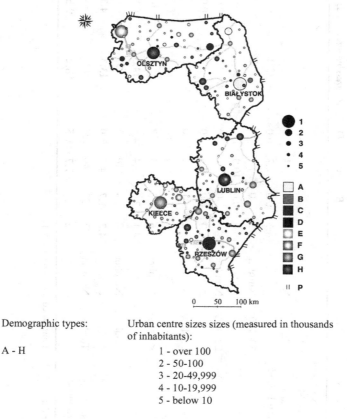

Demographic types:

A - H

Urban centre sizes sizes (measured in thousands of inhabitants):

1 - over 100
2 - 50-100
3 - 20-49,999
4 - 10-19,999
5 - below 10

Figure 12.2 Demographic types of urban centres in Eastern Poland.
Source: Authors' elaboration.

Table 12.2 Demographic types of towns and cities of different sizes in Eastern Poland, compared to the rest of the country in 2019

Demographic types			A	B	C	D	Progressive total	E	F	G	H	Regressive total	Total
Town/city size (in thousands)													
Towns/cities in Eastern Poland	>100	Number	1	0	1	0	2	0	1	1	2	4	6
		%	16.7	0.0	16.7	0.0	33.3	0.0	16.7	16.7	33.3	66.7	100,0
	50–100	Number	1	0	1	0	2	0	1	4	3	8	10
		%	10.0	0.0	10.0	0.0	20.0	0.0	10.0	40.0	30.0	80.0	100,0
	20–49.9	Number	0	0	1	0	1	0	2	18	7	27	28
		%	0.0	0.0	3.6	0.0	3.6	0.0	7.1	64.3	25.0	96.4	100,0
	10–19.9	Number	2	1	2	0	5	0	1	23	12	36	41
		%	4,9	2.4	4,9	0.0	12.2	0.0	2.4	56.1	29.3	87.8	100,0
	<10	Number	6	3	5	7	21	10	29	66	21	126	147
		%	4.1	2.0	3.4	4.8	14.3	6.8	19.7	44.9	14.3	85.7	100,0
	Total	Number	10	4	10	7	31	10	34	112	45	201	232
		%	4.3	1.7	4.3	3.0	13.3	4.3	14.7	48.3	19.4	86.7	100,0
Towns/cities in remaining areas	Total	Number	37	27	40	25	129	67	154	251	107	579	708
		%	5.2	3.8	5.7	3.5	18.2	9.5	21.7	35.5	15.1	81.8	100,0

Source: Authors' calculations, based on www.stat.gov.pl

20–50,000 inhabitants (Table 12.2). In the group of the six largest cities of Eastern Poland, only one—Kielce—represented this particular demographic type (Figure 12.2). In the remaining territory of Poland, type G occurred mostly in small towns inhabited by up to 20,000 people (36–40%).

The considerably larger percentage of urban centres representing type G in Eastern Poland than in the rest of the country was mostly determined by contemporary economic and social factors. The economic factors resulted from the situation on the labour market after 1989. After the crisis due to the system transformation, only some towns/cities in Eastern Poland managed to develop new functions and diversify their labour market. Thus, the migration outflow in many urban centres of type G was rooted in economic regression. Liquidating a large industrial plant, frequently the only place of employment in town, caused disturbances on the local labour market and a drastic increase in unemployment. According to Kantor-Pietraga et al. (2012), the inability to for the people's basic needs related to work made the place less attractive to live and stimulated emigration. Population decline (especially in the productive age group) generated other problems, such as decreasing revenues for the city (CIT, PIT and local taxes).

Economic problems were accompanied by social ones, including unemployment and social apathy. Many settlement units stopped being an attractive space to invest and settle down (Kantor-Pietraga et al., 2012). In addition, the majority of small and medium-sized towns representing type G in Eastern Poland were unfavourably located, region-wise. Most of them were situated on the peripheries of regions, far from the main cities and roads (Domański and Noworól, 2010). Research shows (Kamińska and Mularczyk, 2014b) that such location affected the demographic structures of the urban centres in question.

The situation of towns in the northern part of Eastern Poland (the Warmińsko-Mazurskie Voivodeship), which represented type G, was slightly different. The economic structure in those areas was dominated by the agricultural function, performed primarily in large, state-owned farms (PGR). As agriculture was not a demanding sector as regards the level of education of the workforce, the farms offered employment to many poorly educated inhabitants of towns, especially small ones. The consequence was an uncontrollable increase in unemployment, which most heavily affected the urban centres providing services for the state farming sector. It was a factor stimulating economic migration, especially to Germany (Jażewicz, 2005).

Demographic type G is characterised not only by a negative migration balance, but also natural population decline. As it was mentioned earlier, this type predominated in small and medium-sized towns of Eastern Poland, where agricultural functions are much more advanced than in other regions of the country. The people living there were attached to traditional family and religious values (cf. Kotowska and Giza-Poleszczuk, 2010). It seems to support the assumption that natural population growth in the urban centres of Eastern Poland should be represented by positive values. Positive natural growth was still typical of most towns in Eastern Poland in the first decade of the 21st century (Kamińska and Mularczyk, 2014b). However, positive natural growth lost in importance as an element influencing the number of the population, especially in the smallest towns. Long-lasting negative migration balance was reflected in the decreasing natural growth, because young women (at procreative age) and people at the productive age displayed greater mobility. The outflow of this social group, as well as the changing moral, family, procreative and religious patterns, resulted in lower fertility rates (Kurek and Lange, 2013). As a consequence, in the second decade of the 21st century, many towns in eastern, peripheral voivodeships displayed not only a negative migration balance, but also a natural population decline. This process could not be hampered despite numerous social programmes, offered by the government. They had no influence on increasing the fertility rate.

As regards the number of population, the second position among the towns and cities of Eastern Poland, was occupied by those representing type H, which continued to dominate in the first decade of the 21st century (Kamińska and Mularczyk, 2014b). In type H, the negative migration balance was not compensated by a positive natural growth. This means that the number of population decreased mainly due to the migration loss. In this group of towns, it ranged from −17.05 ‰ to −0.22‰, and the natural growth rate from 0.01 ‰ to 5.78‰. Type H included nearly 20% of towns in Eastern Poland and was less common among urban centres in the remaining part of the country (ca. 15%). The largest percentages of type H towns in Eastern Poland were represented by urban centres populated by over 10,000 inhabitants (25%–30%) (Table 12.2). Two out of six largest cities in Eastern Poland, Lublin and Olsztyn (Figure 12.2), also belonged to this type. In most urban centres from this group, the negative migration balance of the earlier period was smaller and did not cause a significant decrease in natural population growth. Despite the fact that their populations were shrinking because of the high rate of negative migration balance, a large percentage of type H towns displayed positive natural growth, mainly due to historical reasons. As demonstrated by numerous studies (Rosner 2007; Szukalski, 2018), belonging to foreign countries left a permanent imprint on the demographic structures in Poland. The inhabitants of the former Russian sector (Lubelskie and Podlaskie Voivodeships) and Austrian sector (Podkarpackie Voivodeship) were strongly attached to traditional family and procreation values. It was reflected in the high rate of marriages entered at a relatively young age, a significantly larger number of church weddings than civil ones, rare cases of cohabitation if not married and infrequent divorces (Szukalski, 2017). As regards procreation, the traditional values were reflected by the high fertility rate, a small percentage of childless people, a higher average age at childbirth (an effect of the higher fertility rate), as well as rare cases of bearing children out of wedlock (Szukalski, 2018).

As regards the number of population in the urban settlements of Eastern Poland, the third position was occupied by the towns and cities representing type F, which were characterised by a negative natural growth rate, higher than the negative migration balance. This means that the number of inhabitants was decreasing mainly due to natural population decline, which, in this group of towns, ranged from −12.64‰ to −0.42‰, while the migration balance rate stood at −7.4‰ to −0.05‰. Type F towns and cities made up 15% of all the urban centres in Eastern Poland. The largest percentage of them (nearly 20%) were the smallest towns, inhabited by less than 10,000 people. Among the largest cities, with populations exceeding 100,000, this particular demographic type was represented by only one urban centre—Elbląg (Figure 12.2). A significantly larger number of type F settlement units was found outside the peripheral area of Eastern Poland (22%) (Table 12.2). The much smaller percentage of type F towns in Eastern Poland, compared to the settlement units in the remaining part of the country, where natural population decrease was the predominant cause of the population decline, was determined, first of all, by the moral patterns concerning family, procreation and religion (cf. Szukalski, 2009; Kurek and Lange, 2013). In 2019, the regressive types described above (G, H and F) jointly included nearly 84% of the settlement units located in the peripheral areas of Eastern Poland, and 72.3% in the remaining areas of the country (Table 12.2).

Urban centres representing other demographic types were in the minority, both in Eastern Poland and in the whole country. In the peripheral areas, regressive type E was represented by 4.3% and in the remaining parts of Poland—by 9.5% of towns and cities (Table 12.2). Progressive urban centres were also in a minority, except the largest cities, where the number of inhabitants exceeded 100,000. Of these, 2 out of 6 were in Eastern Poland: Białystok (demographic type A) and Rzeszów (type C) (Figure 12.2). The situation was similar in the remaining parts of the country, where progressive type C was represented by 6 out of 32 largest settlement units. Among smaller towns, their percentage was smaller.

The analysis presented above shows that urban centres representing regressive types, where a population decrease was observed, were of greater significance in the structure of the settlement network in the peripheral areas of Eastern Poland than in the remaining areas of the country. The predominant depopulation factor in Eastern Poland towns and cities was the negative migration balance, ranging from $-4.75‰$ in towns inhabited by 10–20,000 people to $-0.25‰$ in the largest cities (Table 12.3). The importance of the negative migration balance in the depopulation of urban centres in the remaining areas was smaller, ranging from $-2.96‰$ in the smallest settlement units, inhabited by less than 10,000 people, to $-0.27‰$ in the largest centres, populated by over 100,000 people (Table 12.3).

The impact of natural growth on the depopulation of towns located in the peripheral areas of Poland was much weaker than the influence of migration loss. It was diversified, ranging from $-2.14‰$ in the smallest towns to $0.65‰$ in the largest ones. In the towns located in the remaining area of the country, natural loss was observed in all size categories of towns. It ranged from $-1.92‰$ in the largest urban centres to $-0.63‰$ in medium-sized towns, populated by 20–50,000 inhabitants (Table 12.3).

In the territory of Eastern Poland, the smallest population decrease due to migration ($-0.25‰$) and the largest natural growth ($0.65‰$) were observed in the largest cities, inhabited by over 100,000 people, where the effects of the economic crisis were less acute as a result of their functional diversity. The adverse effects of the crisis were felt most strongly in small and medium-sized towns, which usually performed central functions for the agricultural environment. Negative population trends were reinforced there by natural population loss, caused by emigration, mostly of women and young people. Traditional family and religious values were insufficient to prevent the negative outcomes of depopulation.

Table 12.3 Mean values of natural growth and negative migration balance in towns and cities of different sizes in Eastern Poland, compared to other parts of Poland, in 2019

Size of town (number of inhabitants)	Towns/cities in Eastern Poland		Towns/cities in remaining areas		Towns/cities in Poland	
	Mean value					
	Natural growth (‰)	Migration balance (‰)	Natural growth (‰)	Migration balance (‰)	Natural growth (‰)	Migration balance (‰)
>100,000	0.65	−0.25	−1.92	−0.27	−1.51	−0.27
50,000–100,000	−0.43	−4.15	−1.48	−2.40	−1.25	−2.79
20,000–49,999	−1.49	−4.58	−0.63	−0.84	−0.81	−1.62
10,000–19,999	−0.37	−4.75	−1.54	−2.35	−1.27	−2.90
<10,000	−2.14	−3.73	−1.75	−2.96	−1.86	−3.17
Total	−1.60	−3.94	−1.53	−2.37	−1.55	−2.76

Source: Authors' calculations, based on www.stat.gov.pl

The predominant role of emigration in the depopulation of towns and cities in Eastern Poland resulted from several factors. One of them was the much more attractive labour market in the metropolitan centres of the neighbouring voivodeships. Emigrants from Warmińsko-Mazurskie Voivodeship settled mostly in Gdańsk (Pomeranian Voivodeship) and Warsaw (Masovian Voivodeship), those from Podlaskie and Lubelskie Voivodeships—in Warsaw, from Podkarpackie – in Krakow (Małopolskie Voivodeship), from Świętokrzyskie—in Warsaw and Krakow (Celińska-Janowicz et al., 2010). People abandoning the towns of Eastern Poland also looked for better living conditions abroad. Moving to another country offered them opportunities to improve their financial situation or develop a career, but it was also rooted from the emigrational tradition, developing in the peripheral areas of Poland for many years (Cieślińska, 2012). Both before and after Poland joined the European Union, the inhabitants of those areas migrated mostly to Great Britain, Germany and the United States. The high rate of negative migration balance indicates that in the urban centres of Eastern Poland, strong pushing out factors are still at work, related to the huge differences in the living standard and conditions among individual regions of Poland, as well as between Poland and the countries accepting the emigrants.

The pace of ageing among the inhabitants of urban centres in Eastern Poland

Over the period analysed in this chapter, population ageing was noticeable in all the towns and cities of Eastern Poland (Figure 12.3). It is a stable trend, observed since the turn of the 21st century.

The demographic old age ratio (AR) calculated for the whole set of the urban centres in Eastern Poland at 7.44 and in individual cases, it ranged from 0 (Stalowa Wola in Podkarpackie Voivodeship) to 17.53 (Iwonicz-Zdrój in Podkarpackie Voivodeship). In 2008–2019, the pace of ageing among the inhabitants of towns and cities in the peripheral areas was faster than in the remaining areas of Poland (AR = 6.74) (Table 12.4). The slowest pace of population ageing was observed in the largest cities, inhabited by over 100,000 people. In Eastern Poland, it was lower (AR = 4.99) than in the remaining areas of Poland (AR = 5.29). In Eastern Poland, cities of this size were characterised by natural population growth, while in the remaining areas of the country there was a decrease in natural population growth (Table 12.3). In peripheral areas, the urban population ageing processes were the fastest in medium-sized towns, populated by 20–50,000 inhabitants (AR = 8.08), while in other parts of Poland, a similar process was observed in small towns populated by 10–20,000 people (AR = 7.16) (Table 12.4).

The urban centres in Eastern Poland did not include any towns or cities where the population age structure was rejuvenating (group I) (Table 12.5). In the remaining area of the country, there were 14 such urban centres: Warsaw (−0.45) and 13 smaller towns, located within the area of influence of large cities. That meant that the urban centres in the peripheral areas of Eastern Poland, including the largest ones, were devoid of any factors attracting young migrants who might have rejuvenated the population.

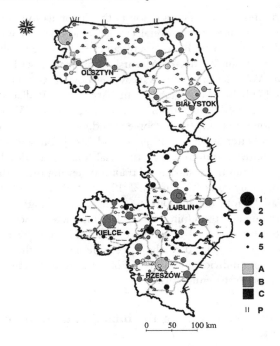

Population ageing pace:

A - low
B - average
C - high

Urban population (in thousands):

1 - over 100
2 - 50-100
3 - 20-49,999
4 - 10-19,000
5 - under 10

Figure 12.3 Pace of population ageing in the towns/cities of Eastern Poland.
Source: Authors' elaboration.

Table 12.4 Mean values of the population ageing ratio (AR) for towns/cities in Eastern Poland and in other areas of Poland, in 2008–2019

Town/city population	Mean values of the population ageing ratio		
	Towns/cities in Eastern Poland	Towns/cities in the remaining areas	Towns/cities in Poland - total
>100,000	4.99	5.29	5.24
50,000–100,000	6.77	6.89	6.87
20,000–49,999	8.08	6.16	6.57
10,000–19,999	8.03	7.16	7.36
<10,000	7.17	6.85	6.93
Total	7.44	6.74	6.89

Source: Authors' calculations based on www.stat.gov.pl

A slow pace of ageing (group II) was observed in 24 towns of Eastern Poland, which made up about 12% of the studied urban centres (Table 12.5). The percentages were similar in the towns of Eastern Poland—11.8%, and in the towns in the remaining areas of Poland—12.2% (Table 12.5). Slow ageing was characteristic of two out of six largest cities, populated by over 100,000 inhabitants: Białystok (2.97) and Rzeszów (2.29). In both, a positive natural growth was observed, which slowed down the pace of population ageing. In the first case, the natural growth compensated the negative migration balance, while in the other it was much less significant. The cities performed the function of regional growth poles; they were characterised by a more diversified labour market compared to the remaining cities of this size.

Among smaller urban centres, the percentage of towns characterised by a slow pace of ageing was smaller than that of the urban centres described above. In Eastern Europe, they mostly included the smallest towns, inhabited by up to 20,000 people (15.3%—the towns of up to 10,000 inhabitants, 7.3%—the towns of 10–20,000 inhabitants). In the remaining areas of the country, the percentage of medium-sized and small towns, characterised by a low pace of ageing was not highly diversified (11.4%–12%). The main cause of a lower than average rate of urban population ageing in peripheral towns was the people's attachment to family and religious values. The relatively slower pace of population ageing was the result of natural growth and a high fertility rate. Only in some small towns in the area in, natural growth was accompanied by a positive migration balance, resulting from a well-developed labour market and a favourable location in relation to roads. The reduction of negative changes in demographic structures may also be due to the location rent in relation to large urban centres. However, due to the fact that the largest urban centres in the studied peripheral areas were of regional significance, at the most, the location rent in relation to them was lower than in relation to centres of supraregional significance in other parts of the country, and they did not play any major role in slowing down the pace of population ageing.

The average ageing rate (group III) was characteristic of over 72% of towns/cities in Poland. Their percentage in the peripheral areas was smaller than in the remaining areas. In Eastern Poland, they stood at 68%, while in the remaining areas of the country—at nearly 74% (Table 12.5). In peripheral areas, medium-sized towns (20–50,000 inhabitants—82%; 50–100,000—80%) constituted the largest percentage of towns representing an average pace of population ageing. Large percentages of this type of urban centres were also observed among towns and cities of other sizes. An average pace of ageing characterised four out of six cities populated by over 100,000 people. Towns representing a similar ageing rate constituted 65% of all urban centres inhabited by up to 10,000 inhabitants and about 63% of those inhabited by 10–20,000 people. The average pace of ageing dominated in towns/cities of all population sizes. It was mostly the result of the economic crisis, i.e. of the liquidation or restructuring of local production plants. Job reductions caused an increase in unemployment, which, in turn, stimulated the inhabitants to migrate. Since it is mainly young people that emigrate, it resulted in a decreasing natural growth. It was possible to maintain the production function in some urban centres, but the scale of employment at local industrial plants generally decreased.

Table 12.5 Diverse pace of population ageing in Polish towns/cities, in 2008–2019

Pace of population ageing			Rejuvenation Group I	Low Group II	Average Group III	High Group IV	Total
Towns by the number of inhabitants							
Towns/cities in Eastern Poland	>100,000	Number	0	2	4	0	6
		%	0.0	33.3	66.7	0.0	100,0
	50,000–100,000	Number	0	1	8	1	10
		%	0.0	10.0	80.0	10.0	100,0
	20,000–49,999	Number	0	0	23	5	28
		%	0.0	0.0	82.1	17.9	100,0
	10,000–19,999	Number	0	3	26	12	41
		%	0.0	7.3	63.4	29.3	100,0
	<10,000	Number	0	18	77	23	118
		%	0.0	15.2	65.3	19.5	100,0
	Total	Number	0	24	138	41	203
		%	**0.0**	**11.8**	**68.0**	**20.2**	**100,0**
Towns/cities in remaining areas	Total	Number	14	84	509	82	689
		%	**2.0**	**12.2**	**73.9**	**11.9**	**100**

Source: Authors' calculations based on www.stat.gov.pl

Urban centres displaying an average pace of population ageing represented different demographic conditions. Three categories were distinguished: the first one was characterised by both: natural growth decline and negative migration balance, the second one—by either natural growth decline or negative migration balance, and the third one—by natural population growth and positive migration balance. In Eastern Poland, out of the 138 settlements representing an average pace of population ageing, 90 (over 65%) were characterised by both natural growth decline and negative migration balance. This group included urban centres from all size classes. These values demonstrate the large scale of the demographic crisis among the towns and cities in peripheral areas. Many of them were affected not only by the economic crisis related to the restructuring of local industrial plants or the decline of state-owned farms, but also to the limited access to transport and an unfavourable location in relation to large urban centres. The next category of towns presenting an average pace of population ageing was characterised by only one negative phenomenon (natural growth decline or negative migration balance) of considerable intensity, which determined the pace of population ageing. There were about 33 (32%) such towns in peripheral areas (205 = ca. 55% in the remaining areas) and they were located mostly in transitional zones between large urban areas and voivodeship peripheries. Ageing processes were strongly influenced by the economic crisis.

The group of towns with an average pace of population ageing also included those where both natural growth and migration balance displayed positive values. There were three such towns in the analysed area of Eastern Poland: Sędziszów (ca. 12,300 inhabitants), Iława (ca. 33,300 inhabitants) and Ełk (ca. 63,100 inhabitants). Sędziszów developed due to its location next to international road S7 and flourishing plants producing power boilers. Iława and Ełk are destinations where tourism and ecology are heavily invested in. In these towns, senior inhabitants were provided with good living conditions.

The group of urban centres characterised by a fast pace of population ageing (group IV) included 123 settlement units. In Eastern Poland, they made up over 20% of towns and centres, and in the remaining areas of the country nearly 12% (Table 12.5). A rapid pace of population ageing was typical primarily of small towns. In peripheral areas, they constituted nearly 30% of settlements inhabited by 10–20,000 people and nearly 20% of those populated by up to 10,000 inhabitants (Table 12.3). In the remaining parts of the country, this percentage was smaller and stood at slightly over 14% in both population size classes. The majority of towns with a fast pace of population ageing, regardless of their size, were characterised by negative migration balance as well as natural population decline. In Eastern Poland, there were 35 such urban centres in the group of 41 settlements included in the study, and in the remaining area of Poland—65 among 82. Urban centres characterised by positive values of natural growth and migration balance were not identified in the group of settlements where the pace of population ageing was the fastest.

In the case of peripheral areas, the population was ageing primarily due to outward migration. In nearly all the towns and cities of Eastern Poland which were characterised by the highest ageing ratios, negative migration balance was considerably greater than natural population decline or counterbalanced the significance of the small natural growth. In peripheral areas, among the urban centres present-ing a fast pace of population ageing, there were none where increased migration was observed. In the case of towns with rapidly ageing populations, located periph-erally, the basic factor deciding that they belonged to this group was a very high level of negative migration balance. Those centres were located in a farming envi-ronment, with an underdeveloped urban network and poor transport accessibility. Unfavourable demographic processes (a growing percentage of population at the post-productive age and unbalanced sex structure due to the outflow of young women) have been observed there for a long time. Therefore, the areas discussed in this chapter are considered to be population problem areas (Bański, 2002).

Conclusions

The analysis presented above made it possible to conclude that the urban centres located in the peripheral areas of Poland and, at the same time, of the European Union, are affected by negative demographic processes, which are manifested by the fast pace of urban population ageing—higher than in the towns and cities in the remaining area of the country. The pace of population ageing in the studied settlement units of Eastern Poland varied, depending on the size expressed by the number of inhabitants. The ageing processes were the slowest in the largest cities, which performed the function of regional growth centres. Most of them were characterised by natural growth, which could be the result of the inhabitants' attachment to traditional family and religious values. Regrettably, natural growth in nearly all largest urban centres of Eastern Poland was accompanied by a negative migration balance, contributing to the declining number of inhabitants. The ageing processes observed in small and medium-sized towns were more intensive than in the large regional centres of Eastern Poland, mainly due to the considerable migra-tion outflow.

The research also showed that in the peripheral areas of Poland, the percentage of urban centres representing a regressive type with regard to the number of inhab-itants, was higher than in the remaining areas of the country. The decrease in the populations of the urban centres in Eastern Poland was the result of the negative migration balance, to a much larger degree than in the remaining regions. It was the smallest in the largest cities, where the economic crisis was milder, and the larg-est in small and medium-sized towns, where it was most severe. The factors stimu-lating emigration from the towns of Eastern Poland included the labour market, which was less developed than in the neighbouring regions, as well as the histori-cally conditioned emigrational culture. Negative population trends in peripheral areas were reinforced with natural population decline, which undoubtedly resulted from increased emigration, mostly among women and young people. Traditional family and religious values did not prevent contemporary negative demographic phenomena in the urban centres of Eastern Poland.

The unfavourable population trends described in this chapter have a negative impact on nearly all spheres of life. They may lead to decreased national and regional budget revenues, a lower level of entrepreneurship, the erosion of human and social capital and shrinking technical infrastructure. This, in turn, results in a lower supply of jobs and, consequently, growing unemployment, causing subsequent waves of emigration. Breaking this vicious circle in the urban centres of Eastern Poland will be very difficult.

The unfavourable demographic processes described above contribute to a deeper peripherality of the region under discussion.

References

Ardener E., 2012. Remote areas: Some theoretical considerations, *Journal of Ethnographic Theory*, 2 (1), pp. 519–533.

Bański J., 2010. Granica w badaniach geograficznych – definicja i próby klasyfikacji, *Przegląd Geograficzny*, 82 (4), pp. 489–508.

Bański J., 2005. Suburban and peripheral rural areas in Poland: The balance of development in the transformation period, *Geografický časopis*, 57 (2), pp. 117–130.

Bański J., 2002. Typy ludnościowych obszarów problemowych, [in:] J. Bański i E. Rydz (eds.), *Społeczne problemy wsi*, PTG, IGiPZ PAN, Warszawa, pp. 41–52.

Celińska-Janowicz D., Miszczuk A., Płoszaj A., Smętkowski M., 2010. Aktualne problemy demograficzne Polski Wschodniej, *Raporty i analizy EUROREG*, 5.

Cieślińska B., 2012. *Emigracje bliskie i dalekie. Studium współczesnych emigracji zarobkowych na przykładzie województwa podlaskiego*, Wydawnictwo Uniwersytetu w Białymstoku, Białystok.

Copus A., 1999. Peripherality and peripherality indicators, *North-The Journal of Nordregio*, 10 (1), pp. 11–15.

Davies S., Michie R., 2011. *Peripheral regions: A marginal concern?*, Paper presented at *European Regional Policy Research Consortium*, Scotland, UK, 2/10/11–4/10/11.

Długosz Z., 1997. Stan i dynamika starzenia się ludności Polski, *Czasopismo Geograficzne*, 68, pp. 227–232.

Domański B., Noworól A., (eds.), 2010. *Badanie funkcji, potencjału i trendów rozwojowych miast w województwie małopolskim*, Małopolskie Obserwatorium Polityki Rozwoju, Kraków.

Ferrao J., Lopes, R., 2004. Understanding peripheral rural areas as contexts for economic development, [in:] W.L. Labrianidis (ed.), *The Future of Europe's Rural Peripheries*, Ashgate Publishing Ltd, Aldershot, Hampshire, pp. 31–61.

Gorzelak G., 2007. Strategiczne kierunki rozwoju Polski Wschodniej, *Ekspertyzy wykonane na zamówienie Ministerstwa Rozwoju Regionalnego na potrzeby opracowania Strategii Rozwoju Społeczno-Gospodarczego Polski Wschodniej do roku 2020, t. I.*, Ministerstwo Rozwoju Regionalnego, Warszawa.

Gren J., 2003. Reaching the peripheral regional growth centres, centre-periphery convergence through the Structural Funds' transport infrastructure actions and the evolution of the centre-periphery paradigm, *European Journal of Spatial Development*, 3, https://archive. nordregio.se/Global/Publications/Publications%202017/Refereed_3_Gren(2003).pdf [access March 5, 2020].

Hall C.M., Harrison D.W., 2015. Vanishing peripheries: Does tourism consume places? *Tourism Recreation Research*, 38 (1), pp. 71–92.

Hoekveld J.J., 2012. Time-space relations and the differences between shrinking cities, *Built Environment*, 38 (2), pp. 179–195. https://stat.gov.pl [accessed March 12, 2020].

Jażewicz I., 2005. Przemiany społeczno-demograficzne i gospodarcze w małych miastach Pomorza środkowego w okresie transformacji gospodarczej, *Słupskie Prace Geograficzne*, 2, pp. 71–79.

Jezierski A. (ed.), 1994. *Historia Polski w liczbach*, Głowny Urząd Statystyczny, Warszawa.

Kamińska W., Mularczyk M., 2014a. Assessment of the economic cohesion of rural areas in Poland. A dynamic and spatial approach, [in:] W. Kamińska, K. Heffner (eds.), *Rural Development and EU Cohesion Policy*, Studia Regionalia KPZK PAN, 39, pp. 41–67.

Kamińska W., Mularczyk M., 2014b. Demographic types of small cities in Poland, *Miscellanea Geographica*, 18 (4), pp. 24–33.

Kantor-Pietraga I., Krzysztofik R., Runge J., 2012. Kontekst geograficzny i funkcjonalny kurczenia się małych miast w Polsce południowej, [in:] K. Heffner, A. Halama, (eds.), *Ewolucja funkcji małych miast w Polsce*, Studia Ekonomiczne, Zeszyty Naukowe Wydziałowe, pp. 9–24.

Kotowska I. E., Giza-Poleszczuk A., 2010. Zmiany demograficzno-społeczne i ich wpływ na rekonceptualizację polityki rodzinnej w kierunku równowagi w zakresie ochrony praw rodziny i poszczególnych jej członków. Polska na tle Europy, [in:] E. Leś, S. Bernini (eds.), *Przemiany rodziny w Polsce i we Włoszech i ich implikacje dla polityki rodzinnej*, Wydawnictwa Uniwersytetu Warszawskiego, Warszawa, pp. 31–68.

Kurek S., Lange M., 2013. *Zmiany zachowań prokreacyjnych w Polsce w ujęciu przestrzennym*, Wydawnictwo Uniwersytetu Pedagogicznego, Kraków.

Miszczuk A., 2013. *Uwarunkowania peryferyjności regionu przygranicznego*, Norbertinum, Lublin.

Okólski M., 2010. Wyzwania demograficzne Europy i Polski, *Studia Socjologiczne*, 4 (199), pp. 37–78.

Olechnicka A., 2004. *Regiony peryferyjne w gospodarce informacyjnej*, Wydawnictwo Naukowe Scholar, Warszawa.

Pezzi M.G., Urso G., 2016. Peripheral areas: conceptualizations and policies. Introduction and editorial note, *Italian Journal of Planning Practice*, 6 (1), https://www.researchgate.net/publication/311934914_Peripheral_areas_Conceptualizations_and_policies_Introduction_and_editorial_note [accessed March 12, 2020].

Population trends 1950–2100. Globally and within Europe, 2020. https://www.eea.europa.eu/data-and-maps/indicators/total-population-outlook-from-unstat-3/assessment-1, [accessed March 8, 2020].

Program Polska Wschodnia 2014–2020. Załącznik nr 4. Diagnoza wyzwań, potrzeb i potencjałów obszarów objętych Programem Operacyjnym Polska Wschodnia 2014–2020, 2014, Ministerstwo Funduszy i Polityki Regionalnej, https://www.polskawschodnia.gov.pl/strony/o-programie/dokumenty/program-polska-wschodnia-2014-2020/, [accessed March 06, 2020].

Proniewski M., 2014. Polityka rozwoju regionów peryferyjnych, Optimum, *Studia Ekonomiczne*, 6 (72), pp. 79–90.

Rakowska J., 2016. Zróżnicowanie poziomu starości demograficznej Polski w ujęciu lokalnym, *Economic and Regional Studies*, 9 (2), pp. 13–23.

Rosenbers M.W., Wilson K., 2018. Population Geographies of Older People, [in:] M.W. Skinner, G.J. Andrews, M.P. Cutchin (eds.), *Geographical Gerontology. Perspectives, Concepts, Approaches*, vol. 8, Routledge, London, pp. 56–67.

Rosner A. (ed.), 2007. *Zróżnicowanie poziomu rozwoju społeczno-gospodarczego obszarów wiejskich a zróżnicowanie dynamiki przemian*, IRWiR PAN, Warszawa.

Rosset E., 1978. *Démographie de la vielliesse*, PAN Ossolineum, Wrocław.

Schmidt J. (ed.), 2007. *Granica*, Awel, Poznań.

Strategia rozwoju Polski Wschodniej, 2020. *Synteza, 2013*, Ministerstwo Rozwoju Regionalnego, Warszawa.

Swianiewicz P., 2007. Strategiczna analiza stanu spójności ekonomicznej i społecznej przeprowadzona na poziomie obszarów Nuts 4 i Nuts 5. *Ekspertyza przygotowana na potrzeby aktualizacji "Strategii rozwoju społeczno-gospodarczego Polski Wschodniej do roku 2020"*, MRR, Warszawa.

Szukalski P., 2018. Wielodzietność we współczesnej Polsce w świetle statystyk urodzeń, *Demografia i Gerontologia Społeczna - Biuletyn Informacyjny*, 3, pp. 1–6.

Szukalski P., 2017. Późne macierzyństwo – nowe zjawisko demograficzne w Polsce, *Demografia i Gerontologia Społeczna – Biuletyn Informacyjny*, 1, pp. 2–6.

Szukalski P., 2009. Przygotowanie do starości jako zadanie dla jednostek i zbiorowości, [in:] P. Szukalski (ed.), *Przygotowanie do starości. Polacy wobec starzenia się*, Instytut Spraw Publicznych, Warszawa, pp. 39–55.

Szul R., 1991. *Przestrzeń, gospodarka, państwo*, Wydawnictwo Naukowe Jan Szumacher, Warszawa.

Talarowski A., 2017. *O różnicach między zaborami*. Teologia Polityczna, https://teologiapolityczna.pl/adam-talarowski-o-roznicach-miedzy-zaborami [accessed March 10, 2020].

Wilczek-Karczewska M., 2015. Rozwój kolei żelaznych na ziemiach polskich w ujęciu historycznoprawnym, *Internetowy Kwartalnik Antymonopolowy i Regulacyjny*, 4 (1), pp. 100–124.

Wilkin J., 2003. Peryferyjność i marginalizacja w świetle nowych teorii rozwoju (nowa geografia ekonomiczna, teoria wzrostu endogennego, instytucjonalizm), [in:] A. Bołtromiuk, (ed.), *Regiony peryferyjne w perspektywie polityki strukturalnej Unii Europejskiej*, Wydawnictwo Uniwersytetu w Białymstoku, Białystok, pp. 44–52.

Wójcik M., 2009. Społeczna geografia wsi, *Czasopismo Geograficzne*, 80 (1–2), pp. 42–62.

Zarębski P., 2012. Obszary peryferyjne i ich potencjał rozwojowy, *Roczniki Naukowe Stowarzyszenia Ekonomistów Rolnictwa i Agrobiznesu*, 14 (4), pp. 131–136.

Zarycki T., 2007. Interdyscyplinarny model stosunków centro-peryferyjnych, *Studia Regionalne i Lokalne*, 1 (27), pp. 5–26.

13 Hallmark features of post-socialist urban development in Central Europe

Josef Kunc and Waldemar Cudny

This book is an edited volume presenting an analysis of socio-economic, functional and spatial transformation that took place after the fall of communism in post-socialist cities located in the region of Central Europe. The region according to the division made by Lewis (2014) includes Eastern Germany, Poland, the Czech Republic, Slovakia and Hungary. The volume encompasses a wide range of case studies presenting cities of different sizes categories and statuses from the afore-mentioned countries. Urban areas from Eastern Germany (Chemnitz, Leipzig), Poland (Łódź, Kielce, Katowice), the Czech Republic (Olomouc, Brno), Slovakia (Bratislava, Nitra) and Hungary (Pécs) were accepted for research. The book presents an in-depth analysis of the structural socio-economic transformation ongoing in post-socialist cities of Central Europe. The main research aim is to present a comprehensive summary of the processes of socio-economic and demographic changes taking place in the studied cities after the fall of communism in the region.

The rise of post-socialist cities corresponds in large part to the fall of the Berlin Wall and then the Soviet Union. At the same time, a political, economic, social and cultural transformation in Central and Eastern Europe was triggered. The term "post-socialist" seems most often used in the literature as a chronological marker, denoting the time after the Wall fell. Where socialist states remain in power, commentators and even scholars often use the term "post-socialist" to describe the contemporary context of policymaking with strong neoliberal overtones (Drummond and Young, 2020). The countries of the former Eastern bloc were forced to very quickly turn 40 years of centrally planned economies into economies with market principles, what brought considerable problems (Hamilton et al., 2005; Žídek, 2019).

With reference to Central and Eastern Europe, Borén and Gentile (2007) assume that the post-socialist transformation can be understood as the economic, political, institutional, and ideological changes associated with the abandoning of communism or state socialism and the embracing of capitalism. Stanilov (2007) adds that in the aftermath of the demise of the Soviet Union, the cities of Central Europe have undergone "radical transformation" at a rapid pace, and in many ways. In the presented publication, the authors divided the transformation processes basically into economic, political, social (cultural), demographic, functional and spatial.

In emerging post-socialist cities in Europe, Stanilov (2007) characterises the immediate issues of post-socialist transformation as involving housing privatisation,

DOI: 10.4324/9781003039792-13

property restitution, commercialisation of the city centres, decentralisation of housing and retail, growth of automobile ownership, etc. This means that privatisation and the increased concentration of urban decision-making in the hands of non-state actors have become the focal point of post-socialist urban change (Stanilov, 2007; Brade et al., 2009). Sýkora (1999) but points out that the post-socialist cities of the 1990s were not quickly and fully transformed into capitalist cities. Their development after the fall of communism showed several specific features which in those days could not be generalised into the model of a transitional city. Moreover, the transformation was not completed in many aspects even after thirty years.

One of the important features was that in the newly formed administrative and executive structures of public administration, post-socialist cities mostly strengthened their position compared to the period of socialism. At the same time, cities have become very important players not only in the political and institutional field but also in the field of economics, social responsibility and environmental sustainability (Mulíček and Toušek, 2004; Bertaud, 2006).

Except for the introduction and conclusions, the book includes eleven chapters with case studies from Germany (the former German Democratic Republic), Slovakia, the Czech Republic, Hungary and Poland, which will be summarised in the further part of this text.

The transformation of the post-socialist cities of Central Europe had a very similar course in most countries of the region, but certain differences in time and space could be observed. The case of East Germany (former GDR), which was annexed to the Federal Republic of Germany in the early 1990s, is treated differently in the discussions, significantly accelerating its transformation and creating certain features not inherent in other former Eastern Bloc countries (Forsyth, 2005; Turok and Mykhnenko, 2007). Re-integration into the capitalist world market had already been formally completed with the economic and monetary union in July 1990. This led to a shock transformation after 1990; within a short period of a few years, profound economic, political, social and cultural changes took place. The transformation was associated with extensive and widespread de-industrialisation, high and persistent unemployment, and massive shrinkage of population numbers (Rink et al., 2012; Haase et al., 2013).

Like many other post-socialist cities, Leipzig (ca. 600,000 inhabitants) and Chemnitz (ca. 250,000 inhabitants) had to cope with a severe demographic and economic decline after the political change and initial difficulties to find their place in the new political and economic system. After the change of systems, both German cities had considerable difficulties in abandoning the outdated development trajectory associated with industrial production and in modernising their economic, urban and governance structures. It was only after 2000 that there was a certain economic and population stabilisation (in the case of Leipzig also significant growth), investment began to manifest itself in the establishment of new businesses, and unemployment decreased.

Leipzig suffered from a de-industrialisation and shrinkage during the 1990s. However, in the 2000s, the city experienced re-growth and, in the 2010s, dynamic re-growth. The redevelopment was so intense that in the 2010s Leipzig became the fastest-growing city in Germany. Massive public investments, subsidies and, support programmes from central and regional governments were the basis for the revitalisation of the urban economy and space in Leipzig. These programmes also mobilised private capital, which joined the investment processes. Leipzig has thus caught up to the level of urban development evident elsewhere in Europe; its re-growth is comparable to that of other European cities of this size, such as Gdansk, Liverpool or Lyon.

However, despite the advantages of re-growth, some unsuspected disadvantages appeared. They included, among others, a shortage of affordable housing and schools or growing traffic congestions. The analysis showed that re-growth affected the afore-mentioned three crucial municipal policies (the housing market, public schools and public transportation) and created pressure for a reaction. The further prospects for Leipzig's population development foresee further re-growth, but it remains to be seen whether the high level of in-migration of recent years will continue. Re-growth solves some shrinkage-related problems, but brings with it new problems and challenges. Urban policymakers should generally be more open to varying future developments and react more quickly to both shrinkage and re-growth.

As a city with industrial traditions, Chemnitz was predestined for the typical accumulation regimes and production methods of Fordism, which were characterised by hierarchical regulation regimes, a strong state, and little societal differentiation. After the system change, Chemnitz had significant difficulties leaving this outdated development path and modernising its economic, urban and governance structures. This affected demographic and economic development, as well as urban structures.

For many years during the transformation period, the city remained focused on top-down management processes. Only very recently have new modes of governance been introduced, such as participatory approaches, the management of plurality, and the support of bottom-up processes. This can be interpreted as the adoption of the concept of "transformation", emphasising the unpredictability of social and spatial contexts. However, the riots and demonstrations of extremist groups following the deadly attack on Chemnitz residents by immigrants/asylum seekers have shown the fragility of this concept. Reconciling different parts of society and negotiating diversity will be a crucial task. A large part of society is afraid of social diversity and is therefore particularly vulnerable to authoritarian ideas. Thus, for years to come, the reconciliation of the different parts of society and the negotiation of diversity will be a major task. This includes the activation of the passive parts of the population and the introduction of participative formats of urban and social development.

Case studies from Polish cities and urban regions (Łódź, Kielce, the Katowice Conurbation and a group of peripheral cities in eastern Poland) address transformation challenges related, in particular, to the previous concentration and diversification of the industrial base and the current de-industrialisation, depopulation, and shrinking of cities and dynamic population ageing. Łódź, the third-largest Polish city (ca. 685,000), was the leading Polish centre of textile production until 1989 and after the fall of communism underwent intensive de-industrialisation and urban shrinkage due to the deep economic crisis and subsequent transformation.

Due to the increasing academic function Łódź gained well-educated specialists in a variety of domains, which allowed the city to develop new industries. Moreover, a special economic zone was also created as one of the most important means for regenerating the city. Another advantage is the rapid development of higher-order functions, related to higher education, business service, logistics and finance. Experts in these domains are attracted to Łódź given its competitive prices for renting modern commercial spaces or apartments. Łódź is increasingly perceived as an interesting tourist city, which is also helped by its central location and convenient road connections (also with Warsaw). A great chance for the further development of Łódź is efficient territorial self-government, successfully applying for EU funds and making rational use of them. On the initiative of the Łódź authorities is the organisation of the Green EXPO 2024 exhibition.

Developing the IT and BPO sectors, creating the financial centre as well as developing tourism in Łódź contribute to the modernisation of the city's economy, which for decades was based on the monoculture of the textile industry. As regards an improvement of the critical demographic situation, a certain chance lies in encouraging Łódź university students to stay in the city after graduating. The incentives include the interesting work market, the rich cultural offer of the city, and the low cost of living, which substantially compensates for earnings, which are lower than in other large Polish urban centres.

Despite the advantages presented above, Łódź still faces many socio-economic problems. For years, the city's main issue was the underdevelopment of the service sector. It is not long since the situation changed radically. Other serious problems are the inefficient city transport system, due to a large number of vehicles and traffic congestion and the growing debt of the city, resulting from the huge investments made by the city authorities. However, the most problematic issue is the demographic condition of the city. Since 1988, depopulation has caused the loss of almost 170,000 inhabitants, making Łódź the fastest-shrinking large city in Poland. According to demographic forecasts, without a substantial migration inflow, the population of the city will fall below 500,000 in the middle of the 21st century. The problem of urban shrinkage requires adopting a proper strategy. Łódź is an example of implementing an active strategy, which involves maintaining the existing resources, and, at the same time, a passive strategy, which involves control over the shrinking process and qualitative development.

Since the beginning of the transformation period, Kielce has been losing its population potential. In 1988–2019, the number of residents decreased by about 16,000 (currently ca. 195,000) and can be called a typical shrinking city. Since the late 1980s, the rate of demographic processes has declined considerably, which is manifested by natural population loss and negative migration balance. Negative trends are also confirmed by future forecasts: it is predicted that in 2050, Kielce will be inhabited by 138,000 people, which is about one-third less than today. The phenomena related to the second demographic transition, i.e. the decreasing fertility rate, falling number of new marriages, and the increasing number of divorces, may aggravate the population decline. The depopulation of the city will bring negative economic consequences, such as declining entrepreneurship and problems on the local labour market.

The results of the survey conducted among Kielce students lead to the conclusion that the shrinking of the city is unlikely to stop. Nearly half of the respondents declared their wish to leave the city in the near future. Implementing these plans would not only increase the negative migration balance, but also affect the population growth. It would also lead to decreasing the quality of the city's social capital. A more optimistic picture of Kielce's demography arises from the students' plans regarding the size and structure of their potential families. Most young people are planning to have at least two children, which would ensure a simple generational replacement.

Many changes characteristic of the second demographic transition were evaluated by the students negatively. Nevertheless, they are becoming increasingly common among the Polish society (divorces, single parenthood). Transformation and modernisation processes have led to a considerable cultural transformation of metropolitan communities. The study described herein indicates that in medium-sized, shrinking cities, these processes are less intense.

The transformation of the Katowice conurbation (ca. 2.4 million inhabitants) as one of the largest urban regions in Poland had at least three basic dimensions: functional, state and political, and geographical. The determining factor for the first of them was the specific monofunctional structure based on coal mining and other sectors of heavy industry. In practice, this factor makes the region similar to other areas of this type in Central Europe; nevertheless, due to the size of the Katowice region, the scale of the changes and related challenges was immeasurably larger.

An important distinguishing feature of the region's transformation is related to the clear morphological and administrative polycentricity of the Katowice conurbation. This seemingly uniform, territorially integrated area is really a collection of settlements characterised by often-differing policies or visions for development and additionally competing for investments that could form a new basis for their development.

Due to the depletion of resources and competition in a globalising world, many mining and industrial plants were sentenced to closure or complete restructuring. Both these processes took place; at the same time, however, it was possible to ease the negative consequences of economic transformation via an inflow of new investments, including direct foreign investments. As mentioned, this was not unaccompanied by problems and was not an unequivocally positive process. Nevertheless, from a perspective of 30 years of transformation, the process should be seen as generally positive. Another positive element is the limiting of industrial pollution and the expansion of green spaces.

Nevertheless, the transformation of the Katowice conurbation at the turn of the 20th and 21st centuries also caused several changes whose negative effects remain a great challenge for local authorities and residents. Undoubtedly the greatest challenge is slowing depopulation and reducing its consequences. In demographic terms, the region is also characterised by growing ageing. Both factors are currently a fundamental challenge for local and regional policy. Over the last 30 years the Katowice conurbation, currently one of the largest urban regions in Central Europe, has undergone probably the deepest transformation in the region. Due to its specific nature, this process is not only unfinished but will most likely last the longest in this part of Europe. Also from this perspective, it is apparent that the "built to suit" concept implemented through careful municipal planning and sustainability should be the basis for further transformation.

The last case study from Poland differs from the previous ones because it is devoted to the group of peripheral cities and towns in eastern Poland. The authors presented the relationship between population growth and migration balance. The analysis made it possible to conclude that the urban centres located in the peripheral areas of Poland and, at the same time, of the European Union, are affected by negative demographic processes, which are manifested by the fast pace of urban population ageing – higher than in the towns and cities in the remaining area of the country.

The pace of population ageing in the studied settlement units of Eastern Poland varied, depending on the size expressed by the number of inhabitants. The ageing processes were the slowest in the largest cities, which performed the function of regional growth centres. Most of them were characterised by natural growth, which could be the result of the inhabitants' attachment to traditional family and religious values. Regrettably, natural growth in nearly all the largest urban centres of Eastern Poland was accompanied by a negative migration balance, contributing to the declining number of inhabitants. The ageing processes observed in small and medium-sized towns were more intensive than in the large regional centres of Eastern Poland, mainly due to the considerable migration outflow.

The factors stimulating emigration from the towns of Eastern Poland included the labour market, which was less developed than in the neighbouring regions, as well as the historically conditioned emigrational culture. Negative population trends in peripheral areas were reinforced with natural population decline, which undoubtedly resulted from increased emigration, mostly among women and young people. The unfavourable population trends have a negative impact on nearly all spheres of life and the change of this negative development in the urban centres of Eastern Poland will be very difficult.

Hungary's Pécs (142,000 inhabitants) is another city in the Central European region that has been facing dynamic population decline and massive de-industrialisation in recent decades. In the example of Pécs, reducing the "idea" of a city into a pure concept (the former "mining city" vs. the new "city of culture"), and failing to account for sectors which did not fit this ideal picture has reduced the diversity of the local economy. Whereas the abandonment of a previously diversified industrial mix under state socialism had created a monofunctional city, the abandonment of manufacturing and the milieus which would sustain it during post-socialism has contributed to a loss of economic potential and processes of urban shrinkage.

Yet there is no easy return of the city of Pécs to the "typical", more successful growth path followed by the majority of Central European industrial regions. In the peripheral context, FDI inflows are limited, and they mostly just reinforce peripherality. It is more likely that the way forward lies in the exploitation of endogenous growth potential found mainly in domestic companies, and strongly localised sources of development. The city representatives must rediscover that its former successes as a peripheral city were once found in high-quality products in niche markets; and that its cultural functions had once been closely intertwined with its industrial innovation and industrial production – not merely publicly funded cultural consumption. Therein lies a path forward, towards re-specialisation and future prosperity in a high value-added niche role, which will, again, be different from the globally integrated spaces of mass production and mass employment.

Pécs must accept that its destiny may be that of a minor city in the 100–110,000 inhabitant range, and that being again a successful city means to accept aspiration and image of a diversified city, where culture (a culture-based economy) and industry are not rival future images, but complementary parts of a unified vision.

The second-largest city in the Czech Republic, Brno (approx. 380,000 inhabitants), is one of the other traditional industrial centres in the region under study. Like many other Central European cities, it has undergone a major economic and social transformation in the last thirty years, characterised by a process of de-industrialisation. Compared to other cities, however, Brno has never had a problem with a sharp and long-term decline in population. For more than 200 years, the industry in Brno not only played a major role in the economic production system, but also served as the system for social reproduction and establishing the cultural forms. Industrial production was, in the case of Brno, the major decisive factor concerning the spatial organisation and the functional-spatial layout of the city.

Transitional development in the transformation period also brought important changes in the forms of city corporatism. The status of traditional industrial branch representatives, operating in groups of lobbyists influencing decision-making processes in cities, gradually weakened in favour of new players. Moreover, since the 1990s both the economically developed and the post-socialist city were viewed as an entrepreneurial concept and Brno city started to fulfill their new role of a manager, who is fully responsible for the economic development. Nevertheless, the influence of the structures connected with the industry was still strong during the transformation period, or, more precisely, it was stronger than it should have been compared to the declining importance of the traditional industrial branches.

In the last 20 years, the industrial, but also commercial, service and institutional functions of the inner city have been gradually weakening at the expense of new economic structures located on the outskirts of the city. The transfer of production activities to newly formed industrial zones, technological and innovation parks connected to administrative and logistics centres, and the construction of large-scale retail entities have become an economic and spatial urban phenomenon. Besides the modern infrastructure, there emerged new substantial spatial elements in the city, the brownfields. The city representatives and residents perceive these localities ambivalent, as a certain threat to the quality of life, however, on the other hand, they represent the potential and the possibility to reshape the abandoned areas, often located in the vicinity of the city centre, into a lively city organism.

Despite the considerable manifestation of post-socialist de-industrialisation which culminated with the loss of economic dominance, the industrial heritage in Brno still leaves a highly significant trace that shapes the basic form of the city. Despite this, Brno has managed to break free from the more or less rigid industrial legacy in recent decades and has become a city with a modern urban environment. The city representatives build current and future development of the city primarily on the university environment and highly educated labour force, high-tech production, cutting-edge technologies in industry and services, culture and entertainment, and last but not least on quality infrastructure and connections with Europe.

The second case study from the Czech Republic is the example of the Moravian city of Olomouc (100,000 inhabitants), which presents the transformation of the university environment, which has often become the driving force of creative regional economies. There are various forms and types of creative centres at Palacký University in Olomouc, more than twenty have been identified. The analysis showed that the creative activities are in close relation to the needs of the city and region. The university works with a wide range of creative spaces, most of them established after the year 1989.

During the last three decades, the university in Olomouc gained a vibrant mosaic of its creativity for a regional creative economy, which is the ultimate goal. Cooperation with employers has been initiated and they find their business solutions at the university. This concerned organisations from the region and from across the whole Czech Republic. Top domestic and foreign scientists were employed in the university's hubs and laboratories. The gradual commercialisation of research results is obvious; however, its intensity has remained low so far. This is a certain handicap for the future operation of the university's prestigious scientific infrastructure. The executive university authorities approved the establishment of the "super-centre" called CATRIN (the Czech Advanced Technology and Research Institute) and this new institute has formally commenced its operation at the beginning of 2021. Time will tell whether this venture is a success.

An example of a successful economic transformation of the urban environment is Bratislava (ca. 440,000 inhabitants). The role of the city grew significantly after the division of Czechoslovakia and the establishment of the Slovak Republic as an independent state in 1993. Newly acquired capital city status and extraordinary development momentum merge into a unique trajectory known as the "Bratislava type". Between 1948 and 1989, Bratislava had a vulnerable geographical location right on the outer border of the Eastern bloc of Europe. This peculiarity blocked any economically sustainable development and geopolitical interventions remained effective.

For the first time since 1989, the city is free to develop, there is also a connection between revitalising inner-city neighborhoods and the growing integration of these places into the global economy. Attractiveness for the development of economic activities has increased during the social and economic transition. Corporate headquarters, business and commercial centres, hotels, and tourist facilities have flooded the central city of Bratislava similarly to metropolises in Central Europe. Revitalisation, economic and social upgrading of areas near the city centre had special prerequisites in Bratislava. There were three qualitatively different areas close to the centre, which were seemingly suitable for revitalisation. Large business centres with residential functions quickly mushroomed, overlooking all of these attractive property locations.

The city and its metropolitan area have started to dominate in the national regional structure. The unique heritage of earlier decades changed slowly under the influence of developments in the local and regional economy. The growth of a truly modern city has started and continues. Today, the city has a mixture of functions generated mainly by the business service economy represented by global corporations. Today, the Bratislava region represents a growth pole even at the scale of the European Union's east part. Statistically, the Bratislava region is considered the top ten richest within the European Union single market.

Another example of a relatively successful post-socialist transformation is the Slovak city of Nitra (approx. 78,000 inhabitants). Like the capital of Slovakia, Bratislava successfully coped with the transformation of the industrial base and with population shrinkage, which, however, was not as crucial as in other described cities in Central Europe. The spatial structure of post-socialist Nitra contains new elements of urban landscapes; however, significant and relatively large urban sections still resemble a socialist city.

Nitra is a city still in the transformation period. Despite the partial shrinking process it is characterised by dynamic developmental changes. The post-socialist urban development is gaining more important impacts on the overall urban organisation, although there are remaining some socialist patterns in the new urban landscape. The land cover change data showed that a gradual shrinkage of natural and agricultural surfaces took place in the period between 1990 and 2019. This was mainly the result of processes affecting the location of infrastructure, industry and housing.

Nitra as the regional centre has attracted human resources, changed the employment structure, mainly in the tertiary and quaternary sectors, reorganised production and non-production activities, and directed functional and spatial development into a new model of the post-socialist urban environment. To some extent, the impact of globalisation, and to a large extent European integration, and the internationalisation of the Slovak economy and society, together with developmental policies and regulations, have all had profound effects on the inner urban structure. Major policy changes and initiatives are needed in Nitra, which should improve its competitiveness in Slovak and Central-Eastern European city networks.

According to Węcławowicz (2018, 265), countries from Central Europe were functioning for about forty years under the domination of the centrally planned economy and the communist political system. These conditions led to the establishment of the socialist city.

The socialist city was characterised by a number of features, such as the domination of the state-owned industry, social equality, underdevelopment of services, strong specialisation in traditional production, population growth due to a positive birth rate, and positive internal migration balance (mainly from rural to urban areas) (Table 13.1).

The collapse of the communist system at the turn of the 1980s and 1990s was the main factor that led to a series of profound changes in socialist cities. These changes, in turn, resulted in the emergence of a new type of city in Central Europe, i.e. a post-socialist city (see: Liszewski, 2001; Ondoš and Korec, 2008; Sýkora, 2009). In the following decades, the post-socialist cities became similar to the cities of Western Europe, adopting the features of a non-liberal city (Węcławowicz, 2018).

"Neoliberalism assumed that human decisions are based on rationality, individuality, and self-interest. Such decisions based on free choices are the best for

Table 13.1 The main features of transition process from socialist to post-socialist (neo-liberal) city

	Socialism		Post-socialism
Processes	Socialist city features	Transition stage processes	Post-socialist city – a neoliberal city
Industrialisation based on traditional branches of production	Production-based economy and domination of traditional industries	De-industrialisation and development of new (modern) industries	Fall of traditional production and rise of modern industries based on the knowledge economy
Limitation of services	Underdevelopment of service sector	High demand for services	Growth of service-based economy
Strong specialisation	Monofunctional structure of economy mostly based on one type of industrial branches	Diversification of economy	Multifunctional structure of the economy
Dominative role of the state	Domination of state-owned economy	Privatisation	Domination of private capital
Top-down urban management	Inflexible urban socio-economic structure	Regionalisation and bottom-up urban management	A flexible urban socio-economic structure
Social housing development	Domination of unified housing estates (blocks of flats)	Privatisation of housing market	Development of modern housing estates, suburbanisation, revitalisation
Positive birth rate and the balance of internal migrations	Population growth	Second demographic transition	Urban shrinkage
Social and economic egalitarianism	Social equality	Social diversification and external migrations	Diverse social and economic structure

Source: Authors' elaboration.

economic and social development, and therefore state or other public interventions should be avoided, and free entrepreneurship should be strengthened" (Cudny, 2019, 8). The rise of a new, post-socialist city was a result of the implementation of a neoliberal agenda to regional and urban development in the post-socialist states of Central Europe. There were several important processes triggered by the fall of communism and the introduction of neoliberalism leading to the creation of post-socialist cities. These processes encompassed, for instance, de-industrialisation and development of new branches of production, growth of demand for services, diversification of the economy, and massive privatisation, including privatisation of the housing market. Moreover, the system change affected the management of urban areas. More bottom-up approaches were introduced into urban management, and regionalisation increased in importance. Demography and social structure underwent in-depth changes as a result of the second demographic transition and social diversification.

The most important results of the aforementioned processes included the fall of traditional production and the rise of modern industries, growth of services, diversification of the economy, and domination of private capital which replaced state ownership. The urban socio-economic structure became thus more flexible and adjustable to the current conditions. The spatial structure has undergone changes as well; they included, for instance, urban fallows creation, the rise of modern housing estates, and revitalisation. The socio-demographic transition encompassed urban shrinkage and the creation of diversified social and economic structures (Table 13.1).

References

Bertaud, A. (2006). The Spatial Structures of Central and Eastern European Cities. In Tsenkova, S., Nedovic-Budic, Z., eds. *The Urban Mosaic of Post-Socialist Europe Space, Institutions and Policy*, pp. 91–110. New York, Verlag-Springer.

Borén, T., Gentile, M. (2007). Metropolitan Processes in Post-Communist States: An Introduction, *Geografiska Annaler: Series B, Human Geography* (2), 95–110.

Brade, I., Gunter, H., Karin Wiest, K. (2009). Recent Trends and Future Prospects of Socio-spatial Differentiation in Urban Regions of Central and Eastern Europe: A Lull Before the Storm? *Cities* 26, 233–244.

Cudny, W. (2019). *City Branding and Promotion: the Strategic Approach*. London, Routledge.

Drummond, L.B.W., Young, D. (Eds.) (2020). *Socialist and Post-Socialist Urbanism: Critical Reflections from a Global Perspective*. Toronto, Buffalo, London, University of Toronto Press.

Forsyth, T. (2005). *Encyclopedia of International Development*. London, Routledge.

Haase, A., Bernt, M., Grossmann, K., Mykhnenko, G., Rink, D. (2013). Varieties of Shrinkage in European Cities, *European Urban and Regional Studies* 23 (1), 1–17.

Hamilton, I.F.E., Dimitrovska Andrews, K., Pichler-Milanović, N. (2005). *Transformation of Cities in Central and Eastern Europe. Towards Globalization*. Tokyo, New York, Paris, United Nations University Press.

Lewis, P.G. (2014). *Central Europe since 1945*. London, New York, Routledge.

Liszewski, S. (2001). Model przemian przestrzeni miejskiej miasta postsocjalistycznego. In Jażdżewska, I., ed. *Miasto postsocjalistyczne – organizacja przestrzeni miejskiej i jej przemiany: XIV Konwersatorium wiedzy o mieście*, pp. 303–310. Łódź, Wydawnictwo UŁ.

Mulíček, O., Toušek, V. (2004). Changes of Brno Industry and Their Urban Consequences, *Bulletin of Geography (socio-economic series)* 3, 61–70.

Ondoš, S., Korec, P. (2008). The Rediscovered City: A Case Study of Post-socialist Bratislava, *Geografický časopis* 60, 199–213.

Rink, D., Haase, A., Grossmann, K., Couch, C., Cocks, M. (2012). From Long-term Shrinkage to Re-growth? A Comparative Study of Urban Development Trajectories of Liverpool and Leipzig, *Built Environment* 38 2, 162–178.

Stanilov, K. (Ed.) (2007). *The Post-Socialist City Urban Form and Space Transformations in Central and Eastern Europe after Socialism*. Dordrecht, Springer.

Sýkora, L. (1999). Changes in the Internal Spatial Structure of Post-Communist Prague. *GeoJournal* 49(1), 79–89.

Sýkora, L. (2009). Post-socialist Cities. In Kitchin, R., Thrift, N., eds. *International Encyclopedia of Human Geography*, pp. 387–395. Oxford, Elsevier.

Turok, I., Mykhnenko, V. (2007). The Trajectories of European Cities, 1960–2005. *Cities* 24(3), 165–182.

Węcławowicz, G. (2018). *Geografia społeczna Polski*. Warszawa, PWN.

Žídek, L. (2019). *Centrally Planned Economies: Theory and Practice in Socialist Czechoslovakia*. London, New York, Routledge.

Index

Note: Pages in *italics* refers figures; **bold** refers tables and followed by 'n' denotes notes.

abortion in Poland 88–89, **88**
Abramowska-Kmon, A. 71
Academia film Olomouc (AFO) 39
AfD (*"Alternative für Deutschland"* / *"Alternative for Germany"*) 59, 61
anti-asylum demonstrations 61
Apanowicz, J. 6
arable land 23, 25, 126
Arbeitsbeschaffungsmaßnahmen / ABM 52
ArcGIS software 17, 137
Areál Slatina 149
Art Deco buildings 56
Art Nouveau buildings 50
Asea Brown Boveri (ABB) 141
Ashton, D. 30, 33, 37, 39
"asylum background," people with 59
"Aufbrüche. Opening Minds. Creating Spaces" 60
Austria 1, 118, 124, 127–128, 150–151, 225–226, 230
Austrian (Moravian) Manchester 138
Avast 148

bachelors, in Poland 71
Balkans trade networks 97
Baltics 1
Banská Bystrica 20
Bański, J. 5
Berlin Wall 2, 48, 50, 162, 242
birth rate: in Chemnitz 51, 53; in Katowice conurbation 208; in Kielce 74; in Leipzig 169
Borén, T. 242
Bouzarovski, S. 15
"branch-plant" manufacturing companies 98

Bratislava 20, 113, 116, 118; economic transition in 121–124; employment structure in 122, **122**; geographical location of *116*; geography, history and local governance design 117–118; inner spatial structure 124–126; population in 118–121, **119–120**; post-socialist change in 21; privatisation of state-owned economic institutions in 123; socio-cultural zones around 127; spatial structure, special problems of 130–131; suburbanisation in 119–120, 126–130
"Bratislava type" 131, 249
Brno 136, 248; business incubators in 149; changes in the city economy and policy 142–144; cultural life in 152; de-industrialisation in 141, 144–150; education in 152; employment in **141**; industrial tradition in 137–142; interconnection in 150–151; internal transportation 150–151; labour market in 152; location of brownfield sites in *147*; map of *138*; modern urban environment 150; population of 146, **146**; spatial stratification of developing zones in *149*; sports events in 152; transitional development 144; transition period, first years of 137–142
brownfield spaces in Katowice conurbation 200
"Brühl" district 57–58, *59*
Budín 117
Bulgaria 1
Burghardt, A. 198
business incubators: in Brno 149; in Olomouc 37

Business Process Outsourcing (BPO) 187
Byrne, D. 143

Carter, D.K. 159
cartographic visualisation 94
catch-up modernisation 158
CATRIN (Czech Advanced Technology
  and Research Institute) 44, 249
Central and Eastern Europe 4, 16, 31,
  53–54, 67, 69, 113, 146, 242
Central Statistical Office (GUS) 68, 76, 89,
  90n2
*Centre of the Region Haná2 for
  Biotechnological and Agricultural Research
  (CRH)* 42–43
čermáň 25
černovická terasa 149
český technologický park 149
*Chamber of Commerce and Industry of Pécs
  and Baranya (PBKIK)* 105
changes in post-socialist cities 5
Charles University in Prague 35
Chemnitz 244; cultural modernisation of
  61; demographic development in 55;
  economic situation in 55; functional
  reconstruction in 54; housing situation
  improvement in 54; housing vacancies
  in 56–57; industrial production in 50;
  industrial shutdowns in 55; inner-
  city areas with pre-1918 housing 56;
  inner-city reconstruction in 54; Karl-
  Marx-Stadt 49–52; multidimensional
  socio-economic changes in 48; non-
  governmental organisations (NGOs) in
  52; political change and early transition
  period 51–54; population losses in
  51; post-socialist transition in **63**; as a
  shrinking city 54–58; societal ruptures
  61–62; unemployment in 51–52, 59;
  urban development, alternative strategies
  of 57–58; urban re-imaginations and
  socio-cultural modernisation 60–61;
  urban restructuration in 53
Chemnitz incident 59, 61–63
Chemnitz University of Technology 55
childlessness in Poland 70
Chrenová 22, 25
Christian Democratic Union 52
City of Modernism 60
city shrinking 72–73, 79, 90n1
Coale, A. J. 69
co-habitation 71, 81
collectivist city 142
Comenius University in Bratislava 121
commercialisation 5, 7, 40, 104, 106

communism 2, 17, 49
Comunian, R. 30, 33, 37, 39, 43
Constantine the Philosopher University in
  Nitra (UKF) 23
consumption-oriented cultural functions
  99
contemporary development in Central
  European manufacturing 95
contemporary population changes 68
contemporary population reproduction
  trends 69
contraception 86, **87**
co-operation 37, 40, **40**, 44
CORINE Land Cover classes 17, 25
Council of Mutual Economic Assistance
  (CMEA) 136, 139
Covid-19 crisis 127
creative cities 30
creative cultural and artistic hub 39
creative economy 30
creative hub, university as *see* university as
  creative hub
creative platforms 33
creative scientific research hub 39–43
Croatia 1
CTP Invest 148
Cudny, W. 4, 179
cultural life in Brno 152
cultural modernisation of Chemnitz 61
cultural sector 30
cumulative causation 100
Czech Lands 139
Czech Republic 1, **36**, 40, 118, 136
Czechoslovak Academy of Sciences 38
Czecho-Slovak Republic 19
Czechoslovakia 1, 2, 20, 38, 117–118, 121,
  137, 139

Dąbrowa Górnicza 200
Danube river 125
De Socio, M. 104–105
de-industrialisation 5, 55, 72, 141, 144–150
Democratic Oppositional Platform (DOP) 51
demographic and economic stabilisation, in
  Chemnitz 55–56
demographic changes: in Central and
  Eastern Europe 69; in Chemnitz 55; in
  Kielce 76; in Poland 70
depopulation 4
Diely 25
divorces in Poland 71, 85, **85**
D-Mark 52
Dostál, P. 121
Dövényi, Z. 115
DPSIR analyses 17, 26

Drážovce urban area 25
Dresden 50

East Germany 51–52
Eastern Bloc 5
Eastern Germany 1
Eastern Poland 247; demographic types
  of urban centres in 227; location of
  voivodeships (provinces) of 220; pace of
  population ageing among the inhabitants
  of urban centres in 233–238, 234; as
  a peripheral area 224–226; relations
  between natural growth and migration
  balance in the towns 226–233
economy: in Bratislava 123; in Katowice
  conurbation 204–207, 205; in Łódź 187
education in Brno 152
Egedy, T. 31
Elbląg 231
Elcoteq 99
employment: in Brno 141; in Łódź 187
Engels, Friedrich 49
entrepreneurial city concept 144
environmental challenges in Katowice
  conurbation 210–211
Esteve, A. 70
Estonia 1
Etzkowitz, H. 32
European Capital of Culture (ECoC) event
  96, 99
European Regional Development Fund
  and the Cohesion Fund 189
European universities 30
Eurosense 17
Evans, G. 30
extra-marital births: in Kielce 82–83; in
  Poland 71

Faculty of Engineering and Information
  Technology 105
Faculty of Health Sciences 35
Faculty of Law 35
Faculty of Medicine and Dentistry 39
Faculty of Natural Sciences 35
Faculty of Physical Culture 35
Faculty of Science 39
Faculty of Theology 35
fascist Slovak state 121
FDI-based manufacturing activities 101
Federal Republic of Germany (FRG) 49
Federal State of Saxony 168
fertility rate, in Kielce 70, 74, 80
financial situation of students in Kielce
  79, 81
First World War 184

Florida, R. 30, 60
Fordist production schemes 48
foreign direct investment (FDI) networks
  95
fortification zones 34
Freie Presse 62
Frejka, T. 70
Fritz-Heckert-Gebiet 50, 56–57
functional reconstruction in Chemnitz 54
functional structure, transition of 5

Gentile, M. 242
gentrification 5, 131
Geodis s.r.o. Bratislava 17
German Democratic Republic (GDR) 1,
  49, 51, 56, 61
German Reunification 162, 166, 167, 171
Germany 1–2, 49
Germany Democratic Republic (DGR) 1
GGG company 57–58
ghettoization 5
Gilmore, A. 43
Give Culture Space 60
global production networks (GPNs) 96
globalisation 72
Górnośląski Związek Metropolitalny
  (GZM) 203, 210
Gorzelak, G. 224
Great Moravia 117
Grigorescu, I. 5
Großmann, K. 55, 57

Hájek, Z. 141
Hamilton, F. E. I. 16
Hampl, M. 121
Haná 44n2
hard creative services 33
Hauni Hungaria 101
hidden sectors in Pécs 94–96, 106–107,
  107
Hidden unemployment 97–98
Higher Education Act (1990), in Olomouc
  38
Hirt, S. 5
Hoekveld, J.J. 218
Holy Grail 104
Horeczki, R. 101
housing provision in Leipzig 166–168
housing shortages, fight against 50
housing situation improvement in
  Chemnitz 54
housing vacancies, in Chemnitz 56–57
Howkins, J. 30
human-driven changes in land use 17
Hungary 1, 118, 136

IG Metall 52
Illner, M. 115, 143
Immediate Construction Program for Schools 170
immigration policy in Poland 208
industrial development in Nitra City 21–22
industrial manufacturing, spatial structure of 145
industrial paternalism 143
industrial production in Chemnitz 50
industrial shutdowns in Chemnitz 55
industrial tradition in Brno 137–142
industrialisation, in Leipzig 161
industrialism 142
informal partnerships in Kielce 81–82, **82**
informal relationships 71
inner spatial structure in Bratislava 124–126
inner suburbanisation in Katowice conurbation 209
inner-city areas with pre-1918 housing in Chemnitz 56
inner-city reconstruction in Chemnitz 54
Institute of Molecular and Translational Medicine 43
Intelmann, D. 61
interconnection in Brno 150–151
international publishing market 3
inter-urban process 16
intra-urban process 16
intra-urban structure of post-socialist Nitra 20–23
Ira, V. 115
iron curtain 118
Istrochem 130
Italy 1

Jaguar Land Rover (JLR) automobile plant 22
Jan Kochanowski University (UJK) 76
Janiszewska, A. 69–71, 87
Jaroszewska, E. 191
Jesuits 34
Johnson, L. 1
Johnson, L. R. 1
Joint Laboratory of Optics of Palacký University Olomouc and the Institute of Physics of the Czech Academy of Sciences 40

Kaßberg area 56–57
Kaczmarek, S. 4
Kamińska, W. 224
Kantor-Pietraga, I. 229
Karl Marx Monument *50*, 61
Karl Marx Year 49
Karl-Marx-Stadt 49–52

Katowice Central Europe conurbation 208–209
Katowice conurbation 195; after 30 years of transformation 204–211; on the background of region *196*; brownfield spaces in 200; economic transformation process, modelling 204–207, *205*; environmental challenges 210–211; inner suburbanisation in 209; Katowice Special Economic Zone (KSEZ) 201–202; metropolisation of conurbations 203–204; political and economic crisis in the face of traditional economy 199–201; population in the area of **199**; regional crisis into the agenda setting 201; re-industrialisation 201–202; social dimension of changes 202–203; socio-demographic transition 207–209; spatial patterns 209–210; strong monofunctionality, formation of 197–199; unemployment in 200
Katowice conurbation 246
Kielce 72–76, 245; birth rate in 74; demographic changes in **75**, 76; divorce rate in 75, 85; extra-marital births in 82–83; fertility rate in 74, 80; financial situation of students in 79, 81; informal partnerships in 81–82, **82**; location of *73*; marriages in 74, 78–79; number of singles in 82, **83**; opinion about contraception in 86, **87**; parenthood plans in 79, **80**; planning of number of children in **80**, 81; population depression in 76; population growth in 74; postponed parenthood in 79–80; premarital sex in 85–86, **86**; survey respondents, characteristics of 76; unemployment in 73
Kielce students: attitude to family values 81–89; plans regarding place of residence and starting a family 77
Klokočina 25
knowledge exchange hubs for the creative economy 30
knowledge institutions 33
knowledge-based economy 72
Knowledge-based industries in Pécs 104
Kocot-Górecka, K. 70
Kohler, H.-P. 69
Korec, P. 4, 115, **129**, 179
Košice 20
Kotowska, I. E. 70
Kovács, S. Z. 101
Kovács, Z. 5, 115
KRACH program 60

Královopolská strojírna 139
Kubeš, J. 5
Kurek, S. 70
Kurkiewicz, J. 70

Laboratory of Growth Regulators 43
labour immigrants in Nitra 23
labour market in Brno 152
Lambooy, J. G. 32
land-cover changes 7, 17, *19*; in Nitra 26–27
Länder 1
Landry, C. 60
Lange, M. 70
Latvia 1
Leetmaa, K. 5
Leipzig 50, 62, 161; de-industrialisation and shrinkage in 243–244; as a growing city during industrialisation 161; housing provision in 166–168; long-term shrinkage, towards 162; massive shrinkage after 1989 162–163; moderate re-growth in the 2000s 163–164; public transport in 170–172; re-growth in 165–166, 244; school infrastructure in 168–170
Leipzig Monday demonstrations 162
Leipziger Freiheit 166
Leipzig-Halle Airport 164
Leipzig-Halle region 160, *165*
Leipzig-Halle-Bitterfeld conurbation 161
Lengyel, I. *99*
Lesthaeghe, R. 69
Lewis, P.G. 1, 242
Leydesdorff, L. 32
Light and food industries 97
Lisowski, A. 5
Liszewski, S. 4, 179, 190
Lithuania 1
Łódź 115, 178, 245; economy in 187; employment in 187; ethnic groups in 182; future of 182; industries development in 189; location of *181*; New Centre of Łódź (NCŁ) programme in 189; ownership structure of industry in 187; population in 182, 184, **186**; research area, presentation of 180–181; research results after 1989 185–190; research results before 1989 181–185; service function in 189; spatial and population development of *183*; Special Economic Zone in 187, 192n1; symbol of the transformations in 187; textile industry in 181; tourism in 189–190

Lorber 3
Lux, G. *99*

"manhattanisation" process 125
Manufacturing industries in Pécs 105
Manufaktura 188
Marcińczak, S. 4
Mareš, J. 139
Margraviate of Moravia 34
marriages: in Kielce 74, 78–79; in Poland 71, 78, **79**, 87
Marx, Karl 49
Matlovič, R. 16, 179, 190
medium-sized manufacturing firms in Pécs 103
Mendel, Gregor Johann 35
metropolitan development, suburbanisation in 115
mining employment 98
Mlyny Gallery 22
modern urban environment, in Brno 150
modernism 60
Moravia IT 148
Moravia 34, 37, 249
Mularczyk, M. 224
Mulíček, O. **141**, 142, *147*, *149*
multidimensional socio-economic changes in Chemnitz 48
multidimensional transition processes 5
Musil, J. 15, 115, 143
Mykhnenko, V. 146
Mynarska, M. 71

National Biomedical and Biotechnological Park 37
natural growth and migration balance, relations between 226–233
Nazi ideology 61
neoliberal principles 22
neoliberalism 250–251
New Centre of Łódź (NCŁ) programme 189
NGO "New Forum" 51
Nitra City 250; Constantine the Philosopher University in (UKF) 23; industrial development in 21–22; intra-urban structure of post-socialist Nitra 20–23; labour immigrants in 23; land-cover change in 26–27; Old Town Centre 25; post-socialist urban change in 17; retail sector 22; shopping centres in 22; study area of 18–20, *18*; tertiary sector of 22; three areas of post-socialist urban change in **24**; University of Agriculture in 22

non-governmental organisations (NGOs) in Chemnitz 52
Nová Mosilana 142

OECD countries 32
Olomouc 34, 44n1; cultural and creative industries in 38b; Moravian city of 249
Olomouc Fortress 44n1
Ondoš, S. 4, 115, 179
Operational programme for Science and Research for Innovation: "large infrastructure," building 41; science centres at Palacký University Olomouc supported by 42; science centres in the Czech Republic supported by 41
Oracle 148
Ourednek, M. 5
Ozawa, T. 206

Pacione, M. 145
Palacký University Olomouc 31, 34–39, 40, 44
parent generation 55
parenthood plans in Kielce 79, 80
passive revolution 61
Pécs 94, 247–248; hidden sectors in 94–95, 96, 106–107, 107; historical background 96–98; industrial mix of 101; knowledge-based industries in 104; location of 95; manufacturing industries in 105; medium-sized manufacturing firms in 103; post-socialist development processes 98–100; Small and Medium Enterprise (SME) sector in 103; in the spatial structure of Hungarian manufacturing 99; structural indicators of 100; unemployment in 100
peripheral areas 222–224; Eastern Poland as 224–226
peripheral rural municipalities, independence of 20
peripherality 223–224
Pest 117
Phelps, N.A. 206
Philipov, D. 69
Phoenix cities 159
Pichler-Milanovic´, N. 5
Piotrków Trybunalski 184
Plaziak, M. 5
Podogrodzka, M. 70
Póla, P. 105
Poland 1–2, 118, 136, 178, 247; abortion in 88–89, 88; attitude to marriage in 87; bachelors in 71; childlessness in 70; demographic changes in 70; diverse

pace of population ageing in 235, 236; divorces in 71, 85, 85; extra-marital births in 71; fertility rate in 70, 80; immigration policy in 208; marriage age in 71, 78, 79; population changes in 70; single parenthood in 84, 84; size structure of urban centres in 220, 221; unmarried young people in 71
political change, in Chemnitz 51–54
population: in Chemnitz 51; in Katowice conurbation 199; in Kielce 74, 76; in Łódz 182, 184, 186; in Poland 70
population ageing 217; Eastern Poland as a peripheral area 224–226; natural growth and migration balance in Eastern Poland 226–233; pace of ageing among the inhabitants Eastern Poland 233–238; in peripheral areas 222–224
post-communist transformation 180
post-industrial city development 145
post-industrial city, manufacturing in 94, 101–104; institutional context of industrial change 104–106; post-socialist development processes 98–100
postponed parenthood in Kielce 79–80
post-socialism 48, 251
post-socialist cities, defined 4, 179
The Post-Socialist City (Stanilov) (2007) 16
post-socialist housing estates 4
post-socialist transformation 2
Power, A. 159
powiat rights 74
Poznański, Izrael 182, 188
Prague–Brno–Olomouc motorway 150
pre-fabricated housing estates 4
premarital sex in Kielce 85–86, 86
Prešov 20
privatisation of state-owned economic institutions in Bratislava 123
privatisation process 5
procreation 87
První brněnská strojírna 139, 142
PSA Group 206
public transport, in Leipzig 170–172
publicly sponsored jobs 52

qualitative methods 6
quantitative methods 6

Rácz, S. 101
ready-made state 158
Regional Centre of Advanced Technologies and Materials (*RCPTM*) 42
re-growth, challenges and problems of 158, 165–166; background, materials

and methods 159–161; challenges and problems of re-growth 165–166; housing provision 166–168; Leipzig as a growing city during industrialisation 161; long-term shrinkage, towards 162; massive shrinkage after 1989 162–163; moderate re-growth in the 2000s 163–164; public transport in Leipzig 170–172; school infrastructure in Leipzig 168–170
re-industrialisation 33; in Katowice conurbation 201–202; in Pécs 95–96, 105
ReNewTown 4
Research and Enterprise in Arts Creative Technology 30
research centres 40
research infrastructure 36
research problem and aims 1–2
residential suburbanisation 21
retail sector of Nitra City 22
Rink, D. 159
Romania 1, 5
Rosset, E. 219
Round Table Talks 2, 51–52, 178
Rufat, S. 4
Runge, J. 6

Saxony 51
"scattered metropolis" phenomenon 208
Scheibler, Karol 182
Schlesinger, P. 30
schools as an example of social infrastructure 168–170
Science and Technology Park of Palacký University Olomouc 39
scientific method 6
second demographic transition 4; in North and West European countries 69; vs. young Poles' procreative behaviours 68–72
Second World War 180
shopping centres in Nitra City 22
Shrink Smart 160
shrinking city 72–73, 90n1
Siedentop, S. 160
Sikorski, D. 5
Silesian Voivodeship 71–72
single life in Kielce 82, **83**
single parenthood in Poland 84, **84**
Skibiński, A. 71, 87
Šlapanice u Brna 139
Slovak Bank 121
Slovak ethnicity 120–121
Slovak higher education system 22
Slovak National Theatre 121

Slovak Republic 1
Slovak University of Agriculture in Nitra (SUA) 22
Slovakia 1, 20, 21, 116, 118, 123–124, 136
Slovenia 1
Slovnaft 122
Small and Medium Enterprise (SME) sector in Pécs 103
smart city projects 32
Social Atlas of Leipzig 160
social segregation 5
socialism 49, 132, **251**
societal ageing 4
socio-demographic transition in Katowice conurbation 207–209
socioeconomic development in Bratislava during post-socialism 113; economic transition, processes of 121–124; geography, history and local governance design 117–118; inner spatial structure 124–126; main dimensions 114–116; population development, processes of 118–121, **119–120**; spatial structure, special problems of 130–131; suburbanisation 126–130
socio-economic revitalisation 5
socio-spatial restructuration 55
soft creative services 33
Sonnenberg 56–57, 60
Sopianae 96
space fragmentation process 5
spatial patterns in Katowice conurbation 209–210
spatial transformation 1, 6, 15–16, 72, 190, 242
Special Economic Zone, in Katowice conurbation 201–202
Spórna, T. **199**
sports events in Brno 152
Sryjakiewicz, T. 191
Stadtumbau Ost 56, 167
Stanilov, K. 5, 16, 242
state-owned economic institutions, privatisation of 123
Statistical Office of the Slovakian Republic 17
Strategia rozwoju Polski Wschodniej 2020 (2013) 224
studentification 36
Subcarpathian Russia 137
suburbanisation 5, 20–21, 53, 55, 115
super-centre 44, 249
Šveda, M. 5
Sýkora, L. 3, 5, 15, 26, 115, 142, 179, 190, 242

Szafrańska, E. 4
Székesfehérvár 117
Szukalski, P. 70, 74

tertiary education 32, 38
tertiary sector of Nitra City 22, 140–141, 143, 145
textile industry in Łódź 181
third role of universities 33
thirty years of post-socialist transition 3–5
Thomi, Walter 114
tourism in Łódź 189–190
Toušek, V. **141**, 142
training institutions 33
transformation shock 195, 199
transition 5, 16, 48, 62–63
transition period, in Brno 137–142, 144, 248
transportation, in Brno 150–151
Tribeč mountain 23
triple helix 31–32
Tsenkova, S. 15
Turok, I. 146

unemployment 5; in Chemnitz 51–52, 59; in Katowice conurbation 200; in Kielce 73; in Pécs 100
UNESCO Creative Cities Network 190
university as creative hub 30; changing role of universities 32–34; city of Olomouc and its university 34, 37–43
University of Agriculture in Nitra 22
University of Maribor 33
University of Pécs 105–106
university-related start-up companies 32
unmarried young people in Poland 71
Upper Silesian Industrial Region 199
Upper Silesian Metropolitan Union 203
Upper Silesian-Basin Metropolis 203–204
Uranium City 101
urban areas in Central Europe 2

urban changes in Nitra City 15; in Brno 150; in Chemnitz 53, 57–58; intra-urban structure of post-socialist Nitra 20–23; study area 18–20, *18*; three areas of **24**
urban land-cover change analysis in Nitra City: in 1990 23–25; in 2019 25–26
urban planning 50
urban shrinking 54, 72

Van de Kaa, D. J. 69
Vašková, L. **141**
Velvet Revolution 2, 38
vicious circle 218
Visegrad Group 1, 118
Vlněna 148
voivodeship 76
Volkswagen 122

Walde, A. 160
Warminsko-Mazurskie Voivodeship 225
Warsaw 184
Webb's demographic types 226–233
Weberian rationalisation 69
Węclawowicz, G. 4, 250
West Germany 3, 51–52
Western norms, adaptation of 22
Wilhelminian-style inner-city districts 164, 167
Wolaniuk, A. 115
Woodcuttings 23
World Bank 1

young Poles' procreative behaviours vs. second demographic transition theory 68–72
Young, C. 4

Zbrojovka 139
Zetor 139
Žilina 20–21
Zobor hills 25
Zsolnay Porcelain Factory (1853) 97

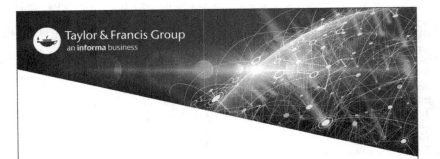

Printed in the United States
by Baker & Taylor Publisher Services